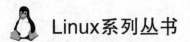 Linux系列丛书

Linux虚拟化和云计算实战指南

吴光科　陈　平　马文涛　杜　明 ◎ 编著

北京理工大学出版社
BEIJING INSTITUTE OF TECHNOLOGY PRESS

内容简介

本书系统地论述了 Linux 运维领域的各种技术，全书分为 21 个章节，主要内容包括：虚拟化概念、虚拟化背景、虚拟化技术意义、虚拟化技术种类及 KVM 虚拟化入门、简介、原理、安装、实战、企业应用；同时重点剖析企业级虚拟化 Docker 入门、简介、原理、安装、实战、企业应用、镜像和容器原理、30+Docker 命令、Docker 网络 None、Host、Bridge、Container 模式剖析、Dockerfile 镜像制作、镜像制作原理、常见参数、Docker Compose 容器编排、编排原理、编排命令、编排企业案例；加入 Docker 集群管理工具：Swarm、入门、简介、原理、部署、安装、配置、实战、排错等实战内容；引入企业级 Openstack 云计算、入门、原理、安装、部署、排错、实战和企业应用、Openstack 各个模块原理、组件整合、云主机创建和管理。

同时还引入最新的 Linux 云计算技术，具体包含：云计算入门、分类、Kubernetes 简介、Kubernetes 组件、Kubernetes Apiserver、Scheduler、Controller-manager 剖析、Kubernetes 原理、Kubernetes Master 功能、Kubernetes Minion 功能、Kubernetes 集群工作原理、工作流程、Kubernetes 证书制作、Kubernetes 各个组件剖析、Kubernetes 最新版本部署实战、Kubernetes 部署环境准备、二进制部署、Kubeadm 部署和管理、Kubernetes 开启 IPVS 代理模块、Kubernetes Service 原理、工作流程、案例演练、Cluster IP、NodePort、Load Balance、容器升级原理、滚动升级过程、参数修改、滚动升级案例、Kubernetes 无状态服务、有状态服务、污点、节点扩容、节点迁移、节点升级、Kubernetes Ingress、Traefik、Kubernetes+NFS 存储持久化、Kubernetes+CEPH-FS 持久化、Kubernetes+CEPH-RBD 模式、挂载云盘、Prometheus 监控 Kubernetes Node、Pod 容器、Granafa 整合、实现邮件、微信报警、ETCD 原理、分布式实战、K-V 备份、扩容、故障切换、Kubernetes 多 Master 节点、管理、添加、实战、Kubernetes 故障排错、Kubernetes 常见面试题、报错解决等实战必备内容。

版权专有 侵权必究

图书在版编目（CIP）数据

Linux 虚拟化和云计算实战指南 / 吴光科等编著.
北京：北京理工大学出版社, 2025.1.
ISBN 978-7-5763-4681-7

Ⅰ. TP316.85；TP393.027

中国国家版本馆 CIP 数据核字第 20253J21S0 号

责任编辑：江 立	文案编辑：江 立
责任校对：周瑞红	责任印制：施胜娟

出版发行 / 北京理工大学出版社有限责任公司
社　　址 / 北京市丰台区四合庄路 6 号
邮　　编 / 100070
电　　话 /（010）68944451（大众售后服务热线）
　　　　　（010）68912824（大众售后服务热线）
网　　址 / http://www.bitpress.com.cn
版 印 次 / 2025 年 1 月第 1 版第 1 次印刷
印　　刷 / 三河市中晟雅豪印务有限公司
开　　本 / 787 mm × 1020 mm　1/16
印　　张 / 21.75
字　　数 / 463 千字
定　　价 / 99.00 元

图书出现印装质量问题，请拨打售后服务热线，负责调换

前言
PREFACE

Linux 是当今三大操作系统（Windows、macOS、Linux）之一。Linux 系统创始人 Linus Benedict Torvalds（林纳斯·本纳第克特·托瓦兹）21 岁的时候，用 4 个月的时间创建了第一个版本 Linux 内核，并于 1991 年 10 月 5 日正式对外发布。该版本继承了 UNIX 以网络为核心的思想，是一个性能稳定的多用户网络操作系统。

随着互联网的飞速发展，IT 技术引领时代潮流，而 Linux 技术又是一切 IT 技术的基石，应用领域包括个人电脑、服务器、嵌入式应用、智能手机、云计算、大数据、人工智能、数字货币、区块链等。

为什么写《Linux 虚拟化和云计算实战指南》这本书呢？这要从我的经历说起。我出生在贵州省一个贫困的小山村，从小经历了山里砍柴、放牛、挑水、做饭、日出而作、日落而归的朴素生活，看到父母一辈子都在小山村里，没有见过大城市，所以从小立志要走出大山，让父母过上幸福的生活。正是这样一个信念让我不断地努力，大学毕业至今，我在"北漂"的 IT 运维路上走过了 10 多年，从初创小公司到国企、机关单位，再到图吧、研修网、京东商城等一线 IT 企业，分别担任过 Linux 运维工程师、Linux 运维架构师、运维经理，到今天创办了京峰教育培训机构。

一路走来，感谢生命中遇到的每一个人，是大家的帮助，让我不断地进步和成长，也让我明白了一个人活着不应该只为自己和自己的家人，而是要为这个社会贡献哪怕只是一点点的价值。

为了帮助更多的人通过技术改变自己的命运，我决定编写《Linux 虚拟化和云计算实战指南》这本书。虽然市面上有很多关于 Linux 的书籍，但是很难找到一本关于 Linux 系统简介、系统原理、系统安装、必备命令、用户权限、磁盘管理、数据库读写分离、Redis、Nginx Web 实战、企业生产环境、企业自动化运维、Docker、Podman 虚拟化、Kubernetes 云计算、Devops 微服务等详细、全面的主流技术的书籍，这是我编写本书的初衷！

本书读者对象包括：系统管理员、网络管理员、在校大学生、Linux 运维工程师、Linux 系统管理人员及从事云计算、网站开发、测试、设计的人员。

尽管笔者花费了大量的时间和精力核对书中的代码和语法，但其中难免还会存在一些纰漏，恳请读者批评指正。

吴光科

2024 年 5 月

致 谢
THANKS

感谢 Linux 之父 Linus Benedict Torvalds，他不仅创造了 Linux 系统，而且影响了整个开源世界，也影响了我的一生。

感谢我亲爱的父母，含辛茹苦地把我们兄弟三人抚养长大，是他们对我无微不至的照顾，让我有更多的精力和动力工作，帮助更多的人。

感谢潘彦伊、周飞、何红敏、周孝坤、杨政平、王帅、李强、刘继刚、常青帅、孙娜、花杨梅、吴俊、李芬伦、陈洪刚、黄宗兴、代敏、杨永琴、姚钗、王志军、谭陈诚、王振、杨浩鹏、张德、刘建波、洛远、谭庆松、王志军、李涛、张强、刘峰、周育佳、谢彩珍、王奇、李建堂、张建潮、佘仕星、潘志付、薛洪波、王中、朱愉、左堰鑫、齐磊、韩刚、舒畅、何新华、朱军鹏、孟希东、黄鑫、陈权志、胡智超、焦伟、曾地长、孙峰、黄超、陈宽、罗正峰、潘禹之、揭长华、姚仑、高玲、陈培元、秦业华、沙伟青、戴永涛、唐秀伦、金鑫、石耀文、梁凯、彭浩、唐彪、郭大德、田文杰、柴宗虎、张馨、佘仕星、赵武星、王永明、何庆强、张镇卿、周聪、周玉海、周泊江、吴啸烈、卫云龙、刘祥胜等挚友多年来对我的信任和支持。

感谢腾讯公司腾讯课堂所有课程经理及平台老师，感谢 51CTO 学院院长一休及全体工作人员对我及京峰教育培训机构的大力支持。

感谢京峰教育培训机构的每位学员对我的支持和鼓励，希望他们都学有所成，最终成为社会的中流砥柱。感谢京峰教育培训机构 COO 蔡正雄，感谢京峰教育培训机构的全体老师和助教，是他们的大力支持，让京峰教育能够帮助更多的学员。

最后感谢我的爱人黄小红，是她一直在背后默默地支持我、鼓励我，让我有更多的精力和时间去完成这本书。

目录 CONTENTS

第 1 章 企业级 KVM 虚拟化实战 1
- 1.1 虚拟化技术概述及简介 1
- 1.2 互联网虚拟化技术种类 1
- 1.3 KVM 虚拟化概念 2
- 1.4 KVM 虚拟化安装 2
- 1.5 KVM 网桥配置实战 3
- 1.6 KVM 虚拟化硬盘扩容 7
- 1.7 KVM 虚拟机批量克隆实战 7
- 1.8 ESXI 虚拟化技术概念 9
- 1.9 XEN 虚拟化技术概念 10

第 2 章 企业级 Docker 虚拟化实战 11
- 2.1 虚拟化技术实现方式 13
- 2.2 Docker LXC 及 Cgroup 原理剖析 13
- 2.3 AUFS 简介 15
- 2.4 Device Mapper 文件系统简介 16
- 2.5 OverlayFS 简介 17
- 2.6 为什么使用 Docker 19
- 2.7 Docker 镜像、容器、仓库 20
- 2.8 Docker 镜像原理剖析 20
- 2.9 CentOS 7.x（7.0+）Linux Docker 平台实战 23
- 2.10 CentOS 8.x（8.0+）Linux Docker 平台实战 25
- 2.11 Ubuntu（16.04+）Linux Docker 平台实战 26
- 2.12 Docker 仓库源更新实战 28

第 3 章 Docker 企业命令实战 30
- 3.1 Docker search 命令实战 30
- 3.2 Docker pull 命令实战 30
- 3.3 Docker images 命令实战 30

- 3.4 Docker run 命令实战 ... 31
- 3.5 Docker ps 命令实战 ... 31
- 3.6 Docker inspect 命令实战 ... 31
- 3.7 Docker exec 命令实战 ... 31
- 3.8 Docker stop|start 命令实战 ... 32
- 3.9 Docker rm 命令实战 ... 32
- 3.10 Docker rmi 命令实战 ... 32
- 3.11 Docker 虚拟化 30 多个命令实战剖析 ... 33

第 4 章 Docker 网络原理实战 ... 35
- 4.1 Host 模式剖析 ... 35
- 4.2 Container 模式剖析 ... 36
- 4.3 None 模式剖析 ... 36
- 4.4 Bridge 模式剖析 ... 36
- 4.5 Bridge 模式原理剖析 ... 36
- 4.6 Bridge 模式实战一 ... 38
- 4.7 Bridge 模式实战二 ... 38
- 4.8 Bridge 模式实战三 ... 39
- 4.9 Bridge 模式实战四 ... 39
- 4.10 Docker 持久化固定容器 IP ... 41
- 4.11 EFK 应用背景剖析 ... 43
- 4.12 EFK 架构原理深入剖析 ... 43
- 4.13 Docker 部署 EFK 日志平台 ... 44
- 4.14 基于 Docker Web 管理 Docker 容器 ... 46

第 5 章 Dockerfile 企业镜像实战 ... 50
- 5.1 Dockerfile 语法命令详解一 ... 50
- 5.2 Dockerfile 语法命令详解二 ... 51
- 5.3 Dockerfile 制作规范及技巧 ... 55
- 5.4 Dockerfile 企业案例一 ... 56
- 5.5 Dockerfile 企业案例二 ... 56
- 5.6 Dockerfile 企业案例三 ... 57
- 5.7 Dockerfile 企业案例四 ... 58

第 6 章 Docker 仓库案例实战 ... 59
- 6.1 Docker 国内源实战 ... 59

6.2	Docker Registry 仓库源实战	60
6.3	Docker Harbor 仓库源实战	62
6.4	Docker 磁盘、内存、CPU 资源实战一	67
6.5	Docker 磁盘、内存、CPU 资源实战二	70
6.6	Docker 资源监控方案和监控实战	74
6.7	Docker stats 监控工具	74
6.8	CAdvisor 监控工具	76
6.9	CAdvisor 部署配置	76
6.10	构建 CAdvisor+InfluxDB+Grafana 平台	77

第 7 章 Docker Compose 容器编排实战 — 81

7.1	Docker Compose 概念剖析	81
7.2	Docker Compose 部署安装	82
7.3	Docker Compose 命令实战	82
7.4	Docker Compose 常见概念	83
7.5	Docker Compose 语法详解	83
7.6	Docker Compose Nginx 案例一	90
7.7	Docker Compose Redis 案例二	92
7.8	Docker Compose Tomcat 案例三	93
7.9	Docker Compose RocketMQ 案例四	95

第 8 章 Docker Swarm 集群案例实战 — 100

8.1	Swarm 概念剖析	100
8.2	Docker Swarm 的优点	101
8.3	Swarm 负载均衡	102
8.4	Swarm 架构图	102
8.5	Swarm 节点及防火墙设置	104
8.6	Docker 虚拟化案例实战	104
8.7	Swarm 集群部署	105
8.8	Swarm 部署 Nginx 服务	106
8.9	Swarm 服务扩容和升级	107
8.10	Manager 和 Node 角色切换	109
8.11	Swarm 数据管理之 volume	109
8.12	Swarm 数据管理之 Bind	110
8.13	Swarm 数据管理之 NFS	111

- 8.14 Docker Swarm 新增节点 ... 113
- 8.15 Docker Swarm 删除节点 ... 116
- 8.16 Docker 自动化部署一 ... 117
- 8.17 Docker 自动化部署二 ... 121

第 9 章 OpenStack+KVM 构建企业级私有云 ... 125

- 9.1 OpenStack 入门简介 ... 125
- 9.2 OpenStack 核心组件 ... 126
- 9.3 OpenStack 准备环境 ... 127
- 9.4 Hosts 及防火墙设置 ... 128
- 9.5 OpenStack 服务安装 ... 128
- 9.6 MQ（消息队列）简介 ... 129
- 9.7 MQ 应用场景 ... 130
- 9.8 安装配置 RabbitMQ ... 132
- 9.9 RabbitMQ 消息测试 ... 134
- 9.10 配置 Keystone 验证服务 ... 135
- 9.11 配置 Glance 镜像服务 ... 139
- 9.12 Nova 控制节点配置 ... 142
- 9.13 Nova 计算节点配置 ... 145
- 9.14 OpenStack 节点测试 ... 148
- 9.15 Neutron 控制节点配置 ... 149
- 9.16 Neutron 计算节点配置 ... 154
- 9.17 OpenStack 控制节点网桥 ... 155
- 9.18 Dashboard 控制节点配置 ... 157
- 9.19 OpenStack GUI 配置 ... 172
- 9.20 OpenStack 核心流程 ... 178

第 10 章 Kubernetes 组件概念 ... 181

- 10.1 云计算概念 ... 181
- 10.2 云计算技术的分类 ... 181
- 10.3 Kubernetes 入门及概念介绍 ... 182
- 10.4 Kubernetes 平台组件概念 ... 183
- 10.5 Kubernetes 工作原理剖析 ... 184
- 10.6 Pod 概念剖析 ... 187
- 10.7 label 概念剖析 ... 187

10.8	Replication Controller 概念剖析	188
10.9	service 概念剖析	188
10.10	node 概念剖析	189
10.11	Kubernetes volume 概念剖析	190
10.12	Deployment 概念剖析	190
10.13	DaemonSet 概念剖析	190
10.14	StatefulSet 概念剖析	191
10.15	ConfigMap 概念剖析	191
10.16	Secrets 概念剖析	192
10.17	CronJob 概念剖析	193
10.18	Kubernetes 证书剖析和制作实战	194

第 11 章 Kubernetes 云计算平台配置实战 … 203

11.1	Kubernetes 节点 hosts 及防火墙设置	203
11.2	Linux 内核参数设置和优化	204
11.3	Docker 虚拟化案例实战	204
11.4	Kubernetes 添加部署源	205
11.5	Kubernetes Kubeadm 案例实战	205
11.6	Kubernetes master 节点实战	207
11.7	Kubernetes 集群节点和删除	208
11.8	Kubernetes 节点网络配置	209
11.9	Kubernetes 开启 IPVS 模式	216
11.10	Kubernetes 集群故障排错	217
11.11	Kubernetes 集群节点移除	217
11.12	etcd 分布式案例操作	217

第 12 章 Kubernetes 企业网络 Flannel 实战 … 219

12.1	Flannel 工作原理	219
12.2	Flannel 架构介绍	220
12.3	Kubernetes Dashboard UI 实战	221
12.4	Kubernetes YAML 文件详解	224
12.5	kubectl 常见指令操作	226
12.6	Kubernetes 本地私有仓库实战	227

第 13 章 Kubernetes 核心组件 service 实战 … 229

13.1	Kubernetes service 概念	229

13.2	Kubernetes service 实现方式	229
13.3	service 实战：ClusterIP 案例演练	230
13.4	service 实战：NodePort 案例演练	231
13.5	service 实战：LoadBalancer 案例演练	232
13.6	service 实战：Ingress 案例演练	233
13.7	Kubernetes Traefik 案例实战	238

第 14 章　Kubernetes 容器升级实战　247

14.1	Kubernetes 容器升级概念	247
14.2	Kubernetes 容器升级实现方式	247
14.3	Kubernetes 容器升级测试	249
14.4	Kubernetes 容器升级验证	250
14.5	Kubernetes 容器升级回滚	251
14.6	Kubernetes 滚动升级和回滚原理	252

第 15 章　Kubernetes+NFS 持久化存储实战　255

15.1	Kubernetes 服务运行状态	255
15.2	Kubernetes 存储系统	255
15.3	Kubernetes 存储绑定的概念	257
15.4	PV 的访问模式	257
15.5	Kubernetes+NFS 静态存储模式	258
15.6	PVC 存储卷创建	259
15.7	Nginx 整合 PV 存储卷	260
15.8	Kubernetes+NFS 动态存储模式	262
15.9	NFS 插件配置实战	262

第 16 章　Kubernetes+CephFS 持久化存储实战　267

16.1	Kubernetes+CephFS 静态存储模式	267
16.2	PV 存储卷创建	267
16.3	PVC 存储卷创建	268
16.4	Nginx 整合 CephFS PV 存储卷	269
16.5	Kubernetes+CephFS 动态存储模式	271
16.6	CephFS 动态插件配置实战	272

第 17 章　Kubernetes+Ceph RBD 持久化存储实战　275

| 17.1 | Kubernetes+Ceph RBD 静态存储模式 | 275 |
| 17.2 | PV 存储卷创建 | 275 |

17.3	PVC 存储卷创建	276
17.4	Nginx 整合 Ceph PV 存储卷	277
17.5	Kubernetes+Ceph RBD 动态存储模式	279
17.6	Ceph RBD 插件配置实战	279

第 18 章 Prometheus 监控 Kubernetes 实战 286

18.1	Prometheus 监控优点	286
18.2	Prometheus 监控特点	286
18.3	Prometheus 组件实战	287
18.4	Prometheus 体系结构	288
18.5	Prometheus 工作流程	289
18.6	Prometheus 和 Kubernetes 背景	289
18.7	Kubernetes 集群部署 node-exporter	289
18.8	Kubernetes 集群部署 Prometheus	290
18.9	Kubernetes 集群部署 Grafana	297
18.10	Kubernetes 配置和整合 Prometheus	299
18.11	Kubernetes+Prometheus 报警设置	301
18.12	Kubernetes Alertmanager 实战	301
18.13	Alertmanager 实战部署	305

第 19 章 Kubernetes etcd 服务实战 310

19.1	etcd 和 ZK 服务概念	310
19.2	etcd 的使用场景	310
19.3	etcd 读写性能	311
19.4	etcd 工作原理	311
19.5	etcd 选主	311
19.6	etcd 日志复制	312
19.7	etcd 安全性	313
19.8	etcd 使用案例	313
19.9	etcd 接口使用	314

第 20 章 Kubernetes+HAProxy 高可用集群 315

20.1	Kubernetes 高可用集群概念	315
20.2	Kubernetes 高可用工作原理	315
20.3	HAProxy 安装配置	316
20.4	配置 Keepalived 服务	319

20.5 Keepalived master 配置实战 ······ 320
20.6 Keepalived Backup 配置实战 ······ 321
20.7 创建 HAProxy 检查脚本 ······ 322
20.8 HAProxy+Keepalived 验证 ······ 323
20.9 初始化 master 集群 ······ 324
20.10 Kubernetes Dashboard UI 实战 ······ 326

第 21 章 Kubernetes 配置故障实战 ······ 328

21.1 etcd 配置中心故障错误一 ······ 328
21.2 etcd 配置中心故障错误二 ······ 328
21.3 Pod infrastructure 故障错误 ······ 329
21.4 Docker 虚拟化故障错误一 ······ 329
21.5 Docker 虚拟化故障错误二 ······ 330
21.6 Dashboard API 故障错误 ······ 330
21.7 Dashboard 网络访问故障错误 ······ 330

第 1 章 企业级 KVM 虚拟化实战

1.1 虚拟化技术概述及简介

IT 行业发展到今天，已经从传统技术、传统运维发展到当下的主流技术、自动化运维虚拟化技术也越来越广泛地应用在企业中，例如百度、阿里巴巴、腾讯、京东、Google 等。

通俗地说，虚拟化就是把物理资源转变为逻辑上可以管理的资源，以打破物理结构之间的壁垒。计算元件运行在虚拟的基础上而不是真实的基础上，可以扩大硬件的容量，简化软件的重新配置过程。

虚拟化技术允许一个平台同时运行多个操作系统，且应用程序都可以在相互独立的空间内运行而互不影响，从而显著提高计算机的工作效率，是一个简化管理、优化资源的解决方案。

虚拟化解决方案的底部是需要进行虚拟化的物理计算机。这台计算机可能直接支持虚拟化，也可能不直接支持虚拟化，后者就需要系统管理程序层的支持。虚拟机管理程序（Virtual Machine Monitor，VMM），可以看作平台硬件和操作系统的抽象化，本质是我们常说的虚拟化技术软件。

通过虚拟化技术软件可以将物理机虚拟生成 N 台虚拟机，应用程序、软件服务（Apache、Nginx、MySQL、Redis、MQ、ZK、Kafka、Ceph、K8S、LVS、Keepalived、Jenkins）可以运行在虚拟机上，而不是直接运行在硬件设备物理机上。

虚拟化技术主要是为了最大化地利用高配硬件设备的资源，提高物理机资源利用率，以实现应用程序、软件服务进程资源隔离，淘汰老旧服务器资源，对老旧服务器资源进行重组、重用，实现企业服务器资源的统一管理和调度。

1.2 互联网虚拟化技术种类

互联网虚拟化技术主要有以下几种。

(1) VMware ESXI。
(2) KVM。
(3) XEN。
(4) Hyper-V。
(5) Open-vz。
(6) Podman。
(7) Docker。

1.3　KVM 虚拟化概念

KVM 虚拟化全称为 Kernel-based Virtual Machine（基于内核的虚拟机），是一个开源的系统虚拟化模块，是针对包含虚拟化扩展（Intel VT 或 AMD-V）的 x86 硬件上 Linux 的完全原生的虚拟化解决方案。

KVM 最早由以色列的公司开发，现在 RedHat 公司斥资 1.07 亿美元收购了 KVM 虚拟化管理程序厂商 Qumranet。严格来讲，KVM 虚拟化技术不是一个软件，而是 Linux 内核里面一种加速虚拟机的功能扩展，自 Linux 2.6.20 之后集成在 Linux 的各个主要发行版本中。KVM 的虚拟化需要硬件支持，如 Intel VT 技术或者 AMD V 技术，属于完全虚拟化。了解 KVM 之前，需要了解 KVM 和 QEMU、Libvirt 之间有什么关系。

KVM 是一款支持虚拟机的技术，是 Linux 内核中的一个功能模块。它在 Linux 2.6.20 之后的任何 Linux 分支中都被支持。但它要求硬件必须达到一定标准。

QEMU 是什么呢。其实它也是一款虚拟化技术，就算不使用 KVM，单纯的 QEMU 也可以完全实现一个虚拟机。

那为何还会有 QEMU-KVM 这个名词呢？虽然 KVM 的技术已经相当成熟，但是在某些方面还是无法虚拟出真实的机器。比如对网卡的虚拟，这时就需要另外的技术做补充，而 QEMU-KVM 正是这样一种技术。它补充了 KVM 技术的不足，且在性能上对 KVM 进行了优化。

Libvirt 又是什么呢？它是一系列库函数，用来管理计算机上的虚拟机。Libvirt 包括各种虚拟机技术，如 KVM、XEN 与 LXC 等，不同虚拟机技术就可以使用不同驱动，但都可以调用 Libvirt 提供的 API 对虚拟机进行管理。我们创建的各种虚拟机都是基于 Libvirt 库及相关命令管理的。

1.4　KVM 虚拟化安装

一台可以运行最新 Linux 内核的 Intel 处理器（含 VT 虚拟化技术）或 AMD 处理器（含 SVM 安全虚拟机技术的 AMD 处理器，又称 AMD-V），需开启 BIOS 虚拟化功能（cpu info ->

Virtualization Technology 选项设置为 Enabled）。执行如下指令：

```
egrep 'vmx|svm' /proc/cpuinfo
```

（1）如果输出结果包含 VMX，它是 Intel 处理器虚拟机技术的标志。

（2）如果输出结果包含 SVM，它是 AMD 处理器虚拟机技术的标志。

（3）如果什么都没有看到，系统不支持虚拟化的处理，不能使用 KVM。

```
yum install -y kvm python-virtinst libvirt bridge-utils virt-manager
qemu-kvm-tools virt-viewer virt-v2v libguestfs-tools
#采用源码安装,安装方法如下
wget ftp://ftp.naist.jp/pub/Linux/momonga/4.1/SOURCES/kvm-33.tar.gz
tar zxf kvm-33.tar.gz
cd kvm-33
./configure --prefix=/usr/local/kvm/
make
make install
#安装完毕,需要如下配置：
ln -s /usr/local/kvm/bin/* /usr/bin/
ln -s /usr/local/kvm/lib/* /usr/lib/
ln -s /usr/local/kvm/lib64/* /usr/lib64/
#若为 x86_64 系统,则执行如下代码
ln -s /usr/local/kvm/lib64/* /usr/lib64/
ln -s /usr/local/kvm/include/kvmctl.h /usr/include/
ln -s /usr/local/kvm/include/linux/* /usr/include/linux/
ln -s /usr/local/kvm/share/qemu /usr/share/
```

1.5　KVM 网桥配置实战

KVM 虚拟机网络配置有以下两种模式。

1）NAT 模式

NAT 模式是让虚拟机访问主机、互联网或本地网络上的资源的简单方法，但是不能从网络或其他客户机访问客户机。

2）Bridge 模式

Bridge 模式下，主机与主机之间、客户机与主机之间的通信都很容易，使虚拟机成为网络中具有独立 IP 的主机，如图 1-1 所示。

其中 Bridge 模式构建步骤如下所述。

（1）新建网卡名称为 br0，即桥接模式。同时在 eth0 网卡中指定桥接网卡名称，在 /etc/sysconfig/network-scripts/ 下修改 ifcfg-eth0，添加如下信息：

```
DEVICE=eth0
HWADDR=00:0C:29:12:4D:30
```

```
TYPE=Ethernet
BRIDGE="br0"
ONBOOT=yes
NM_CONTROLLED=yes
BOOTPROTO=static
IPADDR=192.168.33.10
NETMASK=255.255.255.0
GATEWAY=192.168.33.1
```

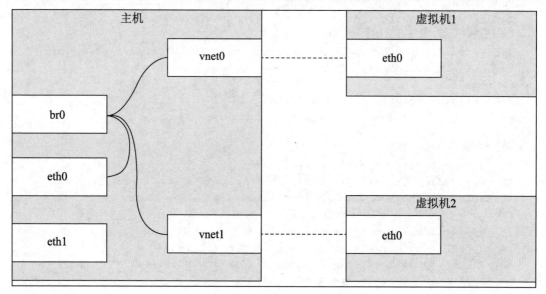

图 1-1　KVM 桥接网络结构图

（2）创建 ifcfg-br0 桥接网卡，内容如下：

```
DEVICE="br0"
HWADDR=00:0C:29:12:4D:30
BOOTPROTO=none
IPV6INIT=no
NM_CONTROLLED=no
ONBOOT=yes
TYPE="Bridge"
USERCTL=no
IPADDR=192.168.33.10
NETMASK=255.255.255.0
GATEWAY=192.168.33.1
```

KVM 虚拟机安装需要选择磁盘镜像的格式，通常有以下两种选择。

① Raw。Raw 格式是原始镜像，会直接作为一个块设备供虚拟机使用，I/O 性能比 Qcow2 高，不支持快照。

② Qcow2。它支持镜像快照、zlib 磁盘压缩、AES 加密等。

无论哪种格式,对磁盘的利用率来说都是一样的,因为实际占用的块数量都是一样的。但是 Raw 的虚拟机会比 Qcow2 的虚拟机 I/O 效率高一些,所以应根据实际应用环境选择磁盘镜像的格式。

(3)查看 KVM 虚拟机配置文件路径,配置如下:

```
[root@localhost ~]# ll /etc/libvirt/qemu/
drwxr-xr-x 2 root root 4096 12月 24 23:10 autostart
-rw------- 1 root root 2464 12月 20 20:11 centos01.xml
-rw------- 1 root root 2386 12月 27 18:42 centos10.xml
drwx------ 3 root root 4096 12月 10 11:11 networks
```

① autostart 目录为真实物理机开机,需要自动启动的各虚拟机的配置文件。
② networks 目录包含-虚拟网络(Virtbr0)用作 NAT 接口。

(4)KVM 创建虚拟机,Raw 格式创建命令如下:

```
virt-install --name=centos01 --ram 300 --vcpus=1 --disk path=/data/kvm/
centos01.img,size=4,bus=virtio --accelerate --cdrom /data/iso/centos7.6.
iso --vnc --vncport=5910 --vnclisten=0.0.0.0 --network bridge=br0,model=
virtio  --noautoconsole
```

(5)安装虚拟机之前先创建 ISO 镜像,最后通过 VNC 客户端连接,进行系统安装。创建方法如下:

```
nohup cp /dev/cdrom /data/iso/centos7.6.iso &
```

(6)KVM 创建虚拟机,Qcow2 格式创建命令如下:

```
qemu-img create -f qcow2 centos02.img 7G
virt-install --name=oelcentos02 --ram 512 --vcpus=1 --disk path=/data/
centos02.img,format=qcow2,size=7,bus=virtio --accelerate --cdrom /data/
iso/centos7.6.iso --vnc --vncport=5910 --vnclisten=0.0.0.0 --network
bridge=br0,model=virtio -noautoconsole  --no-acpi
```

(7)Virt-install 参数说明如下。

--name:指定虚拟机名称。

--ram:分配内存大小。

--vcpus:分配 CPU 核心数,最大与实体机 CPU 核心数相同。

--disk:指定虚拟机镜像,size 指定分配大小单位为 GB。

--network:网络类型,此处用的是默认,一般采用的应该是 Bridge 模式。

--accelerate:加速。

--cdrom:指定安装镜像 ISO。

--vnc:启用 VNC 远程管理,一般安装系统都要启用。

--vncport：指定 VNC 监控端口，默认端口为 5900，端口不能重复。

--vnclisten：指定 VNC 绑定 IP 地址，设置为 0.0.0.0。

--no-acpi：官方推荐使用--no-acpi 参数，原因是 QEMU/KVM 不太支持，可能造成 CPU 的占用偏高。但是 Windows 7 不支持--no-acpi 参数（ACPI 表示高级配置和电源管理接口）。

KVM 虚拟机可以用光盘安装，也可以使用 pxe 网卡安装（推荐），按 Ctrl+B 组合键，执行 autoboot 网卡启动。

（8）重启 Libvirtd 服务，命令如下：

```
/etc/init.d/libvirtd restart
```

（9）查看当前正运行的虚拟机，命令如下：

```
virsh -c qemu:///system list
virsh  list
```

（10）启动、停止、关闭、重启虚拟机，操作指令分别如下：

```
virsh start centos01
virsh stop centos01
virsh shutdown centos01
virsh reboot centos01
```

（11）删除虚拟机，操作指令如下：

```
virsh destroy centos01
virsh undefine centos01
rm -rf /data/kvm/centos01.img
```

（12）KVM 虚拟机克隆，操作指令如下：

```
nohup virt-clone -o centos01 -n centos02 -f /data/kvm/centos02.img&
```

① -o，表示旧的虚拟机名称。

② -n，表示新的虚拟机名称。

③ -f，表示新的虚拟机路径，克隆完毕后需要修改新虚拟机的 MAC 和 UUID。

安装完虚拟机，发现虚拟机竟然是 100MB 网络，传输速率很低，是什么导致的？又该如何解决呢？需要修改 vm01.xml 配置文件网卡段，添加如下代码，改为 e1000，即设置为千兆以太网卡，重新定义 XML 文件，重启虚拟机即可。操作代码如下：

```
<interface type='bridge'>
    <mac address='52:54:00:c2:84:c0'/>
    <source bridge='br0'/>
    <model type='e1000'/>
    <address type='pci' domain='0x0000' bus='0x00' slot='0x03' function='0x0'/>
</interface>
virsh define vm01.xml
```

1.6 KVM 虚拟化硬盘扩容

根据以上方法操作,即可成功创建 KVM 虚拟机,在企业生产环境中,有时需要对现有虚拟机镜像磁盘扩容,操作指令如下:

```
#创建 20GB 扩容镜像
qemu-img create -f raw add_centos01.img 20G
virsh attach-disk centos01 /data/kvm/add_centos01.img vdb --cache none
#查看虚拟机磁盘信息
fdisk -l
#格式化磁盘即可
mkfs.ext3 /dev/vdb
<disk type='file' device='disk'>
    <driver name='qemu' type='raw' cache='none'/>
    <source file='/data/kvm/add_centos01.img'/>
    <target dev='vdb' bus='virtio'/>
</disk>
```

1.7 KVM 虚拟机批量克隆实战

在企业生产环境中,同时需要多个 KVM 虚拟机的时候怎么办?如何快速生成多个虚拟机?最快的办法就是克隆虚拟机。

(1)操作方法一,单个虚拟机克隆命令如下:

```
virt-clone -o vm01 -n vm02 -f /data/kvm/vm02.img
```

(2)操作方法二,多个虚拟机克隆,脚本如下:

```
#!/bin/sh
#Auto Batch Add KVM Virtual
#Author jfedu 2021-6-17
#Define Variables
XML_DIR=/etc/libvirt/qemu/
KVM_DIR=/data/kvm/
COUNT1=`ls $XML_DIR|grep xml|tail -1 |sed 's/[^0-9]//g'`
COUNT2=`expr $COUNT1 + 1`
add_kvm()
{
cat <<eof
#VIR_NAME   RAM{M}     SIZE{g}
centos01    1024       2
centos02    512        3
centos03    1024       4
```

```
centos04    1024         5
centos05    1024         6
eof
echo "================================="
if
    [ ! -e kvm.txt ];then
    echo "The kvm.txt File does not exist,Please create ......"
    exit
fi
#Auto Create machines
if
      [ ! -s kvm.txt ];then
      echo "The kvm.txt is empty file,Please Refer Above Content ......"
      exit
fi
rm -rf kvm.txt.swp;cat kvm.txt|grep -v "#" >>kvm.txt.swp
while read line
do
    NAME=`echo $line |grep -v "^#" |awk '{print $1}'`
    RAM=`echo $line |grep -v "^#" |awk '{print $2}'`
    SIZE=`echo $line |grep -v "^#" |awk '{print $3}'`
    /usr/bin/virt-install --name=${NAME} --ram ${RAM} --vcpus=1 --disk path=/data/kvm/${NAME}.img,size=${SIZE},bus=virtio --accelerate --cdrom /data/iso/centos7.6.iso --vnc --vncport=-1 --vnclisten=0.0.0.0 -network bridge=br0,model-virtio --noautoconsole
done <kvm.txt.swp
}

clone_kvm()
{
    read -p "Please Enter you want create virtual machines count: " COUNT3
    COUNT4=`expr ${COUNT1} + ${COUNT3}`
    for i in `seq ${COUNT2} ${COUNT4}`
    do
    UUID=`/usr/bin/uuidgen`

    MAC=52\\:54\\:$(dd if=/dev/urandom count=1 2>/dev/null | md5sum | sed 's/^\(..\)\(..\)\(..\)\(..\).*$/\1\\:\2\\:\3\\:\4/')
        cd $XML_DIR;cp centos01.xml centos${i}.xml
        cd $XML_DIR;sed -i -e "/uuid/s:.*:<uuid>$UUID</uuid>:g" -e "s/centos01/centos${i}/g" -e "/<mac/s:.*:<mac address='$MAC'/>:g" centos${i}.xml
        cd $KVM_DIR;nohup cp centos01.img centos${i}.img &
        /usr/bin/virsh define $XML_DIR/centos${i}.xml
        echo "The Virtual machines Info:"
        echo "The centos${i} Virtual machines created successfull"
    done
```

```
}
delete_kvm()
{
    read -p "Please Enter you want deleted virtual machines :" name
    for i in 'echo ${name} |sed 's/ /\n/g''
    do
     /usr/bin/virsh destroy $i
     /usr/bin/virsh undefine $i;echo "The $i Virtual machines Deleted Successfully"
    done
}
case $1 in
    add_kvm )
    add_kvm
    ;;
    clone_kvm )
    clone_kvm
    ;;
    delete_kvm )
    delete_kvm
    ;;
    *)
    echo Usage: $0 "{add_kvm|clone_kvm|delete_kvm|help}"
    ;;
esac
```

1.8 ESXI 虚拟化技术概念

VMware 服务器虚拟化第一个产品命名为 ESX，后来 VMware 在第 4 版本时推出了 ESXI。ESXI 和 ESX 的版本最大的技术区别是内核的变化。

从第 4 版本开始 VMware 把 ESX 及 ESXi 产品统称为 vSphere，但是 VMware 从第 5 版本开始取消了原来的 ESX 版本，所以现在 VMware 虚拟化产品 vSphere 都是 ESXI。官方称为 vSphere 虚拟化技术，其实也可以称为 ESXI 虚拟化技术。

VMware、vSphere 是业界领先且最可靠的虚拟化平台。vSphere 将应用程序和操作系统从底层硬件中分离出来，从而简化了操作。现有的应用程序可以看到专有资源，而服务器则可以作为资源池进行管理。因此，具体业务将在简化但恢复能力极强的 IT 环境中运行。

VMware、vSphere、Essentials 和 Essentials Plus 套件专为工作负载不足 20 台服务器的 IT（Internet Technology，互联网技术）环境而设计，只需极少的投资即可通过经济高效的服务器整合和业务连续性为小型企业提供企业级 IT 管理。结合使用 vSphere Essentials Plus 与 vSphere Storage Appliance 软件，无须共享存储硬件即可实现业务连续性。

VMware ESXI 虚拟化有以下特点：
（1）确保业务连续性和始终可用的 IT。
（2）降低 IT 硬件和运营成本。
（3）提高应用程序质量。
（4）增强安全性和数据保护能力。

1.9　XEN 虚拟化技术概念

XEN 是由剑桥大学开发的，是一个基于 X86 架构、发展最快、性能最稳定、占用资源最少的开源虚拟化技术。XEN 可以在一套物理硬件上安全地运行多个虚拟机，与 Linux 是一个完美的开源组合。Novell SUSE Linux Enterprise Server 最先采用了 XEN 虚拟技术。它特别适用于服务器应用整合，可有效节省运营成本，提高设备利用率，最大化利用数据中心的 IT 基础架构。

实际上 XEN 出现的时间要早于 KVM 虚拟化。严格来讲，XEN 是一个开源的虚拟机监视器，属于半虚拟化技术，其架构决定了它注定不是真正的虚拟机，只是自己运行了一个内核的例子。

XEN 虚拟化，同时区分 XEN+pv+ 和 XEN+hvm，其中 pv 只支持 Linux，而 hvm 则支持 Windows 系统。除此之外，XEN 还拥有更好的可用资源、平台支持、可管理性、易实施、支持动态迁移和性能基准等优势。

第 2 章 企业级 Docker 虚拟化实战

Docker 是一款轻量级、高性能的虚拟化技术，是目前互联网使用最多的虚拟化技术。Docker 虚拟化技术的本质类似于集装箱机制。集装箱没有出现的时候，码头上有许多工人在搬运货物；集装箱出现以后搬运模式更加单一、更加高效，码头上更多的不是工人，而是集装箱。

将货物都打包在集装箱里面，可以防止货物之间相互影响。并且如果到了另外一个码头需要转运，有了集装箱以后，直接把它运送到另一个码头即可，完全可以保证里面的货物是整体的搬迁，并且不会损坏货物本身。

Docker 技术机制与集装箱类似。Docker 虚拟化是一个开源的应用容器引擎，让开发者可以打包他们的应用以及依赖包到一个可移植的容器中，然后发布到任何流行的 Linux 机器上，以实现虚拟化。

Docker 容器完全使用沙箱机制，相互之间不会有任何接口，几乎没有性能开销，可以很容易地在机器和数据中心中运行。最重要的是，它们不依赖于任何语言、框架或包括系统。

Docker 自开源后受到广泛的关注和讨论，以至于 dotCloud 公司后来都改名为 Docker Inc。

Redhat 已经在其 RHEL 6.5 中集中支持 Docker；Google 也在其 PaaS 产品中广泛应用，Docker 项目的目标是实现轻量级的操作系统虚拟化解决方案。

Docker 的基础是 Linux 容器（LXC）等技术。在 LXC 的基础上 Docker 进行了进一步的封装，让用户不需要去关心容器的管理，使操作更为简便。用户操作 Docker 的容器就像操作一个快速轻量级的虚拟机一样简单。

Docker 和传统虚拟化（KVM、XEN、Hyper-V、ESXI）结构的不同之处如图 2-1 所示。

（1）Docker 虚拟化技术概念总结如下。

Docker 虚拟化技术是在硬件的基础上，基于现有的操作系统层面上实现虚拟化，直接复用本地主机的操作系统，直接虚拟生成 Docker 容器。而 Docker 容器上部署相关的 App 应用（Apache、MySQL、PHP、Java）。

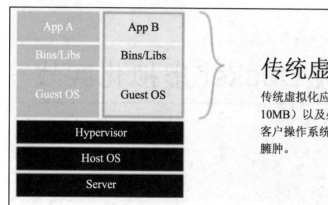

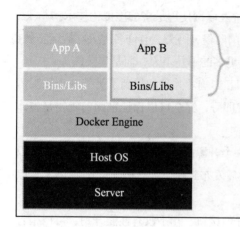

图 2-1 Docker 和传统虚拟化结构的差异

（a）传统虚拟化结构；（b）Docker 虚拟化结构

（2）传统虚拟化技术概念总结如下。

KVM、XEN、ESXI 等传统虚拟化（完全、半虚拟化）技术是在硬件的基础上，基于现有的操作系统实现虚拟化，但是不能复用本地主机的操作系统，而是必须虚拟出自己的 Guest OS 系统，然后在 Guest OS 系统上部署相关的 App 应用（如 Apache、MySQL、PHP、Java 等）。

Docker 虚拟化技术与传统虚拟化技术相比具有以下几种优点。

（1）操作启动快。

运行时性能可以获得极大提升，管理操作（启动、停止、开始、重启等）都是以 s 或 ms 为单位的。

（2）轻量级虚拟化。

可以拥有足够的"操作系统"，仅需添加或减少镜像即可。在一台服务器上可以部署 100～

1 000 个 Containers（容器）。但是利用传统虚拟化技术，虚拟 10～20 个虚拟机就不错了。

（3）开源免费。

Docker 虚拟化技术是开源的、免费的、低成本的，由现代 Linux 内核支持并驱动。注重轻量的 Container，可以在一个物理机上开启更多 "容器"，注定比传统虚拟化技术便宜。

（4）前景及云支持。

Docker 越来越受欢迎，包括各大主流公司都在推动 Docker 的快速发展，因为其性能有很大的优势。随着 Go 语言越来越被人熟知，Docker 的使用也越来越广泛。

2.1 虚拟化技术实现方式

（1）完全拟化技术。

通过软件实现对操作系统的资源再分配，比较成熟。完全虚拟化代表技术有 KVM、ESXI 和 Hyper-V。

（2）半虚拟化技术。

通过代码修改已有的系统，形成一种新的可虚拟化的系统，并调用硬件资源去安装多个系统，整体速度上相对较高。半虚拟化代表技术为 XEN。

（3）轻量级虚拟化。

介于完全虚拟化技术和半虚拟化技术之间。轻量级虚拟化代表技术为 Docker。

2.2 Docker LXC 及 Cgroup 原理剖析

Docker 虚拟化技术结构体系最早为 LXC（Linux Container）和 AUFS（Another Union File System）结构组合。Docker 0.9.0 版本开始引入 LibContainer，可以视作 LXC 的替代品。

LXC 也是一种虚拟化的解决方案。和 KVM、XEN、ESXI 虚拟化技术基于硬件层面虚拟化不同，LXC 是基于内核级的虚拟化技术，Linux 操作系统软件服务进程之所以能够相互独立，且系统能够控制每个服务进程的 CPU 和内存资源，也是得益于 LXC 容器技术。

0.9.0 版本之后的 Docker 虚拟化技术在 LXC 基础上进一步封装，比 LXC 技术更完善，并提供了一系列完整的功能。在 Docker 虚拟化技术中，LXC 主要负责资源管理，AUFS 主要负责镜像管理，而 LXC 又包括 Cgroup、NameSpace、Chroot 等组件，并通过 Cgroup 进行资源管理。从资源管理结构体系上来看，Docker、LXC、Cgroup 三者的关系如下：

Cgroup 在底层落实资源管理，LXC 在 Cgroup 上封装了一层，Docker 又在 LXC 封装了一层。要深入掌握 Docker 虚拟化技术，需要了解负责资源管理的 Cgroup 和 LXC 相关概念和用途。Docker、LXC、Cgroup、AUFS 结构如图 2-2 所示。

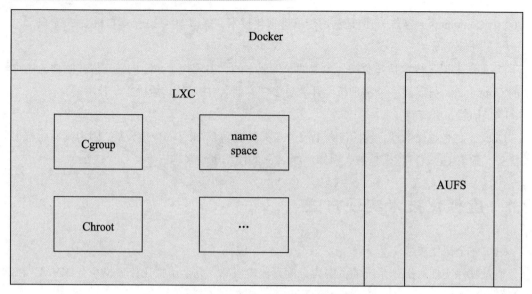

图 2-2 Docker 虚拟化内部结构

Cgroup 又名 Control group，是 Linux 内核提供的一种可以用于限制、记录、隔离进程组的物理资源（CPU、Memory、I/O、NET）的机制。

Cgroup 最初由 Google 的工程师提出，后来被整合进 Linux 内核。Cgroup 也是 LXC 为实现虚拟化所使用的资源管理手段。没有 Cgroup，就没有 LXC；没有 LXC，就没有 Docker。

Cgroup 最初的目标是为资源管理提供一个统一的框架，既整合现有的 Cpuset 等子系统，也为未来开发新的子系统提供接口。现在的 Cgroup 适用于多种应用场景，从单个进程的资源控制，到实现操作系统层次的虚拟化（OS Level Virtualization）。

LXC 可以提供轻量级的虚拟化，以便隔离进程和资源，且不需要提供指令解释机制及全虚拟化的其他复杂性。容器有效地将由单个操作系统管理的资源划分到相互独立的组中，以更好地在组间平衡相互冲突的资源使用需求。

LXC 建立在 Cgroup 基础上，可以理解为 LXC=Cgroup+ namespace + Chroot + veth +用户态控制脚本。

LXC 利用内核的新特性（Cgroup）提供用户空间的对象，用来保证资源的隔离和对应用或系统的资源控制。

典型的 Linux 文件系统由 bootfs 和 rootfs 两部分组成，bootfs（boot file system）主要包含 Bootloader 和 Kernel。Bootloader 主要用于引导加载 Kernel，当 Kernel 被加载到内存中后，bootfs 就被卸载。rootfs（root file system）包含的就是典型 Linux 系统中的/dev、/proc、/bin、/etc 等目录和文件，如图 2-3 所示。

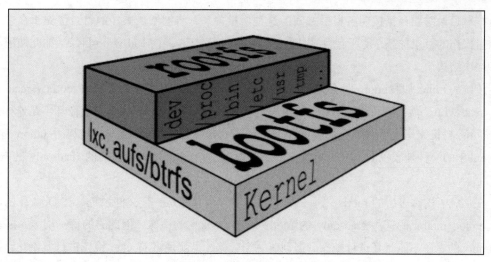

图 2-3　bootfs 和 rootfs 结构图

　　Docker 容器的文件系统最早是建立在 AUFS 基础上的。AUFS 是一种 UnionFS，简单来说就是支持将不同的目录挂载到同一个虚拟文件系统下，并实现一种 layer 的概念。由于 AUFS 未能加入 Linux 内核，考虑到兼容性问题，加入了 devicemapper 的支持。

　　Docker 虚拟化磁盘文件系统，默认使用 devicemapper 方式。Docker 虚拟化目前支持 6 种文件系统：AUFS、Btrfs、Device Mapper、OverlayFS、ZFS、VFS，如图 2-4 所示。

Technology（技术）	Storage driver name（存储驱动程序名）
OverlayFS	overlay
AUFS	aufs
Btrfs	btrfs
Device Mapper	devicemapper
VFS	vfs
ZFS	zfs

图 2-4　Docker 虚拟化支持的文件系统

2.3　AUFS 简介

　　AUFS 将挂载到同一虚拟文件系统下的多个目录分别设置成 read-only（只读）、read-write（读写）以及 whiteout-able 权限，对 read-only 目录只能读，而写操作只能实施在 read-write 目录中。重点在于，写操作是在 read-only 上的一种增量操作，不影响 read-only 目录。

挂载目录时要严格按照各目录之间的这种增量关系，将被增量操作的目录优先于在它基础上增量操作的目录挂载，待所有目录挂载结束后，继续挂载一个 read-write 目录，如此便形成了一种层次结构。

传统的 Linux 加载 bootfs 时会先将 rootfs 设为 read-only，在系统自检之后将 rootfs 从 read-only 改为 read-write，然后就可以在 rootfs 上进行写和读的操作了。但 Docker 的镜像却不是这样，它在 bootfs 自检完毕之后并不会把 rootfs 的 read-only 改为 read-write，而是利用 union mount（UnionFS 的一种挂载机制）将一个或多个 read-only 的 rootfs 加载到之前的 read-only 的 rootfs 层之上。

在加载了这么多层的 rootfs 之后，仍然让它看起来只像是一个文件系统，在 Docker 的体系里把 union mount 的这些 read-only 的 rootfs 叫作 Docker 的镜像。但此时的每一层 rootfs 都是 read-only 的，还不能对其进行操作。当创建一个容器，也就是将 Docker 镜像进行实例化，系统会在一层或多层 read-only 的 rootfs 之上分配一层空的 read-write 的 rootfs。一个完整的容器文件系统层级结构如图 2-5 所示。

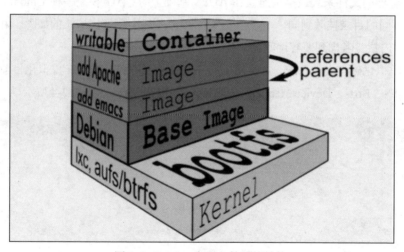

图 2-5　Docker 容器文件系统结构

2.4　Device Mapper 文件系统简介

Device Mapper 是 Linux 2.6 内核中支持逻辑卷管理的通用设备映射机制，它为实现用于存储资源管理的块设备驱动提供了一个高度模块化的内核架构。

Device Mapper 的内核体系架构如图 2-6 所示。

Device Mapper 在内核中通过一个一个模块化的 Target driver 插件实现对 I/O 请求的过滤或者重新定向等工作，当前已经实现的 Target driver 插件包括软 RAID、软加密、逻辑卷条带、多路径、镜像、快照等，图 2-6 中 linear、mirror、snapshot、multipath 表示的是 Target driver。

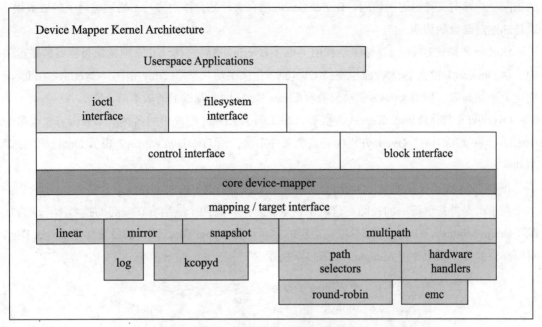

图 2-6　Device Mapper 内核架构

Device Mapper 进一步体现了 Linux 内核设计中策略和机制分离的原则，将所有与策略相关的工作放到用户空间完成，内核中主要提供完成这些策略所需要的机制。

Device Mapper 用户空间相关部分主要负责配置具体的策略和控制逻辑，比如逻辑设备和哪些物理设备建立映射、怎么建立这些映射关系等，而具体过滤和重定向 I/O 请求的工作由内核中相关代码完成。因此，整个 Device Mapper 机制由两部分组成——内核空间的 Device Mapper 驱动、用户空间的 Device Mapper 库及其提供的 Dmsetup 工具。

2.5　OverlayFS 简介

OverlayFS 是目前使用比较广泛的层次文件系统，是一种类似 AUFS 的堆叠文件系统，于 2014 年正式合并入 Linux 3.18 主线内核。OverlayFS 文件系统实现简单，且性能良好，可以充分利用不同或相同 Overlay 文件系统的 Page Cache，具有上下合并、同名遮盖、写时复制等特点。

Page Cache 也称为页缓冲或文件缓冲，由多个磁盘块构成，大小通常为 4KB，在 64 位系统上为 8KB，构成的几个磁盘块在物理磁盘上不一定连续，文件的组织单位为一页，也就是一个 Page Cache 大小，文件读取是由外存上不连续的几个磁盘块到 Buffer Cache，组成 Page Cache，然后供给应用程序。

Page Cache 用于在 Linux 读写文件时缓存文件的逻辑内容，从而加快对磁盘上映像和数据的访问。加速对文件内容的访问，Buffer Cache 缓存文件的具体内容——物理磁盘上的磁盘块，这是加速对磁盘的访问。

Docker 虚拟化 Overlay 存储驱动利用了很多 OverlayFS 特性构建和管理镜像与容器的磁盘结构。从 Docker 1.12 起，Docker 也支持 Overlay2 存储驱动。与 Overlay 相比，Overlay2 在 Inode 优化上更加高效。但 Overlay2 驱动只兼容 Linux Kernel 4.0 及以上的版本。

OverlayFS 加入 Linux Kernel 主线后，在 Linux Kernel 模块中的名称从 OverlayFS 改名为 Overlay。在实际用中，OverlayFS 代表整个文件系统，而 Overlay/Overlay2 表示 Docker 的存储驱动。

在 Docker 虚拟化技术中，OverlayFS 将一个 Linux 主机中的两个目录组合起来，一个在上，一个在下，对外提供统一的视图。这两个目录就是层，将两个层组合在一起的技术称作联合挂载（Union Mount）。在 OverlayFS 中，上层的目录称作 Upper Dir，下层的目录称作 Lower Dir，对外提供的统一视图称作 Merged，如图 2-7 所示。

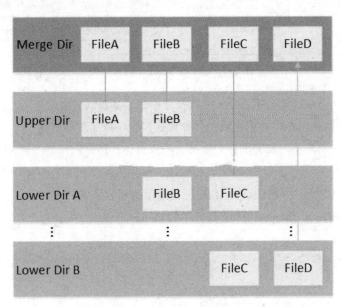

图 2-7　OverlayFS 结构图

当需要修改一个文件时，使用 Copy On Write（写时复制）将文件从只读的 Lower Dir 复制到可写的 Upper Dir 进行修改，结果也保存在 Upper Dir 层。在 Docker 中，底下的只读层是 Image，可写层是 Container。

如果镜像层和容器层可以有相同的文件，则 Upper Dir 中的文件将覆盖 Lower Dir 中的文件。Docker 镜像中的每一层并不对应 OverlayFS 中的层，而是/var/lib/docker/overlay 中的一个文件夹，文件夹以该层的 UUID 命名，然后使用硬链接将下层的文件引用到上层，在一定程度上可以节省磁盘空间。

容器和镜像的层与 OverlayFS 的 Upper Dir、Lower Dir、Merged Dir 之间的对应关系，如图 2-8 所示。

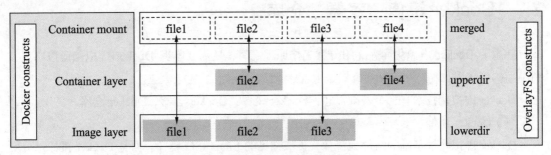

图 2-8　OverlayFS 文件系统结构

2.6　为什么使用 Docker

Docker 在以下几方面具有较大的优势。

（1）更快速的交付和部署。

Docker 在整个开发周期内都可以完美地辅助开发者实现快速交付。Docker 允许开发者在装有应用和服务本地容器做开发。可以直接集成到可持续开发流程中。

开发者可以使用一个标准的镜像构建一套开发容器，开发完成之后，运维人员可以直接使用这个容器部署代码。Docker 可以快速创建容器，快速迭代应用程序，并让整个过程全程可见，使团队中的其他成员更容易理解应用程序是如何创建和工作的。Docker 容器很轻很快，容器的启动时间是秒级的，可有效节约开发、测试、部署的时间。

（2）高效的部署和扩容。

Docker 容器几乎可以在任意平台上运行，包括物理机、虚拟机、公有云、私有云、个人计算机、服务器等。这种兼容性可以让用户把应用程序从一个平台直接迁移到另外一个平台。

Docker 的兼容性和轻量特性可以很轻松地实现负载的动态管理。可以快速扩容或方便地下线应用和服务，这种速度趋近实时。

（3）更高的资源利用率。

Docker 对系统资源的利用率很高，一台主机上可以同时运行数千个 Docker 容器。容器除了运行其中应用外，基本不消耗额外的系统资源，使得应用的性能很高，同时系统的开销尽量小。传统虚拟机方式运行 10 个不同的应用就要建 10 个虚拟机，而 Docker 只需要启动 10 个隔离的应用即可。

（4）更简单的管理。

使用 Docker，只需要小小的修改，就可以替代以往大量的更新工作。所有的修改都以增量的方式被分发和更新，从而实现自动化且高效的管理。

2.7 Docker 镜像、容器、仓库

熟悉了 Docker 虚拟化简介、组件和工作原理之后，还需要掌握 Docker 虚拟化镜像原理、引擎架构等知识。

Docker 虚拟化技术有三个基础概念：Docker 镜像、Docker 容器、Docker 仓库。

（1）Docker 镜像。

Docker 虚拟化最基础的组件为镜像，类似常见的 Linux ISO 镜像。但是 Docker 镜像是分层结构的，由多个层级组成，每个层级分别存储软件实现某个功能。Docker 镜像是静止的、只读的，不能对镜像进行写操作。

（2）Docker 容器。

Docker 容器是 Docker 虚拟化的产物，也是最早在生产环境中使用的对象。Docker 容器的底层是 Docker 镜像，是基于镜像运行，并在镜像最上层添加一层容器层之后的实体。容器层是可读、可写的，容器层如果需用到镜像层中的数据，可以通过 JSON 文件读取镜像层中的软件和数据，对整个容器进行修改。写操作只能作用于容器层，不能直接对镜像层进行写操作。

（3）Docker 仓库。

Docker 仓库是用于存放 Docker 镜像的地方。Docker 仓库分为两类，分别是公共仓库（Public）和私有仓库（Private）。国内和国外有很多默认的公共仓库，对外开放，免费或者付费使用。企业测试环境和生产环境推荐自建私有仓库，私有仓库的特点为安全、可靠、稳定、高效，能够根据自身的业务体系进行灵活升级和管理。

2.8 Docker 镜像原理剖析

完整的 Docker 镜像可以支撑一个 Docker 容器的运行，在 Docker 容器运行过程中主要提供文件系统数据支撑。Docker 镜像是分层结构的，由多个层级组成，每个层级分别存储各种软件实现某个功能。Docker 镜像作为 Docker 中最基本的概念，有以下几个特性。

（1）镜像是分层的，每个镜像都由一个或多个镜像层组成。

（2）可通过在某个镜像加上一定的镜像层得到新镜像。

（3）通过编写 Dockerfile 或基于容器 Commit 实现镜像制作。

（4）每个镜像层拥有唯一镜像 ID，Docker 引擎默认通过镜像 ID 识别镜像。

（5）镜像在存储和使用时，共享相同的镜像层，在 PULL 镜像时，已有的镜像层会自动跳过下载。

（6）每个镜像层都是只读，即使启动成容器，也无法对其进行真正的修改，修改只会作用于最上层的容器层。一个完整的 Docker 容器系统如图 2-9 所示。

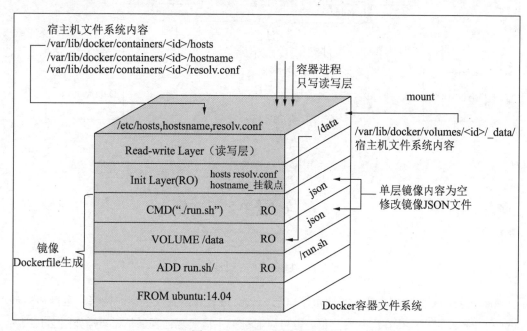

图 2-9　Docker 容器系统结构

Docker 容器是一个或多个运行进程，而这些运行进程将占有相应的内存、CPU 计算资源、虚拟网络设备及文件系统资源。Docker 容器所占用的文件系统资源，则通过 Docker 镜像的镜像层文件提供。基于每个镜像的 JSON 文件，可以通过解析 Docker 镜像的 JSON 的文件获知应该在这个镜像之上运行什么样的进程，应该为进程配置什么样的环境变量，而 Docker 守护进程实现了从静态向动态的转变。

Docker 虚拟化引也是一个 C/S（Client/Server）结构的应用，如图 2-10 所示。

Docker 虚拟化完整体系，包括以下几个组件。

（1）Docker Server 是一个常驻进程。

（2）REST API 实现了 Client 和 Server 之间的交互协议。

（3）Docker CLI 实现容器和镜像的管理，为用户提供统一的操作界面。

（4）Images 为容器提供了统一的软件、文件底层存储。

（5）Container 是 Docker 虚拟化的产物，直接作为生产使用。

（6）Network 为 Docker 容器提供完整网络通信。

（7）Volume 为 Docker 容器提供额外磁盘、文件存储对象。

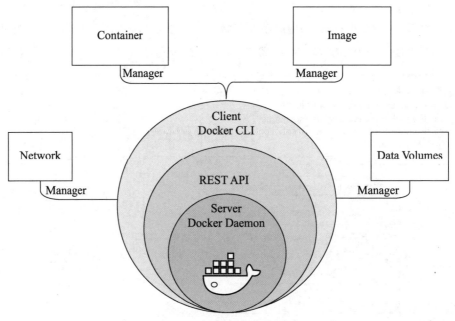

图 2-10　Docker C/S 引擎结构

Docker 使用 C/S 架构，Client 通过接口与 Server 进程通信实现容器的构建、运行和发布。Client 和 Server 可以运行在同一台集群，也可以通过跨主机实现远程通信，架构如图 2-11 所示。

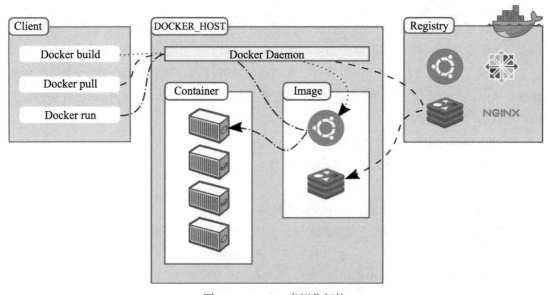

图 2-11　Docker 虚拟化拓扑

由图 2-11 可以清晰地看出 Docker 虚拟化整个生态体系。

（1）基于 Docker Client 客户端-Rest API-操作 Docker Daemon。

（2）Docker Daemon 部署至 Docker 宿主机（通常为硬件物理机）。

（3）基于 Docker pull 可以从 Registry 仓库获取各种镜像至 Docker Host（主机）。

（4）基于 Docker run 可以通过获取的镜像启动 Docker Container（容器）。

（5）基于 Docker build 可以构建满足企业需求的各种 Docker Image（镜像）。

2.9　CentOS 7.x（7.0+）Linux Docker 平台实战

掌握了 Docker 虚拟化概念和原理之后，最重要的就是要在生产环节中落地 Docker。Docker 虚拟化平台最早期只支持 Linux 操作系统，现在最新版 Windows 操作系统也支持 Docker 虚拟化。

本章节将选择不同的发行版本构建 Docker 虚拟化平台。Linux 操作系统主流发行版本包括 Red Hat Linux、CentOS、Ubuntu、SUSE Linux、Fedora Linux 等。以下简要介绍即将部署 Docker 虚拟化平台的两个系统：CentOS 和 Ubuntu。

Docker 官方要求 Linux 内核版本在 3.8+ 以上，生产环境中推荐使用 3.10+ 的 Linux 内核版本。Docker 从 1.13 版本起，采用时间线的方式作为版本号。Docker 版本现在基于 YY.MM，分为社区版（Community Edition）和企业版（Enterprise Edition）。社区版是免费提供给个人开发者和小型团体使用的，而企业版会提供额外的收费服务。

社区版按照 Stable 和 Edge 两种方式发布，每个季度更新 Stable 版本，如 17.06、17.09，每个月份更新 Edge 版本，如 17.09、17.10。

虚拟化和 Docker 虚拟化技术本质的用途：为了最大化地利用高配物理机的资源，提高硬件设备服务器的资源利用率，淘汰一些老、旧服务器，对老、旧服务器进行资源的重组、重用，满足企业快速发展，虚拟化落地实施硬件设备应尽量使用高配物理机资源，参考配置如下。

① 服务器品牌：Dell R730、R820。

② CPU 配置：Intel 至强 E5-2600 系列。

③ MEM 配置：ECC DDR3 256GB。

④ DISK 配置：SAS 12TB（最大支持 24TB）。

⑤ 网络配置：Intel 四端口千兆网卡/双端口万兆网卡。

（1）安装步骤和命令如下：

```
#安装国内阿里源
wget -P /etc/yum.repos.d/ http://mirrors.aliyun.com/docker-ce/linux/
```

```
centos/docker-ce.re po
#安装Docker-CE版本
yum install docker-ce* -y
#检查Docker版本是否安装
rpm -qa|grep -E "docker"
#启动Docker引擎服务
service docker restart
systemctl restart docker.service
#查看Docker服务进程
ps -ef|grep docker
```

（2）安装完成，如图2-12所示。

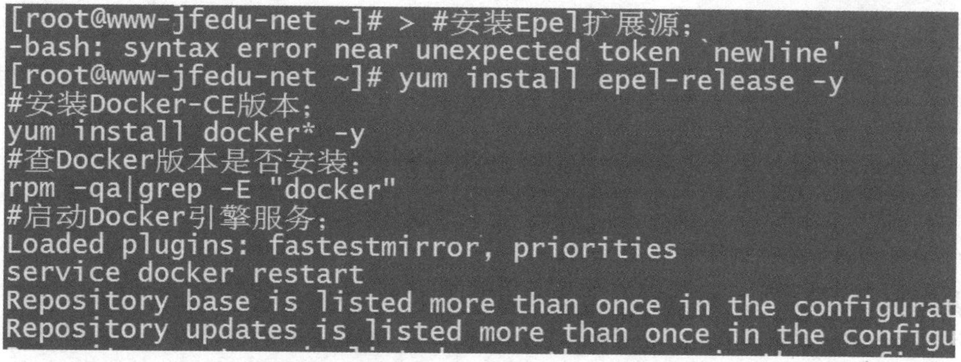

图2-12　Docker安装图解

（3）查看启动进程，如图2-13所示。

图2-13　Docker服务进程

（4）查看Docker基础信息，如图2-14所示。
（5）从Docker仓库下载Nginx镜像，如图2-15所示。

```
[root@jfedu ~]# docker info|more
Client: Docker Engine - Community
 Version:    26.0.0
 Context:    default
 Debug Mode: false
 Plugins:
  buildx: Docker Buildx (Docker Inc.)
```

图 2-14　Docker 基础信息

```
[root@www-jfedu-net ~]#
[root@www-jfedu-net ~]# docker pull docker.io/nginx
Using default tag: latest
Trying to pull repository docker.io/library/nginx ...
latest: Pulling from docker.io/library/nginx
a5a6f2f73cd8: Extracting 6.652 MB/22.49 MB
a5a6f2f73cd8: Pull complete
67da5fbcb7a0: Pull complete
e82455fa5628: Pull complete
Digest: sha256:31b8e90a349d1fce7621f5a5a08e4fc519b634f7
Status: Downloaded newer image for docker.io/nginx:late
[root@www-jfedu-net ~]#
[root@www-jfedu-net ~]#
```

图 2-15　Docker 下载 Nginx 镜像

2.10　CentOS 8.x（8.0+）Linux Docker 平台实战

（1）基于 CentOS 8.x Linux 操作系统，从零开始构建一套 Docker 虚拟化平台，使用二进制 Tar 包方式，部署的方法和步骤如下：

```
#从官网下载Docker软件包
ls -l docker-26.0.0.tgz
#通过Tar工具对其解压缩（-x 表示extract 解压,-z gzip 表示压缩格式,-v verbose 表示
#详细显示,-f file 表示文件属性）
tar -xzvf docker-26.0.0.tgz
#创建Docker程序部署目录/usr/local/docker/
mkdir -p /usr/local/docker/
#将解压后的Docker程序移动至部署目录
\mv docker/* /usr/local/docker/
#查看Docker程序是否部署成功
ls -l /usr/local/docker/
#创建Docker用户和组,同时将Docker部署目录加入PATH环境变量
useradd -s /sbin/nologin docker -M
cat>>/etc/profile<<EOF
export PATH=\$PATH:/usr/local/docker/
```

```
EOF
#使其 PATH 环境变量生效
source /etc/profile
#启动 Docker 引擎服务
nohup /usr/local/docker/dockerd &
#查看 Docker 进程状态
ps -ef|grep -aiE docker
#查看 Docker 版本信息
/usr/local/docker/docker version
docker version
```

（2）根据以上 Docker 平台部署指令，Docker 平台部署成功，查看其版本信息，如图 2-16 所示。

```
[root@jfedu ~]# docker version
Client: Docker Engine - Community
 Version:           26.0.0
 API version:       1.45
 Go version:        go1.21.8
 Git commit:        2ae903e
 Built:             Wed Mar 20 15:21:09 2024
```

图 2-16 Docker 版本查看

2.11 Ubuntu（16.04+）Linux Docker 平台实战

（1）安装步骤和命令如下：

```
#更新 apt 源
apt-get update
#apt 源使用 HTTPS 以确保软件下载过程中不被篡改。需要添加使用 HTTPS 传输的软件包以及
#CA 证书
apt-get install \
apt-transport-https \
ca-certificates \
curl \
software-properties-common
#默认 apt 访问国外源，网络非常慢，此处建议使用国内源，需要添加软件源的 GPG 密钥
curl -fsSL https://mirrors.ustc.edu.cn/docker-ce/linux/ubuntu/gpg | sudo apt-key add -
#curl -fsSL https://download.docker.com/linux/ubuntu/gpg | sudo apt-key
#add -
#向 source.list 中添加 Docker 软件源
add-apt-repository \
"deb [arch=amd64] https://mirrors.ustc.edu.cn/docker-ce/linux/ubuntu
$(lsb_release -cs) stable"
```

第 2 章 企业级 Docker 虚拟化实战

```
#官方源
#add-apt-repository "deb [arch=amd64] https://download.docker.com/linux/
#ubuntu $(lsb_release -cs) stable"
```

（2）Docker 安装操作方法和步骤如下：

```
#更新 apt 软件包缓存
apt-get update
#基于 apt-get 安装 docker-ce 社区版本
apt-get install docker-ce
#查 Docker 版本是否安装
dpkg -s docker-ce
#启动 Docker 引擎服务
service docker restart
#查看 Docker 服务进程
ps -ef|grep docker
```

（3）安装完成，如图 2-17 所示。

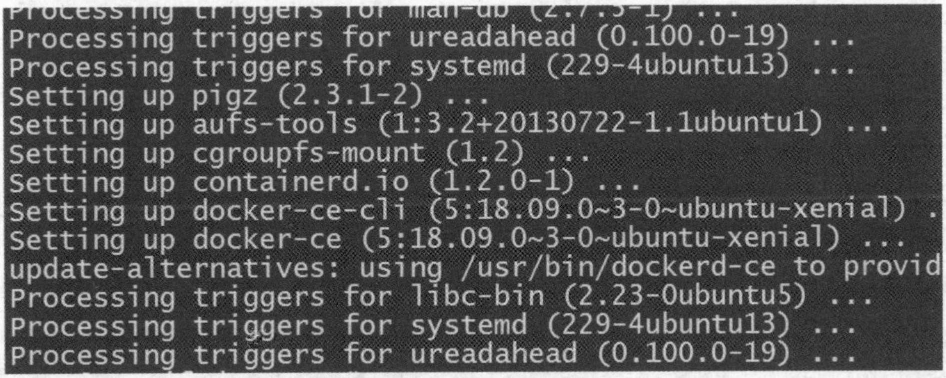

```
Processing triggers for man-db (2.7.5-1) ...
Processing triggers for ureadahead (0.100.0-19) ...
Processing triggers for systemd (229-4ubuntu13) ...
Setting up pigz (2.3.1-2) ...
Setting up aufs-tools (1:3.2+20130722-1.1ubuntu1) ...
Setting up cgroupfs-mount (1.2) ...
Setting up containerd.io (1.2.0-1) ...
Setting up docker-ce-cli (5:18.09.0~3-0~ubuntu-xenial) .
Setting up docker-ce (5:18.09.0~3-0~ubuntu-xenial) ...
update-alternatives: using /usr/bin/dockerd-ce to provid
Processing triggers for libc-bin (2.23-0ubuntu5) ...
Processing triggers for systemd (229-4ubuntu13) ...
Processing triggers for ureadahead (0.100.0-19) ...
```

图 2-17 Ubuntu 安装 Docker

（4）查看启动进程和 Docker 基础信息，如图 2-18 所示。

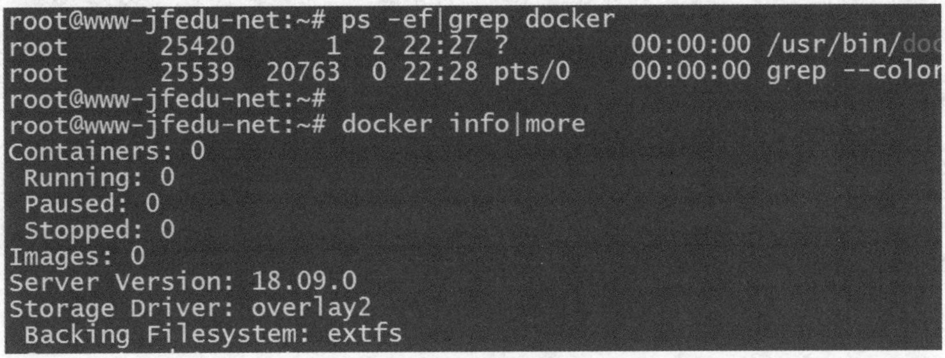

```
root@www-jfedu-net:~# ps -ef|grep docker
root      25420     1  2 22:27 ?        00:00:00 /usr/bin/do
root      25539 20763  0 22:28 pts/0    00:00:00 grep --colo
root@www-jfedu-net:~#
root@www-jfedu-net:~# docker info|more
Containers: 0
 Running: 0
 Paused: 0
 Stopped: 0
Images: 0
Server Version: 18.09.0
Storage Driver: overlay2
 Backing Filesystem: extfs
```

图 2-18 Ubuntu 查看 Docker 信息

（5）从 Docker 仓库下载 Nginx 镜像，如图 2-19 所示。

```
root@www-jfedu-net:~# docker pull docker.io/nginx
Using default tag: latest
latest: Pulling from library/nginx
a5a6f2f73cd8: Pull complete
67da5fbcb7a0: Pull complete
e82455fa5628: Pull complete
Digest: sha256:31b8e90a349d1fce7621f5a5a08e4fc519b634f
Status: Downloaded newer image for nginx:latest
root@www-jfedu-net:~# docker images
REPOSITORY          TAG              IMAGE ID
nginx               latest           e81eb098537d
root@www-jfedu-net:~#
```

图 2-19　Docker 下载 Nginx 镜像

2.12　Docker 仓库源更新实战

Docker 默认连接的国外官方镜像，根据网络情况不同，通常访问时快时慢，大多时候获取速度非常慢，为了提升效率可以自建仓库或先修改为国内仓库源，提升拉取镜像的速度。

Docker 可以配置的国内镜像有很多，例如 Docker 中国区官方镜像、阿里云、网易蜂巢、DaoCloud 等，这些都是国内比较快的镜像仓库。

从国外官网下载 Docker Tomcat 镜像，访问速度慢，如图 2-20 所示。

```
[root@www-jfedu-net ~]# docker pull tomcat
Using default tag: latest
Trying to pull repository docker.io/library/tomcat ...
latest: Pulling from docker.io/library/tomcat
cc1a78bfd46b: Downloading 457.9 kB/45.32 MB
6861473222a6: Pulling fs layer
7e0b9c3b5ae0: Downloading 1.747 MB/4.336 MB
ae14ee39877a: Waiting
8085c1b536f0: Waiting
6e1431e84c0c: Waiting
ca0e3df5a1fd: Waiting
d2cb611ced6c: Waiting
268dc3e43e66: Waiting
79a7e8d254c7: Waiting
5c848af92738: Waiting
789b92e37607: Waiting
```

图 2-20　Docker 下载 Tomcat 镜像

Docker 镜像修改方法为，在命令行输入 vim /etc/docker/daemon.json，执行如下命令：

```
cat>/etc/docker/daemon.json<<EOF
{
"registry-mirrors":["https://registry.docker-cn.com"]
}
EOF
service docker restart
```

重启 Docker 服务即可。修改仓库地址为国内仓库后，获取镜像速度非常快，如图 2-21 所示。

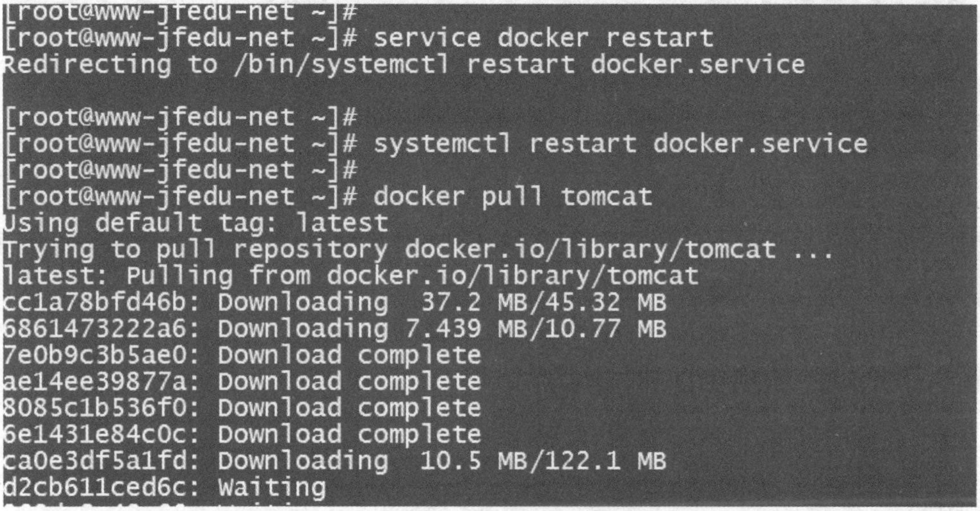

```
[root@www-jfedu-net ~]#
[root@www-jfedu-net ~]# service docker restart
Redirecting to /bin/systemctl restart docker.service
[root@www-jfedu-net ~]#
[root@www-jfedu-net ~]# systemctl restart docker.service
[root@www-jfedu-net ~]#
[root@www-jfedu-net ~]# docker pull tomcat
Using default tag: latest
Trying to pull repository docker.io/library/tomcat ...
latest: Pulling from docker.io/library/tomcat
cc1a78bfd46b: Downloading  37.2 MB/45.32 MB
6861473222a6: Downloading 7.439 MB/10.77 MB
7e0b9c3b5ae0: Download complete
ae14ee39877a: Download complete
8085c1b536f0: Download complete
6e1431e84c0c: Download complete
ca0e3df5a1fd: Downloading  10.5 MB/122.1 MB
d2cb611ced6c: Waiting
```

图 2-21　Docker 下载 Tomcat 镜像

第 3 章　Docker 企业命令实战

Docker 虚拟化平台部署完成后，默认没有图形界面管理，运维人员、测试人员、开发人员需要通过 Docker-Client 命令行操作。以下为 Docker 平台下 30+ 操作指令。熟悉指令的操作能够帮助我们对 Docker 进行高效的管理和维护。

3.1　Docker search 命令实战

Docker search 命令，通常用于从外部仓库或内部仓库中搜索镜像，其后接镜像的名称。命令案例如下：

```
#从 Docker 仓库中搜索 Nginx 镜像
docker search nginx
#从 Docker 仓库中搜索 Tomcat 镜像
docker search tomcat
```

3.2　Docker pull 命令实战

Docker pull 命令，通常用于从外部仓库或内部仓库中下载镜像，根据自身的需求下载，其后接镜像的名称。命令案例如下：

```
#从 Docker 仓库下载 Nginx 镜像
docker pull docker.io/nginx
#从 Docker 仓库下载 Tomcat 镜像
docker pull docker.io/tomcat
```

3.3　Docker images 命令实战

Docker images 命令，通常用于查看 Docker 宿主机本地镜像列表。命令案例如下：

```
#查看已下载的本地 Docker 镜像列表
Docker images
#可以查看具体镜像
Docker images nginx
```

3.4 Docker run 命令实战

Docker run 命令,通常用于创建并启动新容器。命令案例如下:

```
#基于 Docker run 启动 Nginx 镜像
Docker run -itd docker.io/nginx /bin/bash
#i 表示 interactive 交互
#t 表示 tty 终端
#d 表示 daemon 后台启动
#基于 Docker run 启动 Nginx 镜像,映射本地 80 端口至容器 80 端口
Docker run -p 80:80 -itd docker.io/nginx /bin/bash
#-p 端口映射,第一个 80 宿主机监听端口,第二个 80 端口为容器监听
```

3.5 Docker ps 命令实战

Docker ps 命令,通常用于查看已创建容器的运行状态,可以支持查看所有创建的容器。命令案例如下:

```
#查看当前正在运行的容器
docker ps
#查看当前 Linux 系统所有容器,包括运行中的和已经停止的容器
docker ps -a
```

3.6 Docker inspect 命令实战

Docker inspect 命令,通常用于查看已创建容器的详细信息,包括容器的 ID、创建时间、资源配置、网络信息等。命令案例如下:

```
#查看容器详细信息
docker inspect 55e339c80051
#查看容器详细信息,并从信息中过滤 IP 地址
docker inspect 55e339c80051|grep -i ipaddr
```

3.7 Docker exec 命令实战

Docker exec 命令,通常用于进入已创建的容器系统,也可以在 Docker 宿主机远程执行容器

内部命令。命令案例如下：

```
#在 Docker 中容器运行指令 df -h
docker exec 55e339c80051 df -h
#在 Docker 中容器/tmp 目录下创建 jfedu.txt 文件
docker exec 55e339c80051 touch /tmp/test.txt
#进入 Docker 容器/bin/bash 终端,然后执行 df -h 命令
docker exec -it 55e339c80051 /bin/bash
df -h
```

3.8 Docker stop|start 命令实战

Docker stop|start 命令，通常用于停止、启动容器。命令案例如下：

```
#停止正在运行中的容器
docker stop 55e339c80051
#启动已经停止的容器
docker start 55e339c80051
```

3.9 Docker rm 命令实战

Docker rm 命令，通常用于删除已创建的容器，可以删除已经停用的容器，也可以删除正在运行的容器。命令案例如下：

```
#删除某个已经停止的 Docker 容器
docker rm dc455c12ca7d
#强制删除某个运行中的 Docker 容器
docker rm -f 55e339c80051
```

3.10 Docker rmi 命令实战

Docker rmi 命令，通常用于删除已下载的镜像，但是不能删除已创建的容器所需的镜像，除非先删除容器，然后再删除镜像。命令案例如下：

```
#从 Docker images 列表中删除某个镜像
docker rmi 78b258e36eed
#从 Docker images 列表中删除多个镜像
docker rmi e81eb098537d 415381a6cb81
```

3.11 Docker 虚拟化 30 多个命令实战剖析

熟悉命令的操作能够帮助我们对 Docker 进行高效的管理和维护。Docker 平台下 30 多个命令详解如表 3-1 所示。

表 3-1 Docker 命令详解

命 令	详 解
search	在 Docker Hub 中搜索镜像
pull	拉取指定镜像或者库镜像
push	推送指定镜像或者库镜像至 Docker 源服务器
history	展示一个镜像形成历史
images	列出系统当前镜像
run	创建一个新的容器并运行一个命令
start	启动容器
stop	停止容器
attach	当前 Shell 下 attach 连接指定运行镜像
build	通过 Dockerfile 定制镜像
commit	提交当前容器为新的镜像
cp	从容器中复制指定文件或者目录到宿主机中
create	创建一个新的容器,同 run,但不启动容器
diff	查看 Docker 容器变化
events	从 Docker 服务获取容器实时事件
exec	在已存在的容器上运行命令
export	导出容器的内容一个压缩归档文件(对应 import)
import	从压缩文件中的内容创建一个新的文件系统映像(对应 export)
info	显示系统相关信息
inspect	查看容器详细信息
kill	指定 Docker 容器
load	从一个压缩文件中加载一个镜像(对应 save)
login	注册或登录一个 Docker 源服务器
logout	退出登录
logs	输出当前容器日志信息
port	查看映射端口对应的容器内部源端口
pause	暂停容器

（续表）

命　令	详　解
ps	列出容器列表
restart	重启运行的容器
rm	移除一个或多个容器
rmi	移除一个或多个镜像
save	保存一个镜像为一个压缩文件（对应load）
tag	给源中镜像打标签
top	查看容器中运行的进程信息
unpause	取消暂停容器
version	查看Docker版本号
wait	截取容器停止时的退出状态值

第 4 章 Docker 网络原理实战

Docker 虚拟化技术底层基于 LXC+Cgroups+AUFS（Overlay）技术实现，而 Cgroups 是 Linux 内核提供的一种可以限制、记录、隔离进程组（Process Groups）所使用的物理资源的机制。

Docker 虚拟化的产物是 Docker 容器，基于 Docker Engine 启动容器时，默认会给容器指定和分配各种子系统，如 CPU 子系统、Memory 子系统、I/O 子系统、NET 子系统等。

启动一个容器，会为 Network Namespace（子系统）提供一份独立的网络环境，包括网卡、路由、Iptables 规则等，容器与其他容器的 Network Namespace 是相互隔离的。

通过 Docker run 命令创建 Docker 容器时，可以使用 --net 选项指定 Docker 容器的网络模式，Docker 默认有 4 种网络模式。

（1）Host 模式，使用 --net=host 指定。
（2）Container 模式，使用 --net=container:NAME_or_ID 指定。
（3）None 模式，使用 --net=none 指定。
（4）Bridge 模式，使用 --net=bridge 指定，默认设置。

4.1 Host 模式剖析

通常来讲，启动新的 Docker 容器，都会分配独立的 Network Namespace 隔离子系统；如果在运行时指定为 Host 模式，那么 Docker 容器将不会获得一个独立的 Network Namespace，而是和宿主机共用一个 Network Namespace 子系统。

新创建的 Docker 容器不会创建自己的网卡，不会再虚拟出自己的 IP、网关、路由等信息，而是和宿主机共享 IP 和端口等信息，其他软件、目录还是相互独立的。两个容器除了网络方面相同之外，其他如文件系统、进程列表等还是相互隔离的。

4.2 Container 模式剖析

Container 模式是指定新创建的 Docker 容器和已存在的某个 Docker 容器共享一个 Network Namespace 子系统，而不是和宿主机共享。

新创建的 Docker 容器不会创建自己的网卡，不会再虚拟出自己的网卡、IP、网关、路由等信息，而是和指定的 Docker 容器共享 IP 和端口等信息，其他软件、目录还是相互独立的。两个容器除了网络方面相同之外，其他如文件系统、进程列表等还是相互隔离的。如果依附的 Docker 容器关闭，新的 Docker 容器网络也会丢失。

4.3 None 模式剖析

None 模式与其他模式都不同。如果 Docker 容器使用 None 模式，Docker 容器会拥有自己的 Network Namespace 子系统，但是 Docker 引擎并不会为新启动的 Docker 容器配置任何的网络信息。

即新创建的 Docker 容器不会虚拟出自己的网卡、IP、网关、路由等信息，而是需要手动为 Docker 容器添加这些信息。在企业实战环境中，通常会使用 Pipework 工具为 Docker 容器指定 IP 等信息。

4.4 Bridge 模式剖析

Docker 容器的 Bridge 模式也是 Docker 默认的网络模式。该模式会为每个容器分配 Network Namespace 子系统，会自动给每个容器虚拟出自己的网卡、IP、网关、路由等信息，无须手动添加。

默认创建的 Docker 容器会统一通过一对 veth 虚拟网卡连接到一个虚拟网桥交换机 Docker0 上，所有的容器的网络加入一个二层交换机网络，即同一宿主机的所有容器都是可以相互联通和访问的。

4.5 Bridge 模式原理剖析

默认 Docker 引擎启动会在本地生成一个 Docker0 虚拟网卡。Docker0 是一个标准 Linux 虚拟网桥设备。在 Docker 默认的桥接网络工作模式中，Docker0 网桥起至关重要的作用。物理网桥是标准的二层网络设备，标准物理网桥只有两个网口，可以将两个物理网络连接在一起。

但与物理层设备集线器等相比,网桥具备隔离冲突域的功能。网桥通过 MAC 地址学习和泛洪的方式实现二层相对高效的通信。随着技术的发展,标准网桥设备已经基本被淘汰了,替代网桥的是二层交换机。二层交换机也可以看成一个多口网桥。

Docker 容器采用 Bridge 模式结构,如图 4-1 所示。

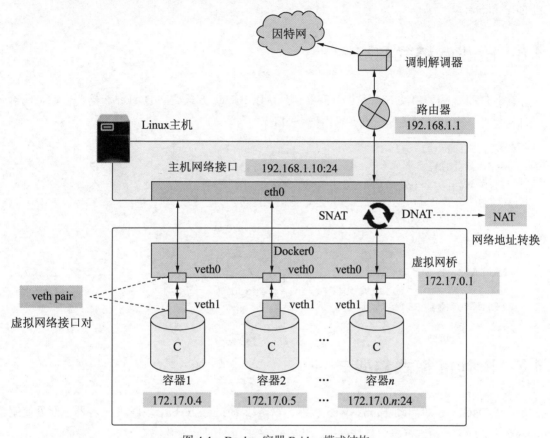

图 4-1　Docker 容器 Bridge 模式结构

Docker Bridge 模式创建过程如下。

(1)启动 Docker 容器,指定模式为桥接模式时(默认模式),Docker 引擎会创建一对虚拟网卡 veth 设备,veth 设备总是成对出现,组成一个数据的通道,数据从一个设备进入,就会从另一个设备出来。veth 设备常用来连接两个网络设备,可以把 veth 接口对认为是虚拟网线的两端。

(2)veth 设备的另外一端放在新创建的容器中,命名为 eth0;然后将另外一块设备放在宿主机中,以类似 vethxxx 的名称命名,并将这个网络设备加入 Docker0 网桥。

(3)Docker 引擎会从 Docker0 子网中动态分配一个新的 IP 给容器使用,并设置 Docker0 的

IP 地址为容器的默认网关。

（4）新创建的容器与宿主机能够通信，宿主机也可以访问容器中的 IP 地址。在 Bridge 模式下，连在同一网桥（交换机）上的容器之间可以相互通信，同时容器也可以访问外网（基于 iptables SNAT）。但是其他物理机不能访问 Docker 容器 IP，需要通过 NAT 将容器 IP 的 port 映射为宿主机的 IP 和 port。

4.6 Bridge 模式实战一

基于 Docker 引擎启动 Nginx Web 容器，默认以 Bridge 方式启动 Docker 容器，会动态地给 Docker 容器分配 IP、网关等信息，操作指令如下：

```
#查看镜像列表
docker images
#运行新的 Nginx 容器
docker run -itd docker.io/nginx:latest
#查看启动的 Nginx 容器
docker ps

#查看 Nginx 容器的 IP 地址
docker inspect 510ea29c39f6|grep -i ipaddr
#访问 Nginx 容器 80 端口服务
curl -I http://172.17.0.2/
```

4.7 Bridge 模式实战二

基于 Docker 引擎启动 Nginx Web 容器，默认以 Bridge 方式启动 Docker 容器，此处使用 pipework 工具手动给容器指定桥接网卡，并手动配置 IP 地址。操作指令如下：

```
#查看镜像列表
docker images
#运行新的 Nginx 容器
docker run -itd --net=none docker.io/nginx:latest
#查看启动的 Nginx 容器
docker ps
#查看 Nginx 容器的 IP 地址（没有 IP 地址）
docker inspect 265a3745752e|grep -i ipaddr
#安装 pipework IP 配置脚本工具,方法如下
#安装 pipework
git clone https://github.com/jpetazzo/pipework
```

```
cp ~/pipework/pipework /usr/local/bin/
#查看pipework工具是否配置正确
pipework -h
#pipework工具手动指定容器的IP,并设置容器为桥接方式上网,命令如下（docker0为网桥
#名称,172.17.0.18/16为容器IP和掩码,172.17.0.1为容器网关）
pipework docker0 265a3745752e 172.17.0.18/16@172.17.0.1
ping 172.17.0.18 -c 2
curl -I http://172.17.0.18/
```

4.8 Bridge 模式实战三

基于 Docker 引擎启动 Nginx Web 容器，默认以 Bridge 方式启动 Docker 容器；Docker0 的网桥 IP 为 172.17.0.0/16 网段，可以通过指令修改 Docker 网桥的 IP 网段。例如，将网桥 IP 段修改为 10.10.0.1/16 段，操作指令如下：

```
#删除原有网络信息
service docker stop
ip link set dev docker0 down
brctl delbr docker0
iptables -t nat -F POSTROUTING
#添加新的Docker0网络信息
brctl addbr docker0
ip addr add 10.10.0.1/16 dev docker0
ip link set dev docker0 up

#配置Docker的文件
cat>/etc/docker/daemon.json<<EOF
{"registry-mirrors": ["http://docker-cn.docker.com"],
"bip": "10.10.0.1/16"
}
EOF
#启动新的Docker容器,查看容器桥接网络IP地址
docker run -itd docker.io/nginx:latest
docker inspect 72fec5ccdf73|grep -i ipaddr
```

4.9 Bridge 模式实战四

基于 Docker 引擎启动 Nginx Web 容器，默认以 Bridge 方式启动 Docker 容器；Docker0 的网桥 IP 为 172.17.0.0/16 网段，默认局域网的其他物理机是不能直接访问 Docker 容器的。

为了实现 Docker 容器与局域网通信，并实现局域网其他的物理机也可以访问容器的 IP（不

配置 NAT 映射），可以自定义桥接网络 br0，将 br0 与物理网卡 eth0 或 ens33 桥接。操作方法如下：

```
#添加 ens33 网卡指定 bridge 桥接网卡名称 br0
cd /etc/sysconfig/network-scripts/
#配置 ifcfg-ens33 网卡
cat>ifcfg-ens33<<EOF
TYPE="Ethernet"
DEVICE="ens33"
ONBOOT="yes"
BRIDGE="br0"
IPADDR=10.0.0.122
NETMASK=255.255.255.0
GATEWAY=192.168.0.1

EOF

#配置 ifcfg-br0 网卡
cat>ifcfg-br0<<EOF
DEVICE="br0"
BOOTPROTO=static
ONBOOT=yes
TYPE="Bridge"
IPADDR=10.0.0.122
NETMASK=255.255.255.0
GATEWAY=192.168.0.1
EOF
#重启 network 网络服务
service network restart
#修改 Docker 引擎,使用 br0 网桥
cat /etc/sysconfig/docker-network
DOCKER_NETWORK_OPTIONS="-b=br0"
#安装&部署 pipework 工具
yum install -y git
git clone https://github.com/jpetazzo/pipework
cp ~/pipework/pipework /usr/local/bin/

#启动 Docker 容器,设置为 none 模式,然后使用 br0 网桥,指令如下
#(br0 为网桥名称,192.168.0.11/24 为容器 IP 和掩码,10.0.0.122 为容器网关)

docker run -itd --net=none --name=nginx-v1 docker.io/nginx
pipework br0 nginx-v1 192.168.0.11/24@10.0.0.122
```

4.10 Docker 持久化固定容器 IP

基于 Docker 引擎创建 Docker 容器，在默认条件下创建容器是 Bridge 模式。启动容器 IP 地址是 DHCP 随机分配且为递增的，容器之间可以互相通信，网段也是固定的。

Docker 容器一旦关闭再次启动，就会导致容器的 IP 地址再次随机分配，而部分容器在部署的时候是不需要互相通信的，所以应使用固态 IP，保证想要通信的容器在同一网段，且容器重启之后 IP 地址也不会随之改变。

根据 4.9 节的 Pipework 脚本可以给 Docker 容器配置固定 IP 地址，但是重启也会丢失 IP 地址，有没有方法实现重启容器 IP 也不丢失呢？持久化固定 IP 地址操作方法如下。

（1）安装桥接工具和 Docker-py 程序，操作指令如下：

```
#安装 Docker-py 程序
#pip install docker-py
yum install python-docker*
#安装桥接扩展包
yum install bridge-utils -y
```

（2）从 Github 仓库下载 Docker-static-ip 固定 IP 的脚本，操作指令如下：

```
#下载 docker-static-ip 脚本
git clone https://github.com/lioncui/docker-static-ip
#部署 docker-static-ip 程序
mv docker-static-ip /usr/local/
#启动 Docker 引擎服务
systemctl start docker.service
#后台启动 duration 脚本
cd /usr/local/docker-static-ip/
python duration.py
#查看 Python 脚本进程
ps -ef|grep -aiE duraion
```

（3）新增配置 br0 桥接网络。

vi ifcfg-ens33 内容修改如下：

```
cat>/etc/sysconfig/network-scripts/ifcfg-ens33 <<EOF
DEVICE=ens33
BOOTPROTO=static
ONBOOT=yes
TYPE=Ethernet
BRIDGE="br0"
IPADDR=192.168.1.151
NETMASK=255.255.255.0
```

```
GATEWAY=192.168.1.1
EOF
```

vi　ifcfg-br0 内容如下：

```
cat>/etc/sysconfig/network-scripts/ifcfg-br0 <<EOF
DEVICE="br0"
BOOTPROTO=static
ONBOOT=yes
TYPE="Bridge"
IPADDR=192.168.1.151
NETMASK=255.255.255.0
GATEWAY=192.168.1.1
EOF
```

启动 Docker 服务，命令操作如下：

```
service docker restart
```

（4）基于本地 CentOS 7 镜像启动 CentOS 云主机，网络模式选择--net=none 即可，操作指令如下：

```
docker run -itd --net=none --privileged --name=jfedu-vm01 centos7-ssh:v1
```

（5）在 /usr/local/docker-static-ip/ 目录下，将需要给 CentOS 容器配置的静态 IP 写入 containers.cfg 文件即可，内容如下：

```
jfedu-vm01,br0,192.168.1.101/24,192.168.1.2
```

（6）查看 Docker 容器的 IP 地址，此时就是 192.168.1.101，命令如下：

```
docker exec jfedu-vm01 ifconfig
```

（7）重启 Docker 容器，再次查看容器的 IP 地址，还是 192.168.1.101，IP 固定成功。

```
docker restart jfedu-vm01
docker exec jfedu-vm01 ifconfig
```

（8）通过 CRT 或者 Xshell 远程登录创建的 CentOS 云主机，命令如图 4-2 所示。

```
[root@localhost docker-static-ip]# ssh -l root 192.168.1.101
root@192.168.1.101's password:
Permission denied, please try again.
root@192.168.1.101's password:
Last failed login: Wed Apr 14 10:06:23 UTC 2021 from 192.168.1.151 on ssh:n
There was 1 failed login attempt since the last successful login.
[root@fd8ffa80c710 ~]#
[root@fd8ffa80c710 ~]# ping -c3 www.baidu.com
PING www.a.shifen.com (14.215.177.38) 56(84) bytes of data.
64 bytes from 14.215.177.38 (14.215.177.38): icmp_seq=1 ttl=128 time=28.7 m
64 bytes from 14.215.177.38 (14.215.177.38): icmp_seq=2 ttl=128 time=40.2 m
64 bytes from 14.215.177.38 (14.215.177.38): icmp_seq=3 ttl=128 time=29.1 m
```

图 4-2　Docker 固定 IP 测试

4.11　EFK 应用背景剖析

运维工程师每天需要对服务器进行故障排错，通过日志可以快速地定位问题。

日志主要包括系统日志、应用程序日志和安全日志。系统运维人员和开发人员可以通过日志了解服务器的软硬件信息、检查配置过程中的错误及错误发生的原因。经常分析日志可以了解服务器的负荷、性能及安全性，从而及时采取措施纠正错误。日志被分散地存储于不同的设备上（每台服务器创建开发普通用户权限，只运行查看日志、查看进程，运维、开发通过命令 tail、head、cat、more、find、awk、grep、sed 统计分析），如果管理数百台服务器，应用登录到每台机器的传统方法查阅日志，这样很烦琐、效率低下。当务之急是使用集中化的日志管理系统，例如开源的 syslog，收集汇总所有服务器上的日志。

集中化管理日志后，日志的统计和检索又成为一件比较麻烦的事情。一般使用 find、grep、awk 和 wc 等 Linux 命令实现检索和统计，但是对于更高的查询、排序和统计等要求以及庞大的机器数量，使用这样的方法难免有点力不从心。

开源实时日志分析 EFK 平台能够完美地解决上述问题。EFK 由 Elasticsearch、Filebeat 和 Kibana 三个开源工具组成。

（1）Elasticsearch 基于 Lucene 全文检索引擎架构，基于 Java 语言编写，对外开源、免费，它的特点有分布式、零配置、自动发现、索引自动分片、索引副本机制、RESTful 风格接口、多数据源、自动搜索负载等。

（2）Filebeat 是一个轻量级日志采集器。Filebeat 属于 Beats 家族的 6 个成员之一。早期的 ELK 架构中使用 Logstash 收集、解析日志并且过滤日志，但是 Logstash 对 CPU、内存、I/O 等资源消耗比较高，相比之下，Filebeat 所占系统的 CPU 和内存几乎可以忽略不计。

（3）Kibana 也是一个开源和免费的工具。它可以为 Logstash 和 Elasticsearch 提供日志分析友好的 Web 界面，可以帮助汇总、分析和搜索重要数据日志。

4.12　EFK 架构原理深入剖析

EFK 架构中也可以使用 Logstash，Logstash 从 Filebeat 获取日志文件。Filebeat 作为 Logstash 的输入对获取到的日志进行处理，然后将处理好的日志文件输出到 Elasticsearch 进行处理，如图 4-3 所示。

EFK 工作流程和原理如下。

（1）使用 Filebeat 获取 Linux 服务器上的日志。当 Filebeat 启动时，它将启动一个或多个 Prospectors（检测者），查找服务器上指定的日志文件，作为日志的源头等待输出到 Logstash。

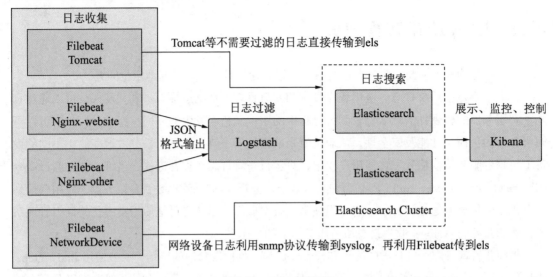

图 4-3 EFK 结构图

（2）Logstash 从 Filebeat 获取日志文件。Filebeat 作为 Logstash 的输入将获取到的日志进行处理，Logstash 将处理好的日志文件输出到 Elasticsearch 进行处理。

（3）Elasticsearch 得到 Logstash 的数据之后进行相应的搜索存储操作。令写入的数据可以被检索和聚合等以便于搜索操作。最后，Kibana 通过 Elasticsearch 提供的 API 将日志信息可视化。

4.13　Docker 部署 EFK 日志平台

（1）提前部署 Docker 虚拟化平台，然后从 Docker 仓库分别获取 EFK 三个镜像包，操作指令如下：

```
docker pull elasticsearch:7.7.1
docker pull kibana:7.7.1
docker pull summerhai/filebeat7.7
```

（2）创建 Elasticsearch 容器，对外监听 9200、9300 端口，命令如下：

```
docker run -d -e ES_JAVA_POTS="-Xms512m -Xmx1024m" -e "discovery.type=single-node" -p 9200:9200 -p 9300:9300 --name es elasticsearch:7.7.1
```

（3）创建 Kibana 容器，对外监听 5601 端口，命令如下：

```
docker run --link es:elasticsearch -p 5601:5601 -d --name kibana kibana:7.7.1
```

采用 "--link" 选项，作用是将两个容器关联到一起以互相通信，因为 Kibana 需要从

Elasticsearch 中获取数据。也可以通过"--network"创建自己的局域网连接各个容器。

进入 Kibana 容器，在 kibana.yml 文件末尾加入代码，支持中文，操作指令如下：

```
#进入 Kibana 容器
docker exec -it kibana /bin/bash
#切换至 config 目录
cd config/
#修改主配置文件 kibana.yml,末尾加入如下代码
i18n.locale: "zh-CN"
#重启 Kibana 容器即可
docker restart kibana
```

（4）创建 Filebeat 容器，并链接 ES 和 Kibana 服务，命令如下：

```
docker run -itd -v /data/filebeat.yml:/etc/filebeat/filebeat.yml -v /data/logs/:/data/logs/ --link es:elasticsearch --link kibana:kibana --name filebeat summerhai/filebeat7.7
```

（5）编写 Filebeat.yml 配置文件，配置文件代码如下：

```
#Filebeat inputs 2021 jfedu.net
filebeat.inputs:
- type: log
  enabled: true
  paths:
    - /var/log/*
#配置多行日志合并规则,以时间为准,一个时间发生的日志为一个事件
  multiline.pattern: '^\d{4}-\d{2}-\d{2}'
  multiline.negate: true
  multiline.match: after
#设置 Kibana 的地址,开始 Filebeat 的可视化
setup.kibana.host: "http://kibana:5601"
setup.dashboards.enabled: true
#------------------------- Elasticsearch output ---------
output.elasticsearch:
    hosts: ["http://elasticsearch:9200"]
    index: "filebeat-%{+yyyy.MM.dd}"
setup.template.name: "my-log"
setup.template.pattern: "my-log-*"
json.keys_under_root: false
json.overwrite_keys: true
#设置解析 JSON 格式日志的规则
processors:
- decode_json_fields:
    fields: [""]
    target: json
```

（6）访问 Kibana Web 界面，同时添加索引 Filebeat，如图 4-4 所示。

（a）

（b）

图 4-4　EFK 日志采集实战

（a）添加索引 Filebeat；（b）查看日志采集柱状统计图

4.14　基于 Docker Web 管理 Docker 容器

默认通过命令行创建及运行 Docker 容器，但 Docker 的 Remote API 可以通过充分利用 REST（代表性状态传输协议）的 API，运行相同的命令。

Docker UI 也是基于 API 方式管理宿主机的 Docker 引擎。Docker UI Web 前端程序让开发人员和管理人员可以通过 Web 浏览器的命令行管理许多任务。

主机上的所有容器都可以通过一条连接处理，该项目几乎没有任何依赖关系。该软件目前仍在大力开发之中，但是它采用麻省理工学院（MIT）许可证，所以可以免费地重复使用。

Docker UI 不包含任何内置的身份验证或安全机制，所以务必将任何公之于众的 Docker UI

连接放在用密码保护的系统中。

基于 Docker Web 管理 Docker 容器，步骤如下：

（1）下载 Docker UI 镜像，在宿主机拉取相关镜像即可，操作指令如下：

```
docker pull uifd/ui-for-docker
docker images
```

（2）启动 Docker UI 服务，并映射 9000 至容器 9000，操作指令如下：

```
docker run -it -d --name docker-web -p 9000:9000 -v /var/run/docker.sock:/var/run/docker.sock docker.io/uifd/ui-for-docker
```

（3）通过 docker ps 命令查看 Docker UI 状态，如图 4-5 所示。

图 4-5　查看 Docker UI 状态

（4）通过浏览器登录 Web 9000 端口访问，如图 4-6 所示。

图 4-6　Docker Web 运行界面

（5）选择 Web 界面 Images 镜像列表，如图 4-7 所示。

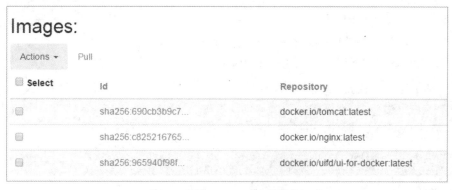

图 4-7 Docker Web 镜像列表

（6）基于镜像启动 Docker 容器虚拟机，并实现端口映射，如图 4-8 所示。

（a）

（b）

图 4-8 Docker Web 启动容器

（a）创建 Docker 容器；（b）查看容器的状态

（7）创建 Docker 容器之后，通过浏览器实现访问 81 端口，如图 4-9 所示。

图 4-9　Docker 容器浏览器访问

（8）Docker 容器资源 Web 监控，如图 4-10 所示。

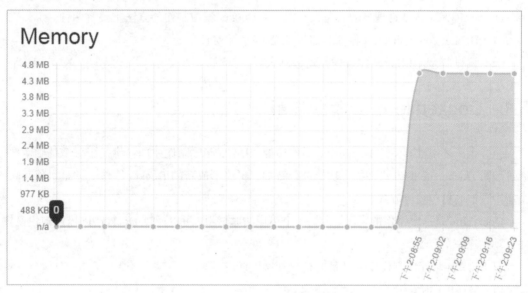

图 4-10　查看 Docker 内存资源

第 5 章 Dockerfile 企业镜像实战

由于 Docker 官网公共仓库镜像大多不完整，无法真正满足企业的生产环境系统，因此需要自行定制或重新打包镜像。

Docker 镜像制作是管理员的必备工作之一。Docker 镜像制作的方法主要有两种。

（1）Docker commit|export 将新容器提交至 images 列表。

（2）编写 Dockerfile，新建镜像至镜像列表。

5.1 Dockerfile 语法命令详解一

企业生产环境推荐使用 Dockerfile 制作镜像。Dockerfile 制作原理：将基于一个基础镜像，通过编写 Dockerfile 的方式将各个功能进行叠加，最终形成新的 Docker 镜像。这是目前互联网企业中首推的打包镜像方式。

Dockerfile 是一个镜像的表示，也是一个镜像的原材料，可以通过 Dockerfile 描述构建镜像，并自动构建一个容器。

以下为 Dockerfile 制作镜像必备的指令和参数详解。

FROM：指定所创建镜像的基础镜像。

MAINTAINER：指定维护者信息。

RUN：运行命令。

CMD：指定启动容器时默认执行的命令。

LABEL：指定生成镜像的元数据标签信息。

EXPOSE：声明镜像内服务所监听的端口。

ENV：指定环境变量。

ADD：赋值指定的<src>路径下的内容到容器中的<dest>路径下，<src>可以为 URL；如果为 tar 文件，会自动解压到<dest>路径下。

COPY：赋值本地主机的<scr>路径下的内容到容器中的<dest>路径下；一般情况下推荐使用 COPY 而不是 ADD。

ENTRYPOINT：指定镜像的默认入口。

VOLUME：创建数据挂载点。

USER：指定运行容器时的用户名或 UID。

WORKDIR：配置工作目录。

ARG：指定镜像内使用的参数（例如版本号信息等）。

ONBUILD：配置当前所创建的镜像作为其他镜像的基础镜像时，所执行的创建操作的命令。

STOPSIGNAL：容器退出的信号。

HEALTHCHECK：如何进行健康检查。

SHELL：指定使用 Shell 时的默认 Shell 类型。

5.2 Dockerfile 语法命令详解二

FROM：指定所创建的镜像的基础镜像，如果本地不存在，则默认会去 Docker Hub 下载指定镜像。

格式为 FROM<image>，或 FROM<image>:<tag>，或 FROM<image>@<digest>。任何 Dockerfile 中的第一条指令必须为 FROM 指令。并且，如果在同一个 Dockerfile 文件中创建多个镜像，可以使用多个 FROM 指令（每个镜像一次）。

MAINTAINER：指定维护者信息，格式为 MAINTAINER<name>，例如：

```
MAINTAINER image_creator@docker.com
```

该信息将会写入生成镜像的 Author 属性域中。

RUN：运行指定命令。

格式：RUN<command>或 RUN ["executable","param1","param2"]。

注意：后一个指令会被解析为 JSON 数组，所以必须使用双引号。

前者默认将在 Shell 终端中运行命令，即/bin/sh -c；后者则使用 exec 执行，不会启动 Shell 环境。

指定使用其他终端类型可以通过第二种方式实现，例如：

```
RUN ["/bin/bash","-c","echo hello"]
```

每条 RUN 指令将在当前镜像的基础上执行指定命令，并提交为新的镜像。当命令较长时可以使用 \ 换行。例如：

```
RUN apt-get update \
&& apt-get install -y libsnappy-dev zliblg-dev libbz2-dev \
&& rm -rf /var/cache/apt
```

CMD：用来指定启动容器时默认执行的命令。它支持三种格式。

- CMD ["executable","param1","param2"] 使用 exec 执行，是推荐使用的方式。
- CMD param1 param2 在/bin/sh 中执行，提供给需要交互的应用。
- CMD ["param1","param2"] 提供给 ENTRYPOINT 的默认参数。

每个 Dockerfile 只能有一条 CMD 命令。如果指定了多条命令，则只有最后一条会被执行。如果用户启动容器时指定了运行的命令（作为 RUN 的参数），则会覆盖 CMD 指定的命令。

LABEL：用来生成用于生成镜像的元数据的标签信息。

格式：LABEL <key>=<value> <key>=<value> <key>=<value> ...。

例如：

```
LABEL version="1.0"
LABEL description="This text illustrates \ that label-values can span
multiple lines."
```

EXPOSE：声明镜像内服务所监听的端口。

格式：EXPOSE <port> [<port>...]。

例如：

```
EXPOSE 22 80 443 3306
```

注意：该命令只起声明作用，并不会自动完成端口映射。在容器启动时需要使用-P(大写)，Docker 主机会自动分配一个宿主机未被使用的临时端口转发到指定的端口；使用-p(小写)，则可以具体指定哪个宿主机的本地端口映射过来。

ENV：指定环境变量，在镜像生成过程中会被后续 RUN 指令使用，在镜像启动的容器中也会存在。

格式：ENV <key><value>或 ENV<key>=<value>...。

例如：

```
ENV GOLANG_VERSION 1.6.3
ENV GOLANG_DOWNLOAD_RUL https://golang.org/dl/go$GOLANG_VERSION.linux-amd64.tar.gz
ENV GOLANG_DOWNLOAD_SHA256 cdd5e08530c0579255d6153b08fdb3b8e47caabbe717bc7bcd7561275a87aeb
RUN curl -fssL "$GOLANG_DOWNLOAD_RUL" -o golang.tar.gz && echo "$GOLANG_DOWNLOAD_SHA256 golang.tar.gz" | sha256sum -c - && tar -C /usr/local -xzf golang.tar.gz && rm golang.tar.gz
ENV GOPATH $GOPATH/bin:/usr/local/go/bin:$PATH
RUN mkdir -p "$GOPATH/bin" && chmod -R 777 "$GOPATH"
```

指令指定的环境变量在运行时可以被覆盖，如 docker run --env <key>=<value> built_image。

ADD：将复制指定的<src>路径下的内容到容器中的<dest>路径下。

格式：ADD <src> <dest>。

其中，<src>可以是Dockerfile所在目录的一个相对路径（文件或目录），也可以是一个URL，还可以是一个 tar 文件（如果是 tar 文件，会自动解压到<dest>路径下）。<dest>可以是镜像内的绝对路径，也可以是工作目录（WORKDIR）的相对路径。路径支持正则表达式，例如：

```
ADD *.c /code/
```

COPY：复制本地主机的<src>（为Dockerfile所在目录的一个相对路径、文件或目录）下的内容到镜像中的<dest>下。目标路径不存在时，会自动创建。路径同样支持正则。

格式：COPY <src> <dest>。

当使用本地目录为源目录时，推荐使用 COPY。

ENTRYPOINT：指定镜像的默认入口命令，该入口命令会在启动容器时作为根命令执行，所有传入值均作为该命令的参数。

支持两种格式：

- ENTRYPOINT ["executable","param1","param2"] (exec 调用执行)。
- ENTRYPOINT command param1 param2(shell 中执行)。

此时，CMD 指令指定值将作为该命令的参数。

每个 Dockerfile 中只能有一个 ENTRYPOINT，当指定多个时，只有最后一个有效。

在运行时可以被--entrypoint参数覆盖，如 docker run --entrypoint。

VOLUME：创建一个数据卷挂载点。

格式：VOLUME ["/data"]。

可以从本地主机或其他容器挂载数据卷，一般用来存放数据库和需要保存的数据等。

```
USER
```

指定运行容器时的用户名或 UID，后续的 RUN 等指令也会使用特定的用户身份。

格式：USER daemon。

当服务不需要管理员权限时，可以通过该指令指定运行用户，并可以在此之前创建所需要的用户。例如，RUN groupadd -r nginx && useradd -r -g nginx nginx 要临时获取管理员权限可以用 gosu 或者 sudo。

WORKDIR：为后续的 RUN、CMD 和 ENTRYPOINT 指令配置工作目录。

格式：WORKDIR /path/to/workdir。

可以使用多个 WORKDIR 指令，后续命令参数如果是相对的，则会基于之前命令指定的路径。例如：

```
WORKDIR /a
WORKDIR b
WORKDIR c
RUN pwd
```

则最终路径为/a/b/c。

ARG 指定一些镜像内使用的参数（例如版本号信息等），这些参数在执行 Docker build 命令时才以--build-arg<varname>=<value>格式传入。

格式：ARG<name>[=<default value>]。

则可以用 Docker build --build-arg<name>=<value>指定参数值。

ONBUILD：配置当所创建的镜像作为其他镜像的基础镜像时，所执行的创建操作指令。

格式：ONBUILD [INSTRUCTION]。

例如，Dockerfile 使用如下内容创建了镜像 image-A：

```
[...]
ONBUILD ADD . /app/src
ONBUILD RUN /usr/local/bin/python-build --dir /app/src
[...]
```

基于 image-A 镜像创建新的镜像时，新的 Dockerfile 中使用 FROM image-A 指定基础镜像，会自动执行 ONBUILD 指令的内容，等价于在后面添加了两条指令。

```
FROM image-A
#Automatically run the following
ONBUILD ADD . /app/src
ONBUILD RUN /usr/local/bin/python-build --dir /app/src
```

使用 ONBUILD 指令的镜像，推荐在标签中注明，例如 ruby:1.9-onbuild。

STOPSIGNAL：指定所创建镜像启动的容器接收退出的信号值。例如：

```
STOPSIGNAL singnal
```

HEALTHCHECK：配置所启动容器如何进行健康检查（判断是否健康），自 Docker 1.12 开始支持。格式有两种：

● HEALTHCHECK [OPTIONS] CMD command：根据所执行命令返回值是否为 0 判断。

● HEALTHCHECK NONE：禁止基础镜像中的健康检查。

[OPTION]支持以下参数：

--interval=DURATION（默认为 30s）：多久检查一次。

--timeout=DURATION（默认为 30s）：每次检查等待结果的超时时间。

--retries=N（默认为 3）：如果失败，重试几次才最终确定失败。

SHELL 指定其他命令使用 Shell 时的默认 Shell 类型。

格式：SHELL ["executable","parameters"]，默认值为 ["bin/sh","-c"]。

注意：对于 Windows 系统，建议在 Dockerfile 开头添加"# escape="指定转移信息。

编写 Dockerfile 之后，可以通过 docker build 命令创建镜像。

docker build [选项]：内容路径，该命令将读取指定路径下（包括子目录）的 Dockerfile，并将该路径下的所有内容发送给 Docker 服务端，由服务端创建镜像。因此，除非生成镜像需要，否则一般建议放置 Dockerfile 的目录为空目录。

如果使用非内容路径下的 Dockerfile，可以通过-f 选项指定其路径；要指定生成镜像的标签信息，可以使用-t 选项。

例如，指定 Dockerfile 所在路径为 /tmp/docker_builder/，并且希望生成镜像标签为 build_repo/first_image，可以使用下面的命令：

```
docker build -t build_repo/first_image /tmp/docker_builder
```

使用.dockerignore 文件可以通过.dockerignore 文件（每一行添加一条匹配模式）让 Docker 忽略匹配模式路径下的目录和文件。例如：

```
#comment
*/tmp*
*/*/tmp*
tmp?
~*
```

5.3 Dockerfile 制作规范及技巧

从企业需求出发，定制适合本企业需求、高效方便的镜像，可以参考官方 Dockerfile 文件，也可以根据自身的需求，在构建中不断优化 Dockerfile 文件。

Dockerfile 制作镜像规范和技巧有以下几点。

（1）精简镜像用途。尽量让每个镜像的用途都比较集中、单一，避免构造大而复杂、多功能的镜像。

（2）选用合适的基础镜像。过大的基础镜像会造成构建出臃肿的镜像，一般推荐比较小巧的镜像作为基础镜像。

（3）提供详细的注释和维护者信息。Dockerfile 也是一种代码，需要考虑方便后续扩展和他人使用。

（4）正确使用版本号。使用明确的具体数字信息的版本号信息，而非 latest，可以统一环境，避免无法确认具体版本号。

（5）减少镜像层数。建议尽量合并 RUN 指令，可以将多条 RUN 指令的内容通过&&连接。

（6）及时删除临时和缓存文件。这样可以避免构造的镜像过于臃肿，并且这些缓存文件并没有实际用途。

（7）提高生产速度。合理使用缓存，减少目录下的使用文件，使用.dockerignore 文件等。

（8）调整合理的指令顺序。在开启缓存的情况下，内容不变的指令尽量放在前面，这样可以提高指令的复用性。

（9）减少外部源的干扰。如果确实要从外部引入数据，需要指定持久的地址，并带有版本信息，让他人可以重复使用而不出错。

5.4　Dockerfile 企业案例一

将启动 Docker 容器，同时开启 Docker 容器对外的 22 端口的监听，实现通过 CRT 或 Xshell 登录。

Docker 服务端创建 Dockerfile 文件，实现容器运行开启 22 端口，内容如下：

```
#设置基本的镜像,后续命令都以这个镜像为基础
FROM centos
#作者信息
MAINTAINER   WWW.JFEDU.NET
#安装依赖工具,删除默认 YUM 源,使用 YUM 源为国内 163 YUM 源
RUN rpm --rebuilddb;yum install make wget tar gzip passwd openssh-server gcc -y
RUN rm -rf /etc/yum.repos.d/*;wget -P /etc/yum.repos.d/ http://mirrors.163.com/.help/CentOS7-Base-163.repo
#配置 SSHD,修改 root 密码为 1qaz@WSX
RUN yes|ssh-keygen -q -t rsa -b 2048 -f /etc/ssh/ssh_host_rsa_key -N ''
RUN yes|ssh-keygen -q -t ecdsa -f /etc/ssh/ssh_host_ecdsa_key -N ''
RUN yes|ssh-keygen -q -t ed25519 -f /etc/ssh/ssh_host_ed25519_key -N ''
RUN echo '1qaz@WSX' | passwd --stdin root
#启动 SSHD 服务进程,对外暴露 22 端口
EXPOSE  22
CMD /usr/sbin/sshd -D
```

基于 Dockerfile，用 docker build 根据 Dockerfile 创建镜像（centos:ssh），命令如下：

```
docker build -t centos:ssh - < Dockerfile
docker build -t centos:ssh .
```

5.5　Dockerfile 企业案例二

开启 SSH 6379 端口，让 Redis 端口对外访问，Dockerfile 内容如下：

```
FROM centos:latest
#作者信息
```

```
MAINTAINER    WWW.JFEDU.NET
#安装依赖工具,删除默认 YUM 源,使用 YUM 源为国内 163 YUM 源
RUN rpm --rebuilddb;yum install make wget tar gzip passwd openssh-server gcc
-y
RUN rm -rf /etc/yum.repos.d/*;wget -P /etc/yum.repos.d/ http://mirrors.163.
com/.help/CentOS7-Base-163.repo
#配置 SSHD,修改 root 密码为 1qaz@WSX
RUN ssh-keygen -q -t rsa -b 2048 -f /etc/ssh/ssh_host_rsa_key -N ''
RUN ssh-keygen -q -t ecdsa -f /etc/ssh/ssh_host_ecdsa_key -N ''
RUN ssh-keygen -q -t ed25519 -f /etc/ssh/ssh_host_ED25519_key -N ''
RUN echo '1qaz@WSX' | passwd --stdin root
#从 Redis 官网下载 Redis 最新版本软件
RUN wget -P /tmp/ http://download.redis.io/releases/redis-5.0.2.tar.gz
#解压 Redis 软件包,并基于源码安装,创建配置文件
RUN cd /tmp/;tar xzf redis-5.0.2.tar.gz;cd redis-5.0.2;make;make PREFIX=
/usr/local/redis install;mkdir -p /usr/local/redis/etc/;cp redis.conf /usr/
local/redis/etc/
#创建用于存储应用数据的目录/data/redis,修改 Redis 配置文件路径
RUN mkdir -p /data/redis/
RUN sed -i 's#^dir.*#dir /data/redis#g' /usr/local/redis/etc/redis.conf
#将应用数据存储目录/data/进行映射,可以实现数据持久化保存
VOLUME ["/data/redis"]
#修改 Redis.conf 监听地址为 bind: 0.0.0.0
RUN sed -i '/^bind/s/127.0.0.1/0.0.0.0/g' /usr/local/redis/etc/redis.conf
#启动 Redis 数据库服务进程,对外暴露 22 和 6379 端口
EXPOSE  22
EXPOSE  6379
CMD /usr/sbin/sshd;/usr/local/redis/bin/redis-server /usr/local/redis/
etc/redis.conf
```

5.6 Dockerfile 企业案例三

基于 Dockerfile 开启 Nginx 80 端口,并远程连接服务器。Dockerfile 内容如下:

```
FROM centos:latest
#作者信息
MAINTAINER    WWW.JFEDU.NET
#安装依赖工具,删除默认 YUM 源,使用 YUM 源为国内 163 YUM 源
RUN rpm --rebuilddb;yum install make wget tar gzip passwd openssh-server gcc
pcre-devel open
ssl-devel net-tools -y
RUN rm -rf /etc/yum.repos.d/*;wget -P /etc/yum.repos.d/ http://mirrors.163.
com/.help/CentOS7-Base-163.repo
#配置 SSHD,修改 root 密码为 1qaz@WSX
```

```
RUN ssh-keygen -q -t rsa -b 2048 -f /etc/ssh/ssh_host_rsa_key -N ''
RUN ssh-keygen -q -t ecdsa -f /etc/ssh/ssh_host_ecdsa_key -N ''
RUN ssh-keygen -q -t ed25519 -f /etc/ssh/ssh_host_ED25519_key -N ''
RUN echo '1qaz@WSX' | passwd --stdin root
#从Nginx官网下载Nginx最新版本软件
RUN wget -P /tmp/ http://nginx.org/download/nginx-1.14.2.tar.gz
#解压Nginx软件包,隐藏Web服务器版本号
RUN cd /tmp/;tar xzf nginx-1.14.2.tar.gz;cd nginx-1.14.2;sed -i -e
's/1.14.2//g' -e 's/nginx\//WS/g' -e 's/"NGINX"/"WS"/g' src/core/nginx.h
#基于源码安装,创建配置文件
RUN cd /tmp/nginx-1.14.2;./configure --prefix=/usr/local/nginx --with-
http_stub_status_module --with-http_ssl_module;make;make install
#启动Nginx服务进程,对外暴露22和80端口
EXPOSE 22
EXPOSE 80
CMD /usr/local/nginx/sbin/nginx;/usr/sbin/sshd -D
```

5.7 Dockerfile 企业案例四

Docker 虚拟化中,如何构建 MySQL 数据库服务器呢? 答案很简单,通过 Dockerfile 生成 MySQL 镜像并启动运行即可,代码如下:

```
FROM centos:v1
RUN groupadd -r mysql && useradd -r -g mysql mysql
RUN rpm --rebuilddb;yum install -y gcc zlib-devel gd-devel
ENV MYSQL_MAJOR 5.6
ENV MYSQL_VERSION 5.6.20
RUN
&& curl -SL "http://dev.mysql.com/get/Downloads/MySQL-$MYSQL_MAJOR/mysql-
$MYSQL_VERSION-linux-glibc2.5-x86_64.tar.gz" -o mysql.tar.gz \
&& curl -SL "http://mysql.he.net/Downloads/MySQL-$MYSQL_MAJOR/mysql-
$MYSQL_VERSION-linux-glibc2.5-x86_64.tar.gz.asc" -o mysql.tar.gz.asc \
&& mkdir /usr/local/mysql \
&& tar -xzf mysql.tar.gz -C /usr/local/mysql \
&& rm mysql.tar.gz* \
ENV PATH $PATH:/usr/local/mysql/bin:/usr/local/mysql/scripts
WORKDIR /usr/local/mysql
VOLUME /var/lib/mysql
EXPOSE 3306
CMD ["mysqld", "--datadir=/var/lib/mysql", "--user=mysql"]
```

第 6 章 Docker 仓库案例实战

Docker 虚拟化有三个基础概念：Docker 镜像、Docker 容器和 Docker 仓库。

1）Docker 镜像

Docker 虚拟化最基础的组件为镜像。与常见的 Linux ISO 镜像类似，但是 Docker 镜像是分层结构的，由多个层级组成，每个层级分别存储各种软件实现某个功能。Docker 镜像是静止的、只读的，不能对镜像进行写操作。

2）Docker 容器

Docker 容器是 Docker 虚拟化的产物，也是最早在生产环境使用的对象。Docker 容器的底层是 Docker 镜像，是基于镜像运行，并在镜像最上层添加一层容器层之后的实体。容器层是可读、可写的，容器层如果需用到镜像层中的数据，可以通过 JSON 文件读取镜像层中的软件和数据，对整个容器进行修改。修改只能作用于容器层，不能直接对镜像层进行写操作。

3）Docker 仓库

Docker 仓库是用于存放 Docker 镜像的地方，Docker 仓库分为两类，分别是公共仓库（Public）和私有仓库（Private），国内和国外有很多默认的公共仓库，对外开放、免费或者付费使用，企业测试环境和生产环境推荐自建私有仓库，私有仓库的特点：安全、可靠、稳定、高效，能够根据自身的业务体系进行灵活升级和管理。

纵观 Docker 镜像、容器、仓库，其中最重要、最基础的当属 Docker 镜像，没有镜像就没有容器，而镜像是静止的、只读的模板文件层，存储在 Docker 仓库中。

6.1 Docker 国内源实战

Docker 默认连接的国外官方镜像，通常根据网络情况不同，访问时快时慢，大多时候获取速度非常慢，为了提升效率可以自建仓库或者先修改为国内仓库源，提升拉取镜像的速度。

Docker 可以配置的国内镜像有很多可供选择，例如，Docker 中国区官方镜像、阿里云、网易蜂巢、DaoCloud 等，这些都是国内比较快的镜像仓库。

（1）修改 Docker 默认镜像源方法，操作指令如下：

```
cat>/etc/docker/daemon.json<<EOF
{
"registry-mirrors":["https://registry.docker-cn.com"]
}
EOF
```

（2）修改完成，重启 Docker 引擎服务，操作指令如下：

```
service docker restart
systemctl restart docker.service
```

6.2 Docker Registry 仓库源实战

Docker 仓库分为公共仓库和私有仓库，在企业测试环境和生产环境中推荐自建内部私有仓库。使用私有仓库有以下优点。

（1）节省网络带宽，针对每个镜像不用去 Docker 官网仓库下载。

（2）可直接从本地私有仓库中下载 Docker 镜像。

（3）组建公司内部私有仓库，方便各部门使用，服务器管理更加统一。

（4）可以基于 GIT 或 SVN、Jenkins 更新本地 Docker 私有仓库镜像版本。

官方提供 Docker Registry 构建本地私有仓库，目前最新版本为 Registry v2，最新版的 Docker 已不再支持 Registry v1，Registry v2 使用 Go 语言编写，在性能和安全性上作了很多优化，重新设计了镜像的存储格式。

（1）登录 Docker 仓库服务器，下载 Docker Registry 镜像，操作指令如下：

```
docker pull registry
```

（2）启动私有仓库容器，操作指令如下：

```
mkdir -p /data/registry/
docker run -itd -p 5000:5000 -v /data/registry:/var/lib/registry docker.io/registry
netstat -tnlp|grep -aiwE 5000
```

（3）默认情况下，会将仓库存放于容器内的/var/lib/registry 目录下，这样如果容器被删除，则存放于容器中的镜像也会丢失，所以一般会指定本地一个目录挂载到容器内的/data/registry 目录下。

（4）客户端上传镜像至本地私有仓库。以下以 busybox 镜像为例，将 busybox 上传至私有仓库服务器。操作指令如下：

```
docker pull busybox
docker tag busybox 192.168.1.123:5000/busybox
docker push 192.168.1.123:5000/busybox
```

（5）默认为 Docker 仓库，报错解决方法：vim /etc/sysconfig/docker 配置文件，注释或者删除以 OPTION 开头的行，操作指令如下：

```
OPTIONS='--selinux-enabled --log-driver=journald --signature-verification
=false --insecure-registry 192.168.1.123:5000'
ADD_REGISTRY='--add-registry 192.168.1.123:5000'
```

（6）检测本地私有仓库，操作指令如下：

```
curl -XGET http://192.168.1.123:5000/v2/_catalog
curl -XGET http://192.168.1.123:5000/v2/busybox/tags/list
```

（7）客户端使用本地私有仓库，登录 Docker 客户端，同样在其/etc/sysconfig/docker 配置文件添加如下代码，同时重启 Docker 服务，获取本地私有仓库。

```
OPTIONS='--selinux-enabled --log-driver=journald --signature-verification
=false --insecure-registry 192.168.1.123:5000'
ADD_REGISTRY='--add-registry 192.168.1.123:5000'
```

（8）重启 Docker 服务，然后从 Docker 仓库下载 busybox 镜像，如图 6-1 所示。

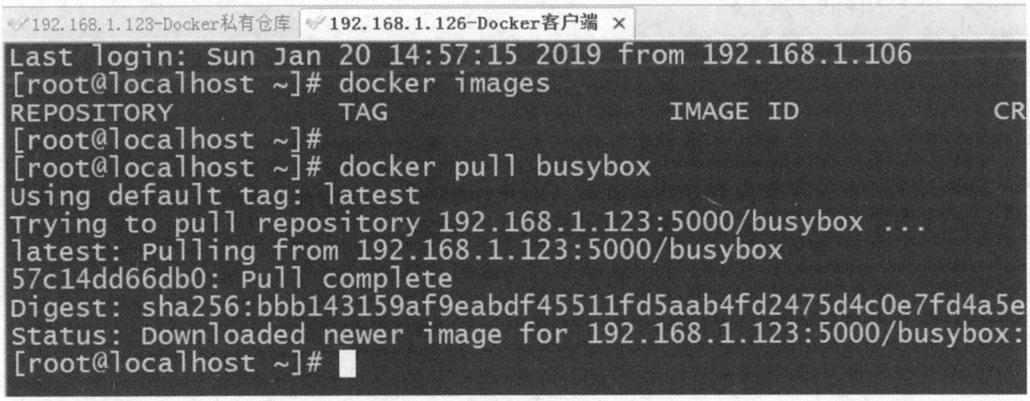

图 6-1 Docker 仓库下载镜像

（9）配置 Docker 仓库第二种方法，在/etc/docker/daemon.json 写入如下内容：

```
{
"insecure-registries":["192.168.1.123:5000"]
}
```

（10）重启 Docker 引擎服务，操作指令如下：

```
service docker restart
```

（11）下载远程仓库镜像，如图 6-2 所示。

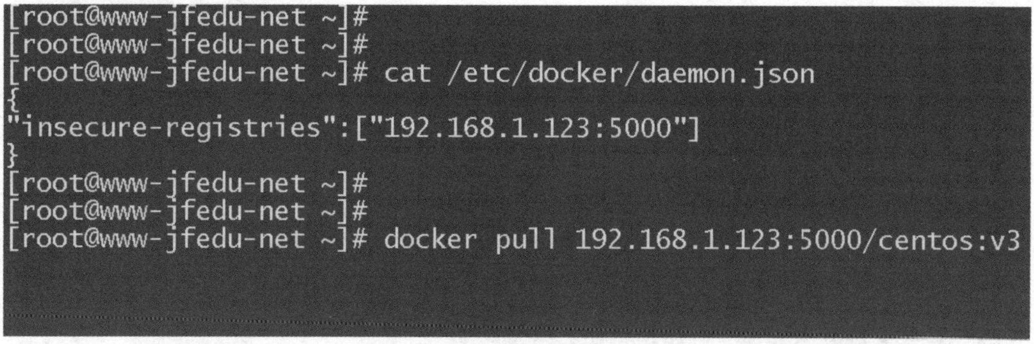

```
[root@www-jfedu-net ~]#
[root@www-jfedu-net ~]#
[root@www-jfedu-net ~]# cat /etc/docker/daemon.json
{
"insecure-registries":["192.168.1.123:5000"]
}
[root@www-jfedu-net ~]#
[root@www-jfedu-net ~]#
[root@www-jfedu-net ~]# docker pull 192.168.1.123:5000/centos:v3
```

图 6-2 下载远程仓库镜像

6.3 Docker Harbor 仓库源实战

构建 Docker 仓库方式除了使用 Registry 之外，还可以使用 Harbor。以下为 Registry 方式的缺点。

（1）缺少认证机制，任何人都可以随意拉取及上传镜像，安全性缺失。

（2）缺乏镜像清理机制，镜像可以拉取却不能删除，日积月累，占用的空间会越来越大。

（3）缺乏相应的扩展机制。

Harbor 仓库可以解决以上几个缺点。Harbor 是一个用于存储和分发 Docker 镜像的企业级 Registry 服务器，通过添加一些企业必需的功能特性，例如安全、标识和管理等，扩展了开源 Docker Distribution。

作为一个企业级私有 Registry 服务器，Harbor 提供了更好的性能和安全性，可提升用户使用 Registry 构建和运行环境传输镜像的效率。Harbor 支持在多个 Registry 节点进行镜像资源复制，镜像全部保存在私有 Registry 中，确保数据和知识产权在公司内部网络中管控。另外，Harbor 也提供了高级的安全特性，诸如用户管理、访问控制和活动审计等。

Harbor 仓库部署有两种方式，一种是 off-line，另一种是 on-line，即离线安装和在线安装。此处选择离线安装。

（1）安装 Docker-Compose 快速编排工具。

```
#安装 Epel-release 扩展源
yum install epel-release -y
#安装 Python pip 工具
yum install python-pip -y
#升级 Python pip 工具
pip install --upgrade pip
#pip install docker-compose
#下载 Docker compose 脚本
```

```
curl -L https://github.com/docker/compose/releases/download/1.18.0/run.sh
> /usr/local/bin/docker-compose
#添加脚本 x 权限
chmod +x /usr/local/bin/docker-compose
#查看其版本信息
docker-compose --version
ln -s /usr/local/bin/docker-compose /usr/bin/docker-compose
```

（2）下载 Harbor 并解压。

```
wget -c https://storage.googleapis.com/harbor-releases/release-1.7.0/
harbor-offline-installer-v1.7.0.tgz
tar -xzf harbor-offline-installer-v1.7.0.tgz
cd harbor
```

（3）修改 Harbor 配置文件 harbor.cfg，修改 hostname 为本机 IP 地址，如图 6-3 所示。

```
[root@jfedu123 harbor]# vim harbor.cfg
## Configuration file of Harbor

#This attribute is for migrator to detect the version of the .cfg
_version = 1.7.0
#The IP address or hostname to access admin UI and registry servic
#DO NOT use localhost or 127.0.0.1, because Harbor needs to be acc
#DO NOT comment out this line, modify the value of "hostname" dire
hostname = 192.168.1.123

#The protocol for accessing the UI and token/notification service,
#It can be set to https if ssl is enabled on nginx.
ui_url_protocol = http

#Maximum number of job workers in job service
max_job_workers = 10
```

图 6-3　Harbor 配置文件修改

（4）执行 Harbor 安装脚本，安装 Harbor，操作指令如图 6-4 所示。

```
./install.sh
```

```
[Step 4]: starting Harbor ...
Creating network "harbor_harbor" with the default driver
Creating harbor-log
Creating redis
Creating registry
Creating registryctl
Creating harbor-adminserver
Creating harbor-db
Creating harbor-core
Creating harbor-portal
Creating harbor-jobservice
Creating nginx
```

（a）

图 6-4　Harbor 仓库平台安装 1

（a）安装 Harbor 仓库服务

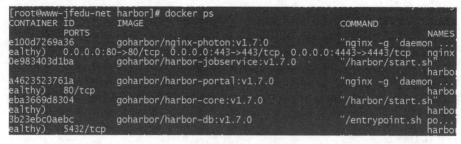

（b）

图 6-4　Harbor 仓库平台安装 1（续）

（b）查看 Harbor 仓库容器状态

（5）登录 Harbor Web 平台，默认用户名为 admin，默认密码为 Harbor12345，可以在 harbor.cnf 中自行设置密码，如图 6-5 所示。

图 6-5　Harbor 仓库平台安装 2

（6）登录 Harbor Web 控制台，可以进行进一步配置，如图 6-6 所示。

图 6-6　Harbor 仓库平台安装 3

（7）创建私有仓库用户名 jfedu，设置密码，并且绑定 library 仓库，如图 6-7 所示。

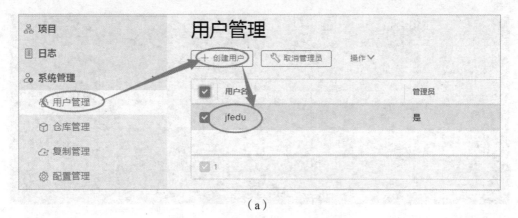

(a)

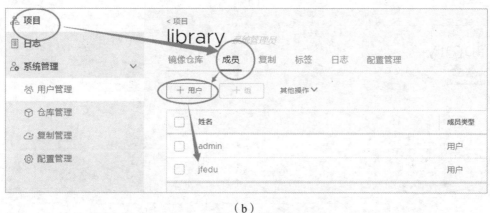

(b)

图 6-7　Harbor 仓库平台安装 4

(a) 创建 Harbor 仓库 jfedu 用户；(b) 查看 Harbor 仓库 jfedu 用户

（8）修改 Docker 客户端仓库地址为 192.168.1.123，同时将 tag 修改为如下格式：

```
192.168.1.123/library/busybox
192.168.1.123/library/nginx
```

（9）docker login 输入创建的用户名和密码，登录成功即可。

```
docker login 192.168.1.123
```

（10）默认访问 Docker 仓库使用 443 端口，要修改为 80 端口仓库地址。

```
cat>/etc/docker/daemon.json<<EOF
{
"insecure-registries":["192.168.1.123"]
}
EOF
service docker restart
```

（11）通过 Docker push 将镜像上传至 Harbor 仓库，如图 6-8 所示。

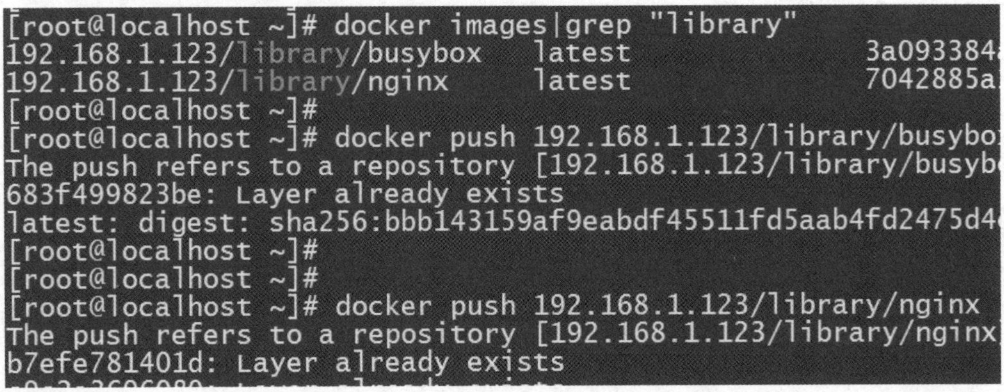

（a）

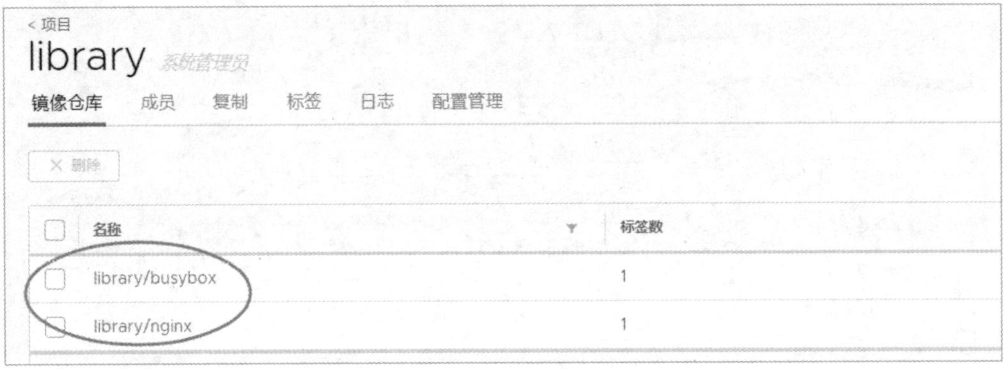

（b）

（c）

图 6-8　Harbor 仓库案例应用

（a）将镜像上传至仓库中；（b）查看仓库中存储的镜像；（c）查看 nginx 镜像的详细信息

6.4 Docker 磁盘、内存、CPU 资源实战一

前面章节介绍了 Docker 引擎启动的容器（虚拟机），其默认会共享宿主机所有的硬件资源（CPU、内存、硬盘等）。

测试环境或生产环境中，如果某一个 Docker 容器非常占资源，又没有对其做任何资源的限制，会导致其他的容器没有资源可用，甚至导致宿主机崩溃。为了防止意外或者错误产生，通常会对 Docker 容器进行资源隔离和限制，默认 Docker 基于 Cgroup 隔离子系统实施资源隔离。

基于 Docker run 启动容器时，可以直接限制 CPU 和内存的资源，但是不能直接限制其对硬盘容量的使用。

1）CPU 和内存资源限制案例一

基于 Docker 引擎启动一台 CentOS 容器，并设置 CPU 为 2 核，内存为 4096MB，启动命令如下：

```
docker run -itd --privileged --cpuset-cpus=0-1 -m 4096m centos:latest
```

2）CPU 和内存资源限制案二

基于 Docker 引擎启动一台 CentOS 容器，并且设置 CPU 为 4 核，内存为 8192MB，启动命令如下：

```
docker run -itd --privileged --cpuset-cpus=2-5 -m 8192m centos:latest
```

以上方法只能限制 Docker 容器 CPU 和内存的资源隔离，如果要实现硬盘容量的限制，则没有默认参数设置，需要通过以下方法实现。

3）基于 Device Mapper 硬盘容量限制

限制 Docker 容器硬盘容量资源，不同的硬盘驱动方式，操作方法不一样。例如，基于 Device Mapper 驱动方式，Docker 容器默认分配硬盘的 rootfs 根分区的容量为 10GB。

可以指定默认容器的大小（在启动容器时指定），也可以在 Docker 配置文件里通过 dm.basesize 参数指定，指定 Docker 容器 rootfs 容量大小为 20GB。

```
docker -d --storage-opt dm.basesize=20G
```

可以修改 Docker 引擎，默认存储配置文件 vim /etc/sysconfig/docker-storage，在 OPTIONS 参数后面加入以下代码：

```
--storage-opt dm.basesize=20G
```

最终 docker-storage 文件，添加之后的代码如下：

```
DOCKER_STORAGE_OPTIONS="--storage-driver devicemapper --storage-opt dm.basesize=20G"
```

除了用第一种默认方法限制 Docker 容器硬盘容量之外，还可以基于现有容器在线扩容，宿主机文件系统类型支持 ext2、ext3、ext4，不支持 XFS。

（1）查看原容器的磁盘空间大小，如图 6-9 所示。

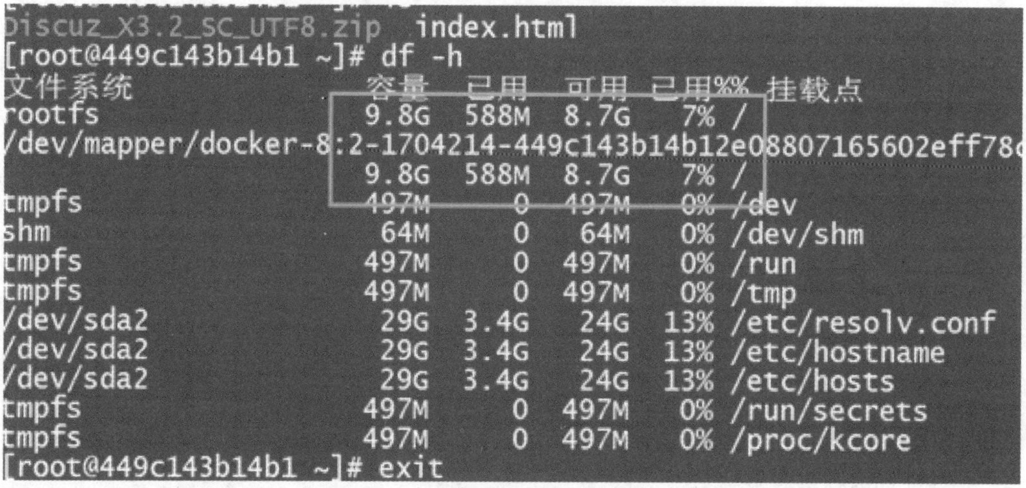

```
Discuz_X3.2_SC_UTF8.zip  index.html
[root@449c143b14b1 ~]# df -h
文件系统                       容量  已用  可用 已用%% 挂载点
rootfs                          9.8G  588M  8.7G    7% /
/dev/mapper/docker-8:2-1704214-449c143b14b12e08807165602eff78c
                                9.8G  588M  8.7G    7% /
tmpfs                           497M     0  497M    0% /dev
shm                              64M     0   64M    0% /dev/shm
tmpfs                           497M     0  497M    0% /run
tmpfs                           497M     0  497M    0% /tmp
/dev/sda2                        29G  3.4G   24G   13% /etc/resolv.conf
/dev/sda2                        29G  3.4G   24G   13% /etc/hostname
/dev/sda2                        29G  3.4G   24G   13% /etc/hosts
tmpfs                           497M     0  497M    0% /run/secrets
tmpfs                           497M     0  497M    0% /proc/kcore
[root@449c143b14b1 ~]# exit
```

图 6-9　查看原容器的磁盘空间大小

（2）查看 Mapper 设备，如图 6-10 所示。

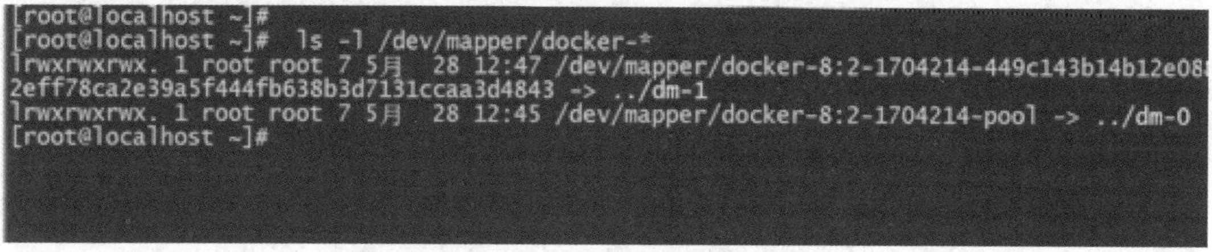

```
[root@localhost ~]#
[root@localhost ~]#  ls -l /dev/mapper/docker-*
lrwxrwxrwx. 1 root root 7 5月  28 12:47 /dev/mapper/docker-8:2-1704214-449c143b14b12e08
2eff78ca2e39a5f444fb638b3d7131ccaa3d4843 -> ../dm-1
lrwxrwxrwx. 1 root root 7 5月  28 12:45 /dev/mapper/docker-8:2-1704214-pool -> ../dm-0
[root@localhost ~]#
```

图 6-10　查看 Mapper 设备

（3）查看卷信息表，如图 6-11 所示。

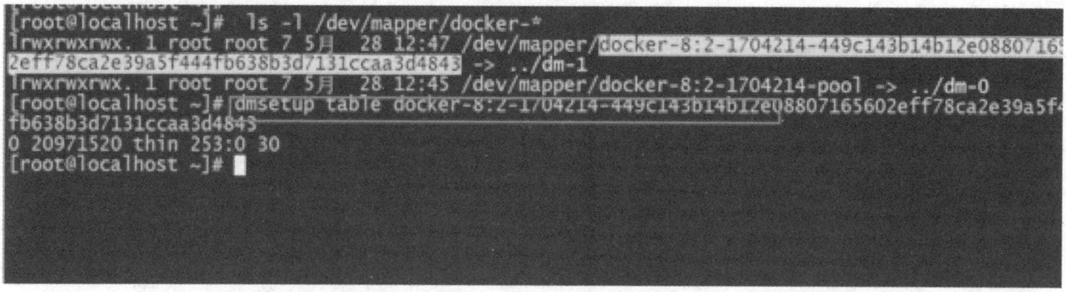

```
[root@localhost ~]#  ls -l /dev/mapper/docker-*
lrwxrwxrwx. 1 root root 7 5月  28 12:47 /dev/mapper/docker-8:2-1704214-449c143b14b12e0880716
2eff78ca2e39a5f444fb638b3d7131ccaa3d4843 -> ../dm-1
lrwxrwxrwx. 1 root root 7 5月  28 12:45 /dev/mapper/docker-8:2-1704214-pool -> ../dm-0
[root@localhost ~]# dmsetup table docker-8:2-1704214-449c143b14b12e08807165602eff78ca2e39a5f4
fb638b3d7131ccaa3d4843
0 20971520 thin 253:0 30
[root@localhost ~]#
```

图 6-11　查看卷信息表

(4)根据要扩展的大小，计算需要多少扇区。第二个数字是设备的大小，表示有多少个 512B 的扇区，这个值略高于 10GB 的大小。计算一个 15GB 的卷需要多少扇区，操作指令如下：

```
echo $((1510241024*1024/512)) 31457280
```

(5) 修改卷信息表，激活并验证（虚线框内 3 部分），如图 6-12 所示。

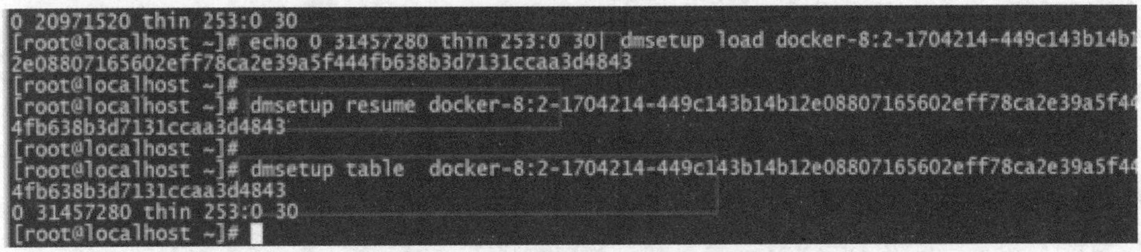

```
0 20971520 thin 253:0 30
[root@localhost ~]# echo 0 31457280 thin 253:0 30| dmsetup load docker-8:2-1704214-449c143b14b1
2e08807165602eff78ca2e39a5f444fb638b3d7131ccaa3d4843
[root@localhost ~]#
[root@localhost ~]# dmsetup resume docker-8:2-1704214-449c143b14b12e08807165602eff78ca2e39a5f44
4fb638b3d7131ccaa3d4843
[root@localhost ~]#
[root@localhost ~]# dmsetup table  docker-8:2-1704214-449c143b14b12e08807165602eff78ca2e39a5f44
4fb638b3d7131ccaa3d4843
0 31457280 thin 253:0 30
[root@localhost ~]#
```

图 6-12　修改卷信息表并激活验证

(6) 修改文件系统大小，如图 6-13 所示。

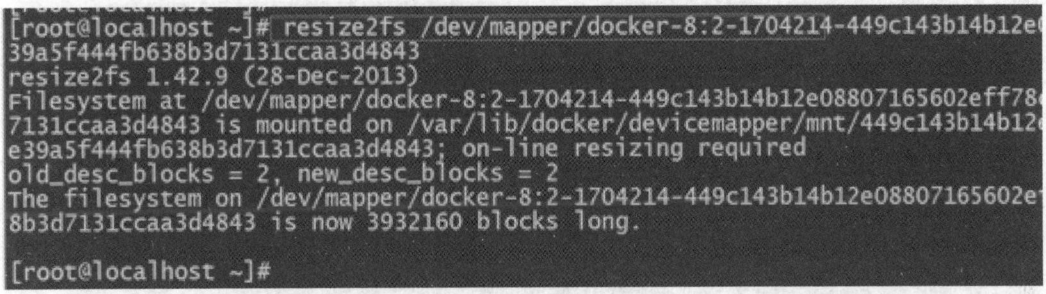

```
[root@localhost ~]# resize2fs /dev/mapper/docker-8:2-1704214-449c143b14b12e
39a5f444fb638b3d7131ccaa3d4843
resize2fs 1.42.9 (28-Dec-2013)
Filesystem at /dev/mapper/docker-8:2-1704214-449c143b14b12e08807165602eff78
7131ccaa3d4843 is mounted on /var/lib/docker/devicemapper/mnt/449c143b14b12
e39a5f444fb638b3d7131ccaa3d4843; on-line resizing required
old_desc_blocks = 2, new_desc_blocks = 2
The filesystem on /dev/mapper/docker-8:2-1704214-449c143b14b12e08807165602e
8b3d7131ccaa3d4843 is now 3932160 blocks long.

[root@localhost ~]#
```

图 6-13　修改文件系统大小

(7) 最后验证磁盘大小，如图 6-14 所示。

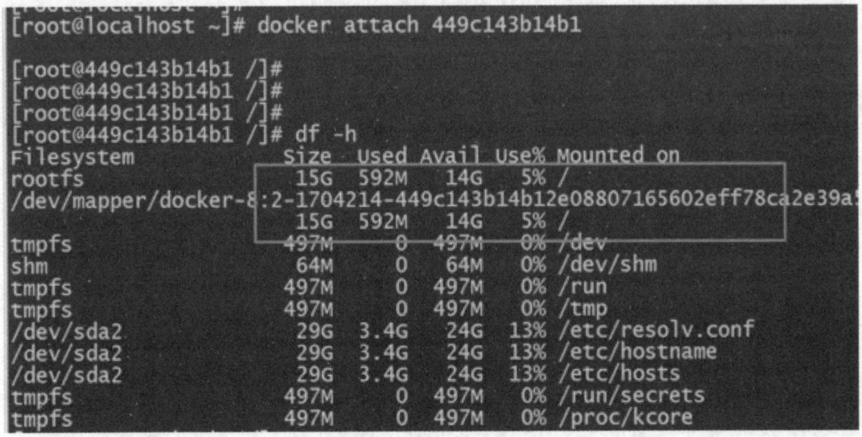

```
[root@localhost ~]# docker attach 449c143b14b1

[root@449c143b14b1 /]#
[root@449c143b14b1 /]#
[root@449c143b14b1 /]#
[root@449c143b14b1 /]# df -h
Filesystem              Size  Used Avail Use% Mounted on
rootfs                   15G  592M   14G   5% /
/dev/mapper/docker-8:2-1704214-449c143b14b12e08807165602eff78ca2e39a5
                         15G  592M   14G   5% /
tmpfs                   497M     0  497M   0% /dev
shm                      64M     0   64M   0% /dev/shm
tmpfs                   497M     0  497M   0% /run
tmpfs                   497M     0  497M   0% /tmp
/dev/sda2                29G  3.4G   24G  13% /etc/resolv.conf
/dev/sda2                29G  3.4G   24G  13% /etc/hostname
/dev/sda2                29G  3.4G   24G  13% /etc/hosts
tmpfs                   497M     0  497M   0% /run/secrets
tmpfs                   497M     0  497M   0% /proc/kcore
```

图 6-14　容器磁盘扩容实战

根据以上所有操作步骤和指令操作，Docker 容器成功扩容。当然，以上步骤也可以写成脚本，然后使用脚本批量扩容分区大小。

6.5 Docker 磁盘、内存、CPU 资源实战二

Docker 容器默认启动的虚拟机，会占用宿主机的资源（CPU、内存、硬盘）。例如，默认 Docker 基于 Overlay2 驱动方式，容器硬盘的 rootfs 根分区空间是整个宿主机的空间大小。

可以指定默认容器的大小（在启动容器时指定），可以在 Docker 配置文件/etc/sysconfig/docker 中，OPTIONS 参数后面添加以下代码，指定 Docker 容器 rootfs 容量大小为 10GB。

```
OPTIONS='--storage-opt overlay2.size=10G'
```

以上方法只适用于新容器生成，且修改后需要重启 Docker，无法做到动态地给正在运行的容器指定大小。默认容器磁盘空间如图 6-15 所示。

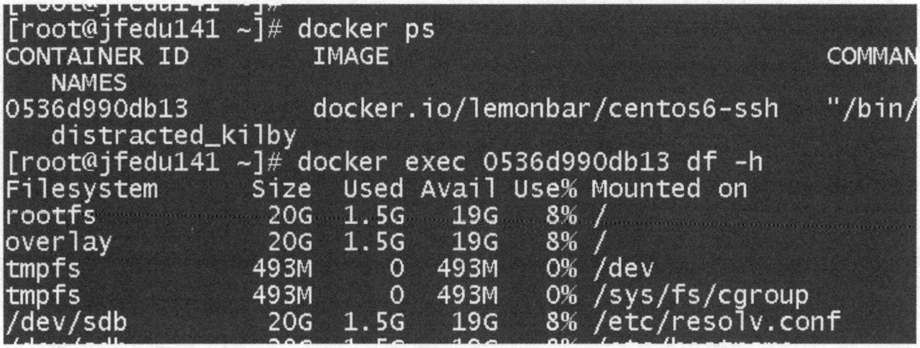
```
[root@jfedu141 ~]# docker ps
CONTAINER ID        IMAGE                              COMMAN
   NAMES
0536d990db13        docker.io/lemonbar/centos6-ssh     "/bin/
   distracted_kilby
[root@jfedu141 ~]# docker exec 0536d990db13 df -h
Filesystem      Size  Used Avail Use% Mounted on
rootfs           20G  1.5G   19G   8% /
overlay          20G  1.5G   19G   8% /
tmpfs           493M     0  493M   0% /dev
tmpfs           493M     0  493M   0% /sys/fs/cgroup
/dev/sdb         20G  1.5G   19G   8% /etc/resolv.conf
```

图 6-15 Docker 容器磁盘空间

修改 Docker 存储配置文件，加入以下代码（默认如果已经为 overlay2，则无须修改）：

```
#修改 Docker 引擎存储配置
vi /etc/sysconfig/docker-storage
DOCKER_STORAGE_OPTIONS="--storage-driver overlay2"
#重启 Docker 引擎服务
service docker restart
```

Overlay2 Docker 磁盘驱动模式下，如果要调整容器大小，通过以上方法会导致 Docker 引擎服务无法启动，还需要让 Linux 文件系统设置为 xfs，并支持目录级别的磁盘配额功能。

CentOS 7.x xfs 磁盘配额配置，新添加一块硬盘，设置磁盘配额方法步骤如下。

（1）添加新的硬盘，如图 6-16 所示。

（2）格式化硬盘为 xfs 文件系统格式，操作指令如下：

```
mkfs.xfs -f /dev/sdb
```

第 6 章 Docker 仓库案例实战

（3）创建 data 目录，后续将作为 Docker 数据目录。

```
mkdir /data/ -p
```

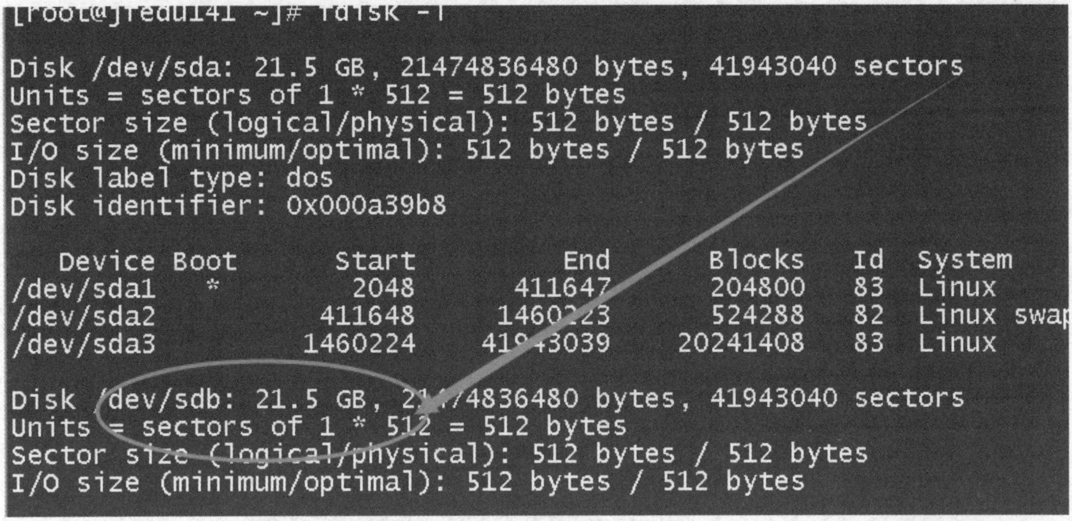

```
[root@jfedu141 ~]# fdisk -l

Disk /dev/sda: 21.5 GB, 21474836480 bytes, 41943040 sectors
Units = sectors of 1 * 512 = 512 bytes
Sector size (logical/physical): 512 bytes / 512 bytes
I/O size (minimum/optimal): 512 bytes / 512 bytes
Disk label type: dos
Disk identifier: 0x000a39b8

   Device Boot      Start         End      Blocks   Id  System
/dev/sda1   *        2048      411647      204800   83  Linux
/dev/sda2         411648     1460223      524288   82  Linux swap
/dev/sda3        1460224    41943039    20241408   83  Linux

Disk /dev/sdb: 21.5 GB, 21474836480 bytes, 41943040 sectors
Units = sectors of 1 * 512 = 512 bytes
Sector size (logical/physical): 512 bytes / 512 bytes
I/O size (minimum/optimal): 512 bytes / 512 bytes
```

图 6-16　添加新的硬盘

（4）挂载 data 目录，并开启磁盘配额功能（默认 xfs 支持配额功能），如图 6-17 所示。

```
mount -o uquota,prjquota /dev/sdb /data/
```

```
[root@jfedu141 ~]#
[root@jfedu141 ~]# mount -o uquota,prjquota /dev/sdb /data/
[root@jfedu141 ~]#
[root@jfedu141 ~]# cd /data/
[root@jfedu141 data]# ls
[root@jfedu141 data]#
[root@jfedu141 data]# ll
total 0
[root@jfedu141 data]# mkdir docker
[root@jfedu141 data]# ls
docker
[root@jfedu141 data]#
[root@jfedu141 data]# mount|grep data
/dev/sdb on /data type xfs (rw,relatime,attr2,inode64,usrquota,prjquota)
[root@jfedu141 data]#
```

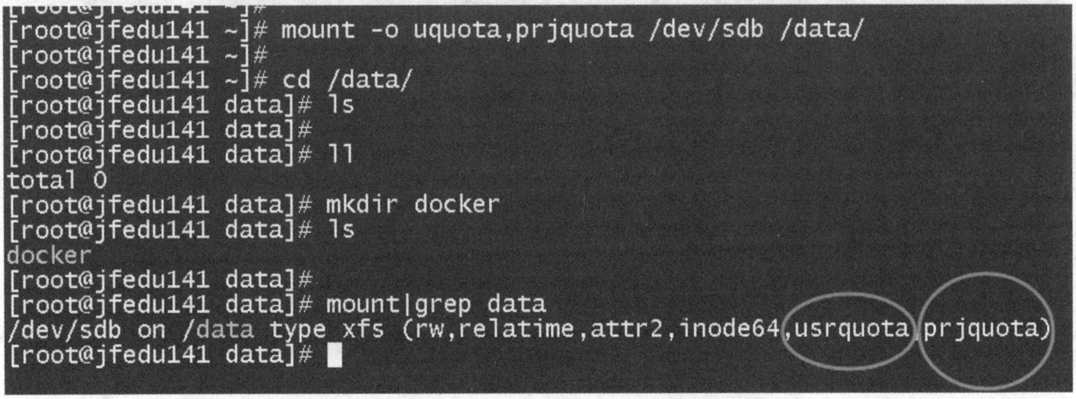

图 6-17　挂载 data 目录并开启磁盘配额功能

挂载配额类型有以下几种。

① 根据用户（uquota/usrquota/quota）。

② 根据组（gquota/grpquota）。

③ 根据目录（pquota/prjquota）（不能与 grpquota 同时设定）。

（5）查看配额配置详情，命令如下，如图 6-18 所示。

```
xfs_quota -x -c 'report' /data/
```

```
[root@jfedu141 ~]# xfs_quota -x -c 'report' /data/
User quota on /data (/dev/sdb)
                        Blocks
User ID      Used       Soft        Hard      Warn/Grace
---------- --------------------------------------------------
root            0          0           0      00 [--------]

Project quota on /data (/dev/sdb)
                        Blocks
Project ID   Used       Soft        Hard      Warn/Grace
---------- --------------------------------------------------
#0              0          0           0      00 [--------]

[root@jfedu141 ~]#
```

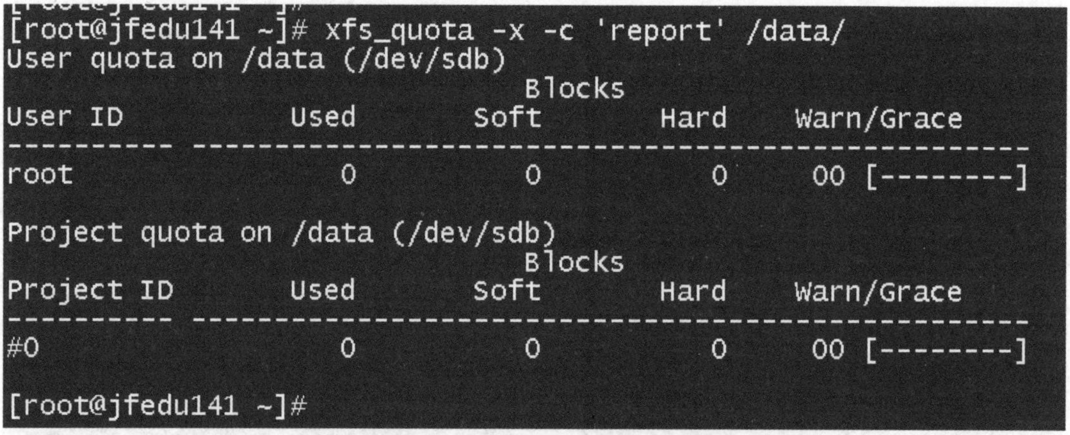

图 6-18　查看配额配置详情

（6）可以通过 xfs_quota 命令为用户和目录分配配额，也可以通过命令查看配额信息，如图 6-19 所示。

```
xfs_quota -x -c 'limit bsoft=10M bhard=10M jfedu' /data
xfs_quota -x -c 'report' /data/
```

```
[root@jfedu141 ~]# xfs_quota -x -c 'limit bsoft=10M bhard=10M jfedu' /data
[root@jfedu141 ~]#
[root@jfedu141 ~]# xfs_quota -x -c 'report' /data/
User quota on /data (/dev/sdb)
                        Blocks
User ID      Used       Soft        Hard      Warn/Grace
---------- --------------------------------------------------
root            0          0           0      00 [--------]
jfedu           0      10240       10240      00 [--------]

Project quota on /data (/dev/sdb)
                        Blocks
Project ID   Used       Soft        Hard      Warn/Grace
```

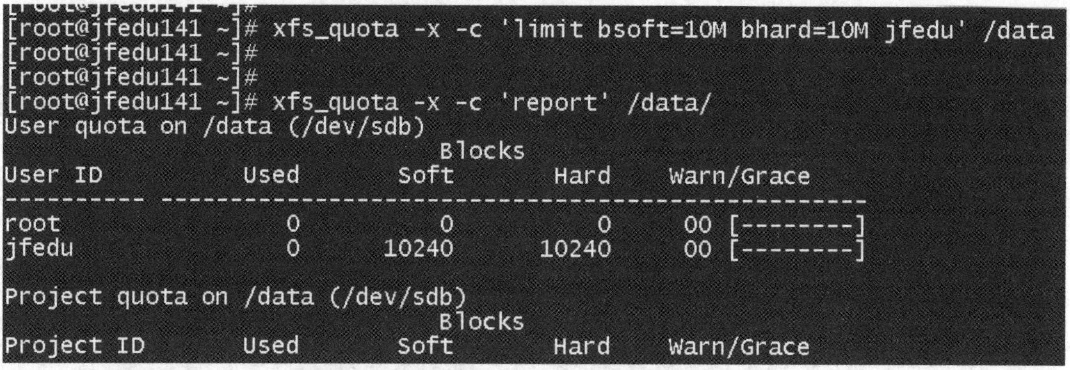

图 6-19　分配和查看配额

（7）将 Docker 引擎默认数据存储于目录/var/lib/docker 下并重命名，然后将/data/docker 目录软链接至/var/lib/下即可。操作指令如下：

```
mkdir -p /data/docker/
cd /var/lib/
mv docker docker.bak
ln -s /data/docker/ ./
```

（8）重启 Docker 服务，并查看进程，可以看到 docker overlay2.size 大小配置，如图 6-20 所示。

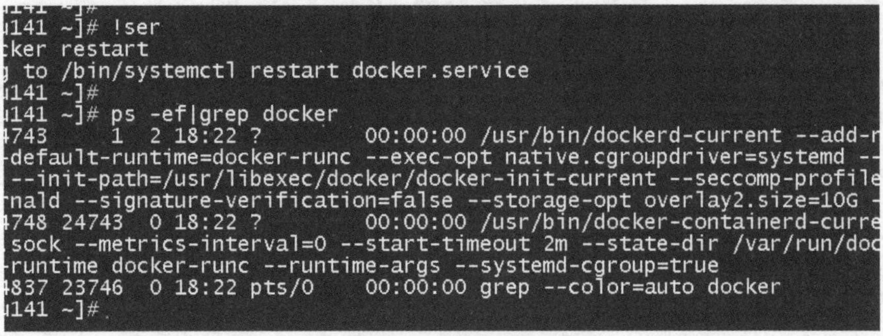

图 6-20　查看进程

（9）基于 Docker 客户端指令启动 Docker 容器，并查看最新容器的磁盘空间为 10GB，则设置容器大小成功，如图 6-21 所示。

（a）

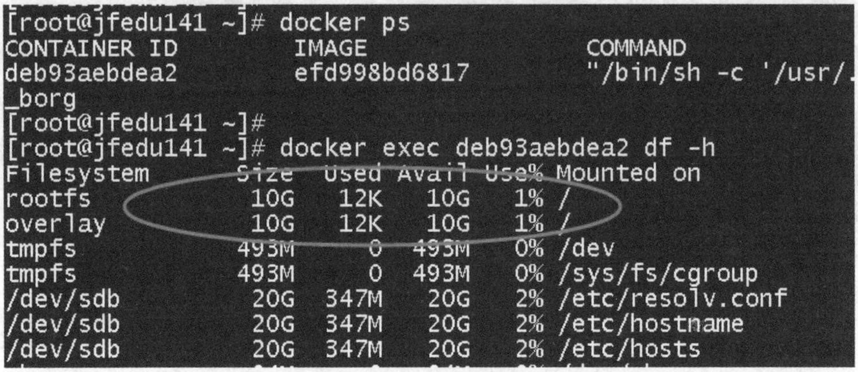

（b）

图 6-21　Docker 容器硬盘扩容

（a）通过 docker run 命令启动新的容器；（b）查看 Docker 容器硬盘容量

6.6 Docker 资源监控方案和监控实战

Docker 虚拟化平台自带了 Docker 容器（虚拟机）资源监控的功能，通过对 Docker 容器的资源监控，用户可以随时掌握容器进行相关的资源性能，对容器性能进行更好的评估。通常，监控容器资源的指标主要包括以下几方面。

（1）容器的 CPU 情况和使用量。

（2）容器的内存情况和使用量。

（3）容器的本地镜像情况。

（4）主机的容器运行情况。

基于 docker ps -a 和 docker images 命令查看容器的本地镜像、容器运行的情况，使用 docker stats 命令可以监控相关容器实例情况。

6.7 Docker stats 监控工具

（1）查看本地镜像和容器列表，命令如下，如图 6-22 所示。

```
docker images
docker ps
```

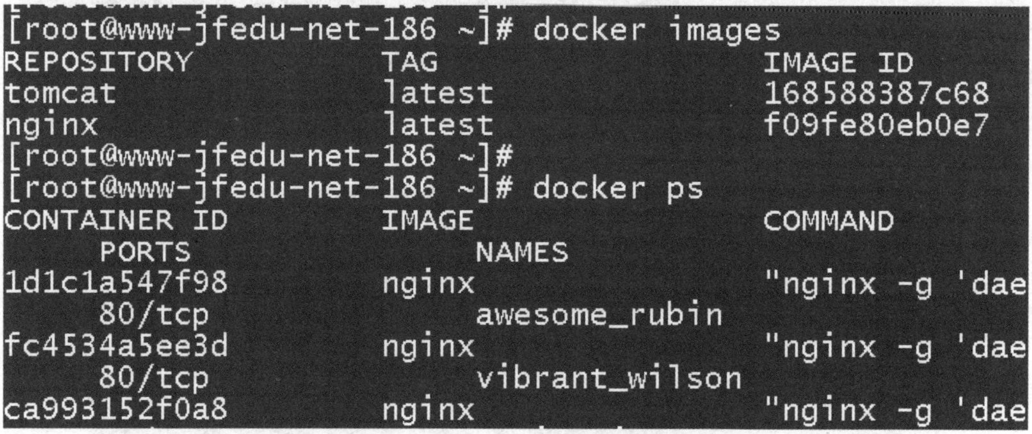

```
[root@www-jfedu-net-186 ~]# docker images
REPOSITORY          TAG                 IMAGE ID
tomcat              latest              168588387c68
nginx               latest              f09fe80eb0e7
[root@www-jfedu-net-186 ~]#
[root@www-jfedu-net-186 ~]# docker ps
CONTAINER ID        IMAGE               COMMAND
     PORTS               NAMES
1d1c1a547f98        nginx               "nginx -g 'dae
     80/tcp              awesome_rubin
fc4534a5ee3d        nginx               "nginx -g 'dae
     80/tcp              vibrant_wilson
ca993152f0a8        nginx               "nginx -g 'dae
```

图 6-22　查看本地镜像和容器列表

（2）通过 docker stats 查看所有运行容器的资源，如图 6-23 所示。

```
docker stats
```

（3）通过 docker stats 容器 ID，查看指定容器的资源，如图 6-24 所示。

```
docker stats 1d1c1a547f98
```

图 6-23 查看所有运行容器的资源

图 6-24 查看指定容器的资源

（4）通过 docker stats 容器 ID，指定参数 --no-stream 非流式查看容器资源，操作指令如下：

```
docker stats 1d1c1a547f98 --no-stream
```

（5）获取容器 CPU 的信息，操作指令如下：

```
docker stats 1d1c1a547f98 --no-stream|awk 'NR&gt;1 {print $1,"CPU:"$3}'
```

（6）获取容器 MEM 的信息，操作指令如下：

```
docker stats 1d1c1a547f98 --no-stream|awk 'NR&gt;1 {print $1,"MEM:"$4}'
```

（7）获取容器 I/O 读写的信息，操作指令如下：

```
docker stats 1d1c1a547f98 --no-stream|awk 'NR&gt;1 {print $1,"IO:"$(NF-1)}'
```

如果想看到更为详细的容器属性，还可以通过 netcat 命令，使用 Docker 远程 API 查看。发送一个 HTTP GET 请求 /containers/[CONTAINER_NAME]，其中 CONTAINER_NAME 是想要统计的容器名称，从这里可以看到一个容器 stats 请求的完整响应信息。

当然这种方式并不令人满意，大家还是希望能够看到非常直观、详细的可视化界面。

6.8 CAdvisor 监控工具

CAdvisor 是 Google 开发的容器监控工具。CAdvisor 是一个易于设置且非常有用的工具，用户无须登录服务器即可查看资源消耗，而且它还可以生成图表。此外，当集群需要额外的资源时，压力表（Pressure Gauges）可提供快速预览。CAdvisor 是免费、开源的。

CAdvisor 的资源消耗也比较低。但是它也有局限性：它只能监控一个 Docker 主机。因此，如果是多节点，那就比较麻烦，需要在所有的主机上都安装一个 CAdvisor。值得注意的是，如果使用的是 Kubernetes，可以使用 heapster 监控多节点集群。在图表中的数据仅仅是时长 1min 的移动窗口，无法查看长期趋势。资源使用率在危险水平时，它也没有生成警告的机制。

如果在 Docker 节点的资源消耗方面没有任何可视化界面，那么 CAdvisor 是一个不错的步入容器监控的开端。然而，如果打算在容器中运行任何关键任务，那就需要一个更强大的工具或者方法。

6.9 CAdvisor 部署配置

（1）下载 CAdvisor 镜像，操作指令如下：

```
docker pull google/cadvisor
```

（2）基于镜像启动 CAdvisor 容器，操作指令如下：

```
docker run -v /var/run:/var/run:rw -v /sys:/sys:ro -v /var/lib/docker:/var/lib/docker:ro -p 8080:8080 -d --name cadvisor google/cadvisor
```

（3）浏览器访问 8080，地址为 http://106.12.133.186:8080/containers/，如图 6-25 所示。

```
root

Docker Containers
Subcontainers

/docker

/system.slice

/user.slice
```

图 6-25　CAdvisor 容器资源监控

（4）查看某个容器的详细资源，如图 6-26 所示。

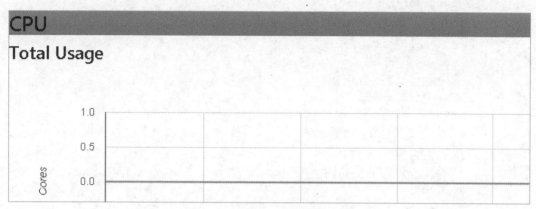

图 6-26　查看某个容器的详细资源

6.10　构建 CAdvisor+InfluxDB+Grafana 平台

1）CAdvisor 简介

CAdvisor 提供的操作界面略显简陋，且需要在不同页面之间跳转，模拟只能监控一台服务器，这不免会让人质疑其实用性。但 CAdvisor 的一个亮点是它可以将监控到的数据导出给第三方工具，由这些工具进一步加工处理。

可以把 CAdvisor 定位为一个监控数据收集器，收集和导出数据是它的强项，而非展示数据。

2）InfluxDB 简介

InfluxDB 是用 Go 语言编写的一个开源分布式时序、事件和指标数据库，无须外部依赖。类似的数据库有 Elasticsearch、Graphite 等。

3）Grafana

Grafana 是一个可视化面板（Dashboard），有着非常漂亮的图表和布局展示、功能齐全的度量仪表盘和图形编辑器，支持 Graphite、Zabbix、InfluxDB、Prometheus 和 OpenTSDB 作为数据源。Grafana 主要特性：灵活丰富的图形化选项；可以混合多种风格；支持白天和夜间模式；多个数据源。

4）CAdvisor+InfluxDB+Grafana 部署

CAdvisor+InfluxDB+Grafana 单独部署方式比较烦琐，此处采用 Docker-compose 方式部署，首先编写 compose.yml 文件，然后启动 compose 相关容器服务即可。

（1）Docker-compose.yml 文件代码如下：

```
version: '3'services:
  influxdb:
```

```yaml
        image: tutum/influxdb:0.9
        container_name: influxdb
        restart: always
        environment:
            - PRE_CREATE_DB=cadvisor
        ports:
            - "8083:8083"
            - "8086:8086"
        expose:
            - "8090"
            - "8099"
        volumes:
            - influxdbData:/data
    cadvisor:
        image: google/cadvisor
        container_name: cadvisor
        links:
            - influxdb:influxsrv
        command: -storage_driver=influxdb -storage_driver_db=cadvisor -storage_driver_host=influxsrv:8086
        restart: always
        ports:
            - "8080:8080"
        volumes:
            - /:/rootfs:ro
            - /var/run:/var/run:rw
            - /sys:/sys:ro
            - /var/lib/docker/:/var/lib/docker:ro
    grafana:
        image: grafana/grafana
        container_name: grafana
        restart: always
        links:
            - influxdb:influxsrv
        ports:
            - "3000:3000"
        environment:
            - HTTP_USER=admin
            - HTTP_PASS=admin
            - INFLUXDB_HOST=influxsrv
            - INFLUXDB_PORT=8086
            - INFLUXDB_NAME=cadvisor
            - INFLUXDB_USER=root
            - INFLUXDB_PASS=root

volumes:
    influxdbData:
```

启动 Docker-compose，命令如下：

```
docker-compose up -d
```

（2）Docker-compose 默认会启动三个类别容器，分别为 Grafana、CAdvisor 和 InfluxDB，对外访问 IP+端口如下：

- Grafana：http://106.12.133.186:3000/。
- CAdvisor：http://106.12.133.186:8080/。
- Influxdb：http://106.12.133.186:8086/。

（3）浏览器访问 Grafana Web 界面，默认用户名和密码为 admin/admin，然后选择 add-database source，填写 InfluxDB 数据库的 IP 和端口，数据库名为 cadvisor，用户名和密码为 admin/admin，访问 URL 地址 http://106.12.133.186:3000/login，如图 6-27 所示。

图 6-27　浏览器访问 Grafana Web 界面

（4）创建 Grafana 图像，设置监控项目，例如添加 MEM 内存使用监控，操作方法如图 6-28 所示。

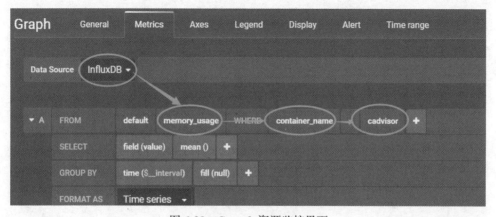

图 6-28　Granafa 资源监控界面

（5）创建 Grafana 图像，设置监控项目，例如添加 CPU 使用监控，操作方法如图 6-29 所示。

图 6-29　添加 CPU 使用监控

第 7 章 Docker Compose 容器编排实战

Docker Compose 是 Docker 官方编排（Orchestration）项目之一，负责快速在集群中部署分布式应用。Compose 定位是 defining and running complex applications with Docker，前身是 Fig，兼容 Fig 的模板文件。

7.1 Docker Compose 概念剖析

学了前面的章节内容，读者应该已经知道 Dockerfile 可以让用户管理一个单独的应用容器，而本章即将学习的 Compose 则是允许用户在一个模板（YAML 格式）中定义一组相关联的应用容器（称为一个 project，即项目）。例如，一台 Web 服务容器关联后端的数据库服务容器等。

Docker Compose 将所管理的容器分为三层，分别是工程（Project）、服务（Service）以及容器（Container）。Docker Compose 运行目录下的所有文件（docker-compose.yml、extends 文件或环境变量文件等）组成一个工程，若无特殊指定工程名即为当前目录名。一个工程当中可包含多个服务，每个服务中定义了容器运行的镜像、参数和依赖。

一个服务当中可包括多个容器实例，Docker Compose 并没有解决负载均衡的问题，因此需要借助其他工具实现服务发现及负载均衡。

Docker Compose 的配置文件默认为"docker-compose.yml"，可通过环境变量 COMPOSE_FILE 或-f 参数自定义。配置文件定义了多个有依赖关系的服务及每个服务运行的容器。

使用 Dockerfile 模板文件，可以让用户很方便地定义一个单独的应用容器。在工作中，经常会碰到需要多个容器相互配合完成某项任务的情况。例如，要实现一个 Web 项目，除了 Web 服务容器本身，往往还需要再加上后端的数据库服务容器，甚至包括负载均衡容器等。

Docker Compose 允许用户通过单独的 docker-compose.yml 模板文件（YAML 格式）定义一组相关联的应用容器为一个项目。

Docker Compose 项目用 Python 语言编写，调用 Docker 服务提供的 API 对容器进行管理。

因此，只要所操作的平台支持 Docker API，就可以在其上利用 Compose 进行编排管理。

7.2 Docker Compose 部署安装

安装 Docker Compose 之前，需要先安装 Docker 引擎服务，此处使用 Compose 镜像方式安装 Compose 项目，然后执行如下命令：

```
#下载 Docker Compose 脚本
curl -L https://github.com/docker/compose/releases/download/1.18.0/run.sh > /usr/local/bin/docker-compose
#添加脚本 x 权限
chmod +x /usr/local/bin/docker-compose
#查看其版本信息
docker-compose --version
ln -s /usr/local/bin/docker-compose /usr/bin/docker-compose
```

7.3 Docker Compose 命令实战

Docker Compose 跟 Docker 客户端一样有很多指令，如表 7-1 所示。

表 7-1　Docker Compose 指令操作

命令名称	命令功能	中文备注
build	Build or rebuild services	构建服务
config	Validate and view the Compose file	配置和查看
create	Create services	创建服务
down	Stop and remove containers	停止和移除
exec	Execute a command in a running container	执行命令
help	Get help on a command	帮助命令
images	List images	查看镜像
kill	Kill containers	杀掉容器
logs	View output from containers	查看日志
pause	Pause services	暂停服务
port	Print the public port for a port binding	打印端口
ps	List containers	列出容器
pull	Pull service images	下载镜像
push	Push service images	上传镜像
restart	Restart services	重启服务

（续表）

命令名称	命令功能	中文备注
rm	Remove stopped containers	删除容器
run	Run a one-off command	运行命令
start	Start services	启动服务
stop	Stop services	停止服务
top	Display the running processes	查看服务进程
up	Create and start containers	启动容器
version	Docker Compose version information	查看版本信息

7.4 Docker Compose 常见概念

1）服务（Service）

一个应用容器，实际上可以运行多个相同镜像的实例。

2）项目（Project）

由一组关联的应用容器组成的一个完整业务单元。

一个项目可以由多个服务（容器）关联而成，Compose 是面向项目进行管理。

7.5 Docker Compose 语法详解

Docker Compose 语法详解如下：

```
version
#指定本 yml 是依从 Compose 哪个版本制定的
build
#指定为构建镜像上下文路径
#例如 webapp 服务,指定为从上下文路径 ./dir/Dockerfile 所构建的镜像
version: "3"
services:
  webapp:
    build: ./dir
#或者,作为具有在上下文指定的路径的对象,以及可选的 Dockerfile 和 args
version: "3"
services:
  webapp:
    build:
      context: ./dir
      dockerfile: Dockerfile-alternate
```

```yaml
      args:
        buildno: 1
      labels:
        - "com.example.description=Accounting webapp"
        - "com.example.department=Finance"
        - "com.example.label-with-empty-value"
      target: prod
```
#context: 上下文路径
#dockerfile: 指定构建镜像的 Dockerfile 文件名
#args: 添加构建参数,这是只能在构建过程中访问的环境变量
#labels: 设置构建镜像的标签
#target: 多层构建,可以指定构建哪一层

cap_add, cap_drop
#添加或删除容器拥有的宿主机的内核功能

```yaml
cap_add:
  - ALL                                              #开启全部权限
cap_drop:
  - SYS_PTRACE                                       #关闭 ptrace 权限
```

cgroup_parent
#为容器指定父 cgroup 组,意味着将继承该组的资源限制

```yaml
cgroup_parent: m-executor-abcd
```

command
#覆盖容器启动的默认命令

```yaml
command: ["bundle", "exec", "thin", "-p", "3000"]
```

container_name
#指定自定义容器名称,而不是生成的默认名称

```yaml
container_name: my-web-container
```

depends_on
#设置依赖关系
#docker-compose up: 以依赖性顺序启动服务。在以下示例中,先启动 DB 和 Redis,才会
#启动 Web
#docker-compose up SERVICE: 自动包含 Service 的依赖项。在以下示例中,docker-
#compose up web 还将创建并启动 db 和 redis
#docker-compose stop: 按依赖关系顺序停止服务。在以下示例中,Web 在 DB 和 Redis 之
#前停止

```yaml
version: "3"
services:
  web:
    build: .
    depends_on:
      - db
      - redis
  redis:
    image: redis
```

```yaml
    db:
      image: postgres
#注意：Web 服务不会等待 Redis DB 完全启动之后才启动
deploy
#指定与服务的部署和运行有关的配置。只在 Swarm 模式下才会有用
version: "3"
services:
  redis:
    image: redis:alpine
    deploy:
      mode: replicated
      replicas: 6
      endpoint_mode: dnsrr
      labels:
        description: "This redis service label"
      resources:
        limits:
          cpus: '0.50'
          memory: 50M
        reservations:
          cpus: '0.25'
          memory: 20M
      restart_policy:
        condition: on-failure
        delay: 5s
        max_attempts: 3
        window: 120s
```

#可以选择以下参数

#endpoint_mode：访问集群服务的方式

endpoint_mode: vip

#Docker 集群服务一个对外的虚拟 IP。所有请求都会通过这个虚拟 IP 到达集群服务内部的机器

endpoint_mode: dnsrr

#DNS 轮询（DNSRR）。所有请求会自动轮询获取到集群 IP 列表中的一个 IP 地址

#labels：在服务上设置标签。可以用容器上的 labels（跟 deploy 同级的配置）覆盖 deploy
#下的 labels

#mode：指定服务提供的模式

#replicated：复制服务，复制指定服务到集群的机器上

#global：全局服务，服务将部署至集群的每个节点

#replicas: mode 为 replicated 时，需要使用此参数配置具体运行的节点数量

#resources：配置服务器资源使用的限制，例如上例中，配置 Redis 集群运行需要的 CPU 的百
#分比和内存的占用，避免占用资源过高出现异常

#restart_policy：配置如何在退出容器时重新启动容器

#condition：可选 none、on-failure 或 any（默认值为 any）

#delay：设置多久之后重启（默认值为 0）

```
#max_attempts：尝试重新启动容器的次数，超出次数则不再尝试（默认值为"一直重试"）
#window：设置容器重启超时时间（默认值为 0）
#rollback_config：配置在更新失败的情况下应如何回滚服务
#parallelism：一次要回滚的容器数。如果设置为 0，则所有容器将同时回滚
#delay：每个容器组回滚之间等待的时间（默认为 0s）
#failure_action：如果回滚失败，该怎么办。可以设为 continue 或者 pause（默认为 pause）
#monitor：每个容器更新后，持续观察是否失败了的时间 (ns|us|ms|s|m|h)（默认为 0s）
#max_failure_ratio：在回滚期间可以容忍的故障率（默认为 0）
#order：回滚期间的操作顺序。其中一个 stop-first（串行回滚），或者 start-first（并行
#回滚）（默认为 stop-first）
#update_config：配置应如何更新服务，对于配置滚动更新很有用
#parallelism：一次更新的容器数
#delay：在更新一组容器之间等待的时间
#failure_action：如果更新失败，该怎么办。可设为 continue、rollback 或 pause （默
#认为 pause）
#monitor：每个容器更新后，持续观察是否失败了的时间 (ns|us|ms|s|m|h)（默认为 0s）
#max_failure_ratio：在更新过程中可以容忍的故障率
#order：回滚期间的操作顺序。可设为 stop-first（串行回滚）或 start-first（并行回滚）
#（默认为 stop-first）
#注：仅支持 V3.4 及更高版本
devices
#指定设备映射列表
devices:
  - "/dev/ttyUSB0:/dev/ttyUSB0"
dns
#自定义 DNS 服务器，可以是单个值或列表的多个值
dns: 8.8.8.8
dns:
  - 8.8.8.8
  - 9.9.9.9
dns_search
#自定义 DNS 搜索域，可以是单个值或列表
dns_search: example.com
dns_search:
  - dc1.example.com
  - dc2.example.com
entrypoint
#覆盖容器默认的 entrypoint
entrypoint: /code/entrypoint.sh
#也可以是以下格式
entrypoint:
    - php
    - -d
```

```yaml
      - zend_extension=/usr/local/lib/php/extensions/no-debug-non-zts-20100525/xdebug.so
      - -d
      - memory_limit=-1
      - vendor/bin/phpunit
env_file
#从文件添加环境变量，可以是单个值或列表的多个值
env_file: .env
#也可以是列表格式
env_file:
  - ./common.env
  - ./apps/web.env
  - /opt/secrets.env
environment
#添加环境变量。可以使用数组或字典、任何布尔值。布尔值需要用引号引起来，以确保 YML 解析
#器不会将其转换为 True 或 False
environment:
  RACK_ENV: development
  SHOW: 'true'
expose
#暴露端口，但不映射到宿主机，只被连接的服务访问
#仅可以指定内部端口为参数
expose:
  - "3000"
  - "8000"
extra_hosts
#添加主机名映射。类似 docker client --add-host
extra_hosts:
  - "somehost:162.242.195.82"
  - "otherhost:50.31.209.229"
#以上会在此服务的内部容器中 /etc/hosts 创建一个具有 IP 地址和主机名的映射关系
162.242.195.82  somehost
50.31.209.229   otherhost
healthcheck
#用于检测 Docker 服务是否健康运行
healthcheck:
  test: ["CMD", "curl", "-f", "http://localhost"]      #设置检测程序
  interval: 1m30s                                       #设置检测间隔
  timeout: 10s                                          #设置检测超时时间
  retries: 3                                            #设置重试次数
  start_period: 40s                                     #启动后，多少秒开始启动检测程序
image
#指定容器运行的镜像，以下格式都可以
image: redis
```

```yaml
    image: ubuntu:14.04
    image: tutum/influxdb
    image: example-registry.com:4000/postgresql
    image: a4bc65fd                                    #镜像ID
    logging
#服务的日志记录配置
#driver: 指定服务容器的日志记录驱动程序,默认值为json-file,有以下三个选项
    driver: "json-file"
    driver: "syslog"
    driver: "none"
#仅在 json-file 驱动程序下,可以使用以下参数,限制日志的数量和大小
    logging:
      driver: json-file
      options:
        max-size: "200K"                               #单个文件大小为200KB
        max-file: "10"                                 #最多10个文件
#当达到文件限制上限,会自动删除旧的文件
#syslog 驱动程序下,可以使用 syslog-address 指定日志接收地址
    logging:
      driver: syslog
      options:
        syslog-address: "tcp://192.168.0.42:123"
    network_mode
#设置网络模式
    network_mode: "bridge"
    network_mode: "host"
    network_mode: "none"
    network_mode: "service:[service name]"
    network_mode: "container:[container name/id]"
    networks
#配置容器连接的网络,引用顶级 networks 下的条目
    services:
      some-service:
        networks:
          some-network:
            aliases:
              - alias1
          other-network:
            aliases:
              - alias2
    networks:
      some-network:
        #Use a custom driver
        driver: custom-driver-1
      other-network:
```

```
    #Use a custom driver which takes special options
    driver: custom-driver-2
#aliases: 同一网络上的其他容器可以使用服务名称或此别名来连接到对应容器的服务
restart
#no: 是默认的重启策略,在任何情况下都不会重启容器
#always: 容器总是重新启动
#on-failure: 在容器非正常退出时(退出状态非0),才会重启容器
#unless-stopped: 在容器退出时总是重启容器,但是不考虑在 Docker 守护进程启动时就已经
#停止了的容器
restart: "no"
restart: always
restart: on-failure
restart: unless-stopped
#注: Swarm 集群模式,请改用 restart_policy
secrets
#存储敏感数据,例如密码
version: "3.1"
services:
mysql:
  image: mysql
  environment:
    MYSQL_ROOT_PASSWORD_FILE: /run/secrets/my_secret
  secrets:
    - my_secret
secrets:
  my_secret:
    file: ./my_secret.txt
security_opt
#修改容器默认的 schema 标签
security-opt:
  - label:user:USER                    #设置容器的用户标签
  - label:role:ROLE                    #设置容器的角色标签
  - label:type:TYPE                    #设置容器的安全策略标签
  - label:level:LEVEL                  #设置容器的安全等级标签
stop_grace_period
#指定在容器无法处理 SIGTERM (或者任何 stop_signal 的信号)时,等待多久后发送
#SIGKILL 信号关闭容器
stop_grace_period: 1s                  #等待 1s
stop_grace_period: 1m30s               #等待 1min30s
#默认的等待时间是 10s
stop_signal
#设置停止容器的替代信号。默认情况下使用 SIGTERM
#以下示例中,使用 SIGUSR1 替代信号 SIGTERM 来停止容器
stop_signal: SIGUSR1
```

```
sysctls
#设置容器中的内核参数,可以使用数组或字典格式
sysctls:
  net.core.somaxconn: 1024
  net.ipv4.tcp_syncookies: 0

sysctls:
  - net.core.somaxconn=1024
  - net.ipv4.tcp_syncookies=0
tmpfs
#在容器内安装一个临时文件系统。可以是单个值或列表的多个值
tmpfs: /run
tmpfs:
  - /run
  - /tmp
ulimits
#覆盖容器默认的 ulimit
ulimits:
  nproc: 65535
  nofile:
    soft: 20000
    hard: 40000
volumes
#将主机的数据卷或文件挂载到容器里
version: "3"
services:
  db:
    image: postgres:latest
    volumes:
      - "/localhost/postgres.sock:/var/run/postgres/postgres.sock"
      - "/localhost/data:/var/lib/postgresql/data"
```

7.6　Docker Compose Nginx 案例一

基于 Docker Compose 构建 Nginx 容器,并实现发布目录映射,通过浏览器实现访问,操作步骤如下:

(1)编写 docker-compose.yml 文件,内容如下:

```
version: "3"
services:
  nginx:
    container_name: www-nginx
    image: nginx:latest
    restart: always
```

```
    ports:
      - 80:80
    volumes:
      - /data/webapps/www/:/usr/share/nginx/html/
```

（2）创建发布目录/data/webapps/www/，并在发布目录新建 index.html 页面，命令如下：

```
mkdir -p /data/webapps/www/
echo "<h1>www.jfedu.net Nginx Test pages.</h1>" >>/data/webapps/www/index.html
```

（3）启动和运行 Docker Compose，启动 Nginx 容器，命令如下，如图 7-1 所示。

```
docker-compose up -d
```

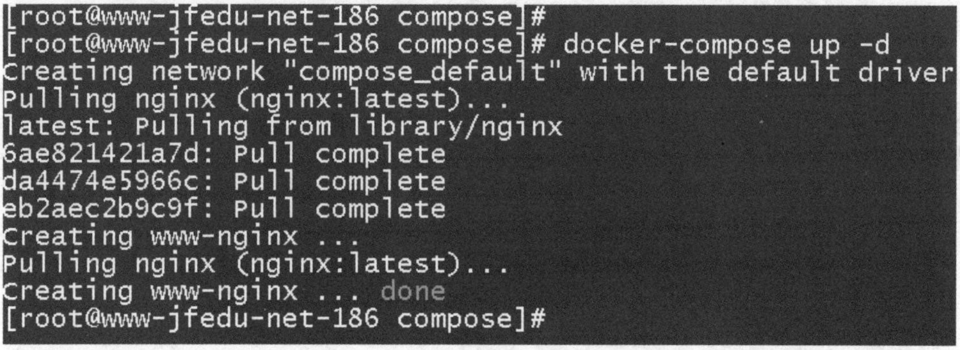

图 7-1　Docker Compose 启动 Nginx 容器

（4）通过浏览器访问宿主机 80 端口，即可访问 Nginx 容器，如图 7-2 所示。

图 7-2　Docker Compose Nginx 案例

（5）docker-compose.yml 内容剖析如下：

version：版本号，通常写 2 和 3 版本。

service：Docker 容器服务名称。

container_name：容器的名称。

restart：设置为 always，容器在停止的情况下总是重启。

image：Docker 官方镜像上找到最新版的镜像。

ports：容器自己运行的端口号和需要暴露的端口号。

volumes：数据卷。表示数据、配置文件等存放的位置。

7.7　Docker Compose Redis 案例二

基于 Docker Compose 构建 Redis 容器，并通过客户端访问 Redis 6379 端口即可，操作步骤如下。

（1）编写 docker-compose.yml 文件，内容如下：

```
version: "3"
services:
 redis:
    image: redis:latest
    container_name: redis
    ports:
      - "16379:6379"
    environment:
      - TZ="Asia/Shanghai"
    volumes:
      - /data:/data
    command: /usr/local/bin/redis-server
```

（2）redis.conf 配置文件代码如下：

```
requirepass 123456
appendonly yes
daemonize no
```

（3）启动和运行 Docker Compose，启动 Redis 容器，命令如下，如图 7-3 所示。

```
docker-compose up -d
```

```
"docker-compose.yml" 12L, 237C written
[root@www-jfedu-net redis]# docker-compose up -d
Recreating redis-test ... done
[root@www-jfedu-net redis]#
[root@www-jfedu-net redis]#
[root@www-jfedu-net redis]# docker ps|grep redis-test
17a6f157ede3           redis:latest
 seconds              0.0.0.0:16379->6379/tcp        redis-test
[root@www-jfedu-net redis]#
[root@www-jfedu-net redis]#
[root@www-jfedu-net redis]#
[root@www-jfedu-net redis]#
```

图 7-3　Docker Compose 启动 Redis 容器

7.8 Docker Compose Tomcat 案例三

基于 Docker Compose 构建 Nginx 容器和 Tomcat 容器，并且实现 Nginx 和 Tomcat 发布目录映射，同时实现 Nginx 均衡 Tomcat 服务，通过浏览器访问 Nginx 80 端口即访问 Tomcat 的 8080 端口，操作步骤如下。

（1）编写 docker-compose.yml 文件，内容如下：

```
version: "3"
services:
  tomcat01:
    container_name: tomcat01
    image: tomcat:latest
    restart: always
    ports:
      - 8080
  tomcat02:
    container_name: tomcat02
    image: tomcat:latest
    restart: always
    ports:
      - 8080
  nginx:
    container_name: www-nginx
    image: nginx:latest
    restart: always
    ports:
      - 80:80
    volumes:
      - /data/webapps/www/:/usr/share/nginx/html/
      - ./default.conf:/etc/nginx/conf.d/default.conf
    links:
      - tomcat01
      - tomcat02
```

（2）创建发布目录 /data/webapps/www/，并在发布目录新建 index.html 页面，命令如下：

```
mkdir -p /data/webapps/www/
echo "<h1>www.jfedu.net Nginx Test pages.</h1>" >>/data/webapps/www/index.html
```

（3）创建 Nginx 默认配置文件 default.conf，内容如下：

```
upstream tomcat_web {
       server tomcat01:8080 max_fails=2 fail_timeout=15;
       server tomcat02:8080 max_fails=2 fail_timeout=15;
}
```

```
server {
    listen       80;
    server_name  localhost;
    location / {
        root   /usr/share/nginx/html;
        index  index.html index.htm;
        proxy_pass http://tomcat_web;
        proxy_set_header host $host;
    }
    error_page   500 502 503 504  /50x.html;
    location=/50x.html {
        root   /usr/share/nginx/html;
    }
}
```

（4）启动 Docker Compose，命令如下，如图 7-4 所示。

```
docker-compose up -d
```

图 7-4　Docker Compose Tomcat 案例 1

（5）通过浏览器直接访问 Nginx 容器，默认访问宿主机的 80 端口即可，如图 7-5 所示。

图 7-5　Docker Compose Tomcat 案例 2

（6）查看运行的 Docker 容器引用的状态和进程信息，命令如下，如图 7-6 所示。

```
docker-compose ps
docker-compose top
```

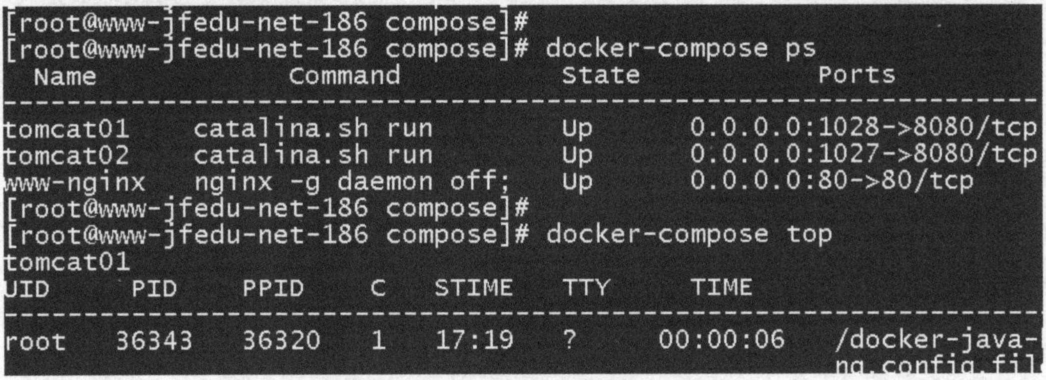

图 7-6 Docker 容器引用状态和进程信息

7.9 Docker Compose RocketMQ 案例四

基于 Docker Compose 构建 MQ 容器，并且通过客户端访问 MQ 8080 Web 端口即可，操作步骤如下。

（1）编写 docker-compose.yml 文件，内容如下：

```
version: "3"
services:
  rmqnamesrv:
    image: foxiswho/rocketmq:server
    container_name: rmqnamesrv
    ports:
      - 9876:9876
    volumes:
      - ./data/logs:/opt/logs
      - ./data/store:/opt/store
  rmqbroker:
    image: foxiswho/rocketmq:broker
    container_name: rmqbroker
    ports:
      - 10909:10909
      - 10911:10911
    volumes:
      - ./data/logs:/opt/logs
```

```yaml
    - ./data/store:/opt/store
    - ./broker.conf:/etc/rocketmq/broker.conf
  environment:
    NAMESRV_ADDR: "rmqnamesrv:9876"
    JAVA_OPTS: " -Duser.home=/opt"
    JAVA_OPT_EXT: "-server -Xms128m -Xmx128m -Xmn128m"
  command: mqbroker -c /etc/rocketmq/broker.conf
  depends_on:
    - rmqnamesrv
rmqconsole:
  image: styletang/rocketmq-console-ng
  container_name: rmqconsole
  ports:
    - 8080:8080
  environment:
    JAVA_OPTS: "-Drocketmq.namesrv.addr=rmqnamesrv:9876 -Dcom.rocketmq.sendMessageWithVIPChannel=false"
  depends_on:
    - rmqnamesrv
```

（2）broker.conf 配置文件代码如下：

```
# Licensed to the Apache Software Foundation (ASF) under one or more
# contributor license agreements. See the NOTICE file distributed with
# this work for additional information regarding copyright ownership.
# The ASF licenses this file to You under the Apache License, Version 2.0
# (the "License"); you may not use this file except in compliance with
# the License.  You may obtain a copy of the License at
#
#    http://www.apache.org/licenses/LICENSE-2.0
#
# Unless required by applicable law or agreed to in writing, software
# distributed under the License is distributed on an "AS IS" BASIS,
# WITHOUT WARRANTIES OR CONDITIONS OF ANY KIND, either express or implied.
# See the License for the specific language governing permissions and
# limitations under the License.


# 所属集群名字
brokerClusterName=DefaultCluster

#broker 名字,注意此处不同的配置文件填写的不一样,如果在 broker-a.properties 使用:
#broker-a,
```

```
#在 broker-b.properties 使用：broker-b
brokerName=broker-a

#0 表示 Master,>0 表示 Slave
brokerId=0

#nameServer 地址,以分号分隔
namesrvAddr=rocketmq-nameserver1:9876;rocketmq-nameserver2:9876

#启动 IP,如果 docker 报如下错误:
com.alibaba.rocketmq.remoting.exception.Remoting
#ConnectException: connect to <192.168.0.120:10909> failed
#解决方式 1 加上一句 producer.setVipChannelEnabled(false);,解决方式 2
#brokerIP1 设置宿主机 IP,不要使用 Docker 内部 IP
#brokerIP1=192.168.0.253

#在发送消息时,自动创建服务器不存在的 Topic,默认创建的队列数
defaultTopicQueueNums=4

#是否允许 Broker 自动创建 Topic,建议线下开启,线上关闭。这里是 false
autoCreateTopicEnable=true

#是否允许 Broker 自动创建订阅组,建议线下开启,线上关闭
autoCreateSubscriptionGroup=true

#Broker 对外服务的监听端口
listenPort=10911

#删除文件时间点,默认为凌晨 4 点
deleteWhen=04

#文件保留时间,默认为 48 小时
fileReservedTime=48

#commitLog 每个文件的大小默认为 1GB
mapedFileSizeCommitLog=1073741824

#ConsumeQueue 每个文件默认存 30 万条,根据业务情况调整
mapedFileSizeConsumeQueue=300000

#destroyMapedFileIntervalForcibly=120000
```

```
#redeleteHangedFileInterval=120000
#检测物理文件磁盘空间
diskMaxUsedSpaceRatio=88
#存储路径
#storePathRootDir=/home/ztztdata/rocketmq-all-4.1.0-incubating/store
#commitLog 存储路径
#storePathCommitLog=/home/ztztdata/rocketmq-all-4.1.0-incubating/store/
#commitLog
#消费队列存储
#storePathConsumeQueue=/home/ztztdata/rocketmq-all-4.1.0-incubating/
#store/consumequeue
#消息索引存储路径
#storePathIndex=/home/ztztdata/rocketmq-all-4.1.0-incubating/store/index
#checkpoint 文件存储路径
#storeCheckpoint=/home/ztztdata/rocketmq-all-4.1.0-incubating/store/
#checkpoint
#abort 文件存储路径
#abortFile=/home/ztztdata/rocketmq-all-4.1.0-incubating/store/abort
#限制的消息大小
maxMessageSize=65536

#flushCommitLogLeastPages=4
#flushConsumeQueueLeastPages=2
#flushCommitLogThoroughInterval=10000
#flushConsumeQueueThoroughInterval=60000

#Broker 的角色
#- ASYNC_MASTER 异步复制 Master
#- SYNC_MASTER 同步双写 Master
#- SLAVE
brokerRole=ASYNC_MASTER

#刷盘方式
#- ASYNC_FLUSH 异步刷盘
#- SYNC_FLUSH 同步刷盘
flushDiskType=ASYNC_FLUSH

#发消息线程池数量
#sendMessageThreadPoolNums=128
#拉消息线程池数量
#pullMessageThreadPoolNums=128
```

（3）启动和运行 Docker Compose，启动 Nginx 容器，如图 7-7 所示。

```
docker-compose up -d
```

(a)

(b)

图 7-7　Docker Compose RocketMQ 案例

（a）通过 docker-compose 命令启动 MQ 容器服务；（b）访问 RocketMq 控制台

第 8 章 Docker Swarm 集群案例实战

Docker Swarm 和 Docker Compose 一样，都是 Docker 官方容器编排项目。不同的是，Docker Compose 是一个在单个服务器或主机上创建多个容器的工具，而 Docker Swarm 则可以在多个服务器或主机上创建容器集群服务，对于微服务的部署，显然 Docker Swarm 会更加适合。

8.1 Swarm 概念剖析

Swarm 是 Docker 公司自主研发的容器集群管理系统。Swarm 在早期是作为一个独立服务存在，在 Docker Engine v1.12 中集成了 Swarm 的集群管理和编排功能，可以通过初始化 Swarm 或加入现有 Swarm 启用 Docker 引擎的 Swarm 模式。

Docker Engine CLI 和 API 包括了管理 Swarm 节点命令，比如添加、删除节点，以及在 Swarm 中部署和编排服务，也增加了服务栈（Stack）、服务（Service）、任务（Task）的概念。

Swarm 集群管理和任务编排功能已经集成到了 Docker 引擎中，称为 SwarmKit。SwarmKit 是一个独立的、专门用于 Docker 容器编排的项目，可以直接在 Docker 上使用。

Swarm 集群由多个运行 Swarm 模式的 Docker 主机组成，关键的是，Docker 默认集成了 Swarm mode。Swarm 集群中有 Manager（管理成员关系和选举）、Worker（运行 Swarm Service）。

一个 Docker 主机可以是 Manager，也可以是 Worker 角色，当然，还可以既是 Manager，同时也是 Worker。

当创建一个 Service 时，就定义了它的理想状态（副本数、网络、存储资源、对外暴露的端口等）。Docker 会维持它的状态，例如，如果一个 Worker 节点不可用了，Docker 会调度不可用节点的任务到其他节点上。

运行在容器中的一个任务，是 Swarm Service 的一部分，且通过 Swarm Manager 进行管理和调度，和独立的容器是截然不同的。

Swarm Service 相比单容器的一个最大优势就是，用户能够修改一个服务的配置，包括网络和数据卷，不需要手动重启服务。Docker 会更新配置，把过期配置的任务停掉，重新创建一个新配置的容器。

当然，某些场景下 Docker Compose 也能做与 Swarm 类似的事情，但是 Swarm 相比 Docker Compose，功能更加丰富，比如可以自动扩容、缩容，分配任务到不同的节点等。

一个节点是 Swarm 集群中的一个 Docker 引擎实例。也可以认为这就是一个 Docker 节点。可以在单台物理机或云服务器上运行一个或多个节点，但是在生产环境中，典型的部署方式是：Docker 节点交叉分布式部署在多台物理机或云主机上。

通过 Swarm 部署一个应用，向 Manager 节点提交一个 Service，然后 Manager 节点分发工作（Task）给 Worker Node。Manager 节点同时也会容器编排和集群管理功能，它会选举出一个 Leader 指挥编排任务。Worker 节点接收和执行从 Manager 分发过来的任务。

一般地，Manager 节点同时也是 Worker 节点，但是，也可以将 Manager 节点配置成只进行管理的功能。Agent 则运行在每个 Worker 节点上，时刻准备接收任务。Worker 节点会上报 Manager 节点分配给它的任务当前状态，这样 Manager 节点才能维持每个 Worker 的工作状态。

Service 就是在 Manager 或 Worker 节点上定义的 Tasks。Service 是 Swarm 系统最核心的架构，同时也是和 Swarm 最主要的交互者。

在副本集模式下，Swarm Manager 将会基于扩容的需求，把任务分发到各个节点。对于全局 Service，Swarm 会在每个可用节点上运行一个任务。

任务携带 Docker 引擎和一组命令让其运行在容器中。它是 Swarm 的原子调度单元。Manager 节点在扩容的时候会交叉分配任务到各个节点上，一旦一个任务分配到一个节点，它就不能再移动到其他节点。

8.2 Docker Swarm 的优点

1）Docker Engine 集成集群管理

使用 Docker Engine CLI 创建一个 Docker Engine 的 Swarm 模式，在集群中部署应用程序服务。

2）去中心化设计

Swarm 角色分为 Manager 和 Worker 节点，Manager 节点故障不影响应用使用。

3）扩容缩容

可以声明每个服务运行的容器数量，通过添加或删除容器数自动调整期望的状态。

4）期望状态协调

Swarm Manager 节点不断监视集群状态，并调整当前状态与期望状态之间的差异。

5）多主机网络

可以为服务指定 overlay 网络。当初始化或更新应用程序时，Swarm Manager 会自动为

overlay 网络上的容器分配 IP 地址。

6）服务发现

Swarm Manager 节点为集群中的每个服务分配唯一的 DNS 记录和负载均衡 VIP。可以通过 Swarm 内置的 DNS 服务器查询集群中每个运行的容器。

7）负载均衡

实现服务副本负载均衡，提供入口访问。

8）安全传输

Swarm 中的每个节点使用 TLS 相互验证和加密，确保与安全的其他节点通信。

9）滚动更新

升级时，逐步将应用服务更新到节点，如果出现问题，可以将任务回滚到先前版本。

8.3 Swarm 负载均衡

Swarm Manager 使用 ingress 负载均衡暴露需要让外部访问的服务。Swarm Manager 能够自动分配一个外部端口到 Service，当然，用户也能够配置一个外部端口，可以指定任意没有使用的端口，如果不指定，那么 Swarm Manager 会给 Service 指定 30000～32767 中任意一个端口。

Swarm 模式有一个内部的 DNS 组件，它能够在 Swarm 里面自动分发每个服务。Swarm Manager 使用内部负载均衡机制接收集群中节点的请求，基于 DNS 名字解析实现。

8.4 Swarm 架构图

Docker Swarm 结构如图 8-1 所示。

Swarm Manager 各部分功能详解如下。

（1）API：接收命令，创建一个 Service（API 输入）。

（2）Orchestrator：Service 对象创建的 Task 进行编排工作（编排）。

（3）Allocater：为各个 Task 分配 IP 地址（分配 IP）。

（4）Dispatcher：将 Task 分发到 Node（分发任务）。

（5）Scheduler：安排一个 Worker 运行 Task（运行任务）。

Worker Node 功能详解如下。

（1）Worker：连接到分发器接收指定的 Task。

（2）Executor：将 Task 指派到对应的 Worker 节点。

Swarm Manager 创建 1 个拥有 3 个 Nginx 副本集的 Service，它会将 Task 分配到对应的 Node，如图 8-2 所示。

第 8 章 Docker Swarm 集群案例实战

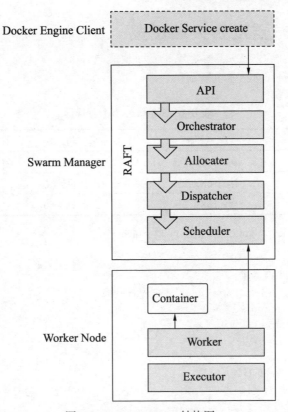

图 8-1 Docker Swarm 结构图

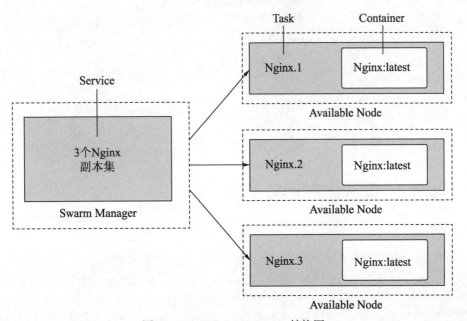

图 8-2 Docker Swarm Nginx 结构图

8.5 Swarm 节点及防火墙设置

Swarm 集群配置环境如表 8-1 所示。

表 8-1 Swarm集群配置环境

集群角色	宿主机IP	宿主机系统	Docker版本
Manager	192.168.1.145	CentOS 7.6	20.10.6
Node1	192.168.1.123	CentOS 7.6	20.10.6
Node2	192.168.1.147	CentOS 7.6	20.10.6

对 Manager、Node1、Node2 节点进行如下配置：

```
#添加hosts解析
cat >/etc/hosts<<EOF
127.0.0.1 localhost localhost.localdomain
192.168.1.145 manager
192.168.1.123 node1
192.168.1.147 node2
EOF
#临时关闭SELinux和防火墙
sed -i '/SELINUX/s/enforcing/disabled/g' /etc/sysconfig/selinux
setenforce 0
systemctl stop firewalld.service
systemctl disable firewalld.service
#同步节点时间
yum install ntpdate -y
ntpdate pool.ntp.org
#修改对应节点主机名
hostname 'cat /etc/hosts|grep $(ifconfig|grep broadcast|awk '{print $2}')|awk '{print $2}'';su
#关闭swapoff
swapoff -a
```

8.6 Docker 虚拟化案例实战

相关代码如下：

```
#安装依赖软件包
yum install -y yum-utils device-mapper-persistent-data lvm2
#添加Docker repository,这里使用国内阿里云yum源
yum-config-manager --add-repo http://mirrors.aliyun.com/docker-ce/linux/centos/docker-ce.repo
```

```
#安装docker-ce,这里直接安装最新版本
yum install -y docker-ce
#修改 Docker 配置文件
mkdir /etc/docker
cat > /etc/docker/daemon.json <<EOF
{
  "exec-opts": ["native.cgroupdriver=systemd"],
  "log-driver": "json-file",
  "log-opts": {
    "max-size": "100m"
  },
  "storage-driver": "ov
  "storage-opts": [
    "overlay2.override
  ],
  "registry-mirrors":                                          cs.com"]
}
EOF
#注意,由于国内拉取镜像                                              rs
mkdir -p /etc/syst
sed -i '/^ExecStar                                     .0.0:2375/g' /usr/lib/
systemd/system/do
#重启 Docker 服务
systemctl daemon
systemctl enable
systemctl rest
ps -ef|grep -
```

8.7 Swarm

（1）根据以上步骤   部署 Swarm 集群，在 Manager 节点初始化集群，操作

```
docker swa                                          1.145
[root@manager
[root@manager
[root@manager                  92.168.1.145
Swarm initial                                 w2ii) is now a manager.
To add a worker to this s                                 and:
    docker swarm join --token SWMTKN-1-              yctzj6g8npam11rntiotpkzw1se1o5kpk7h0t-
To add a manager to this swarm, run 'docker swarm join-token manager' and follow the instr
[root@manager ~]#
[root@manager ~]#
[root@manager ~]#
```

图 8-3 Docker Swarm 初始化集群

（2）将 Node1 节点加入 Swarm 集群，操作指令如下：

```
docker swarm join --token SWMTKN-1-5lioovnms9y4zyctzj6g8npam11rntiotpkz
wlselo5kpk7h0t-09et4wl0hpr5zphkjnmecx470 192.168.1.145:2377
```

（3）将 Node2 节点加入 Swarm 集群，操作指令如下：

```
docker swarm join --token SWMTKN-1-5lioovnms9y4zyctzj6g8npam11rntiotpkzw
lselo5kpk7h0t-09et4wl0hpr5zphkjnmecx470 192.168.1.145:2377
```

（4）查看 Swarm 集群 Node 状态，操作指令如下，如图 8-4 所示。

```
docker node ls
```

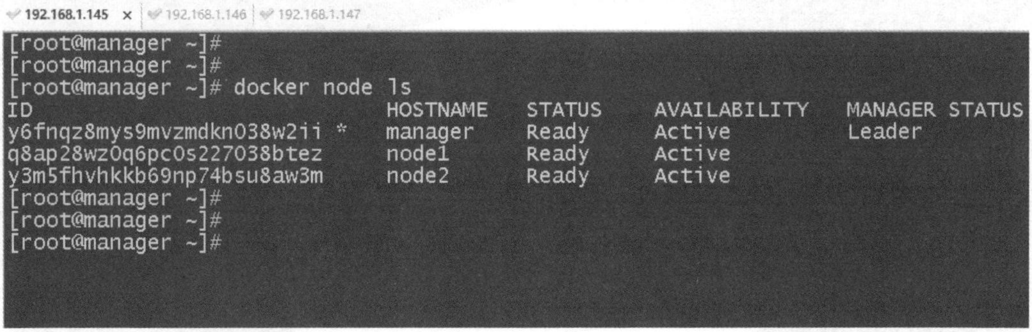

图 8-4　Docker Swarm 查看 Node 状态

8.8　Swarm 部署 Nginx 服务

（1）基于 Swarm 集群创建 Nginx Web 服务，操作指令如下：

```
docker service create --replicas 1 --name nginx-test nginx:latest
```

其中，--replicas 表示副本集数，--name 表示服务名称。

（2）查看 Nginx 服务，当前运行在 Node2 上，操作指令如下，如图 8-5 所示。

```
docker service ps nginx-test
```

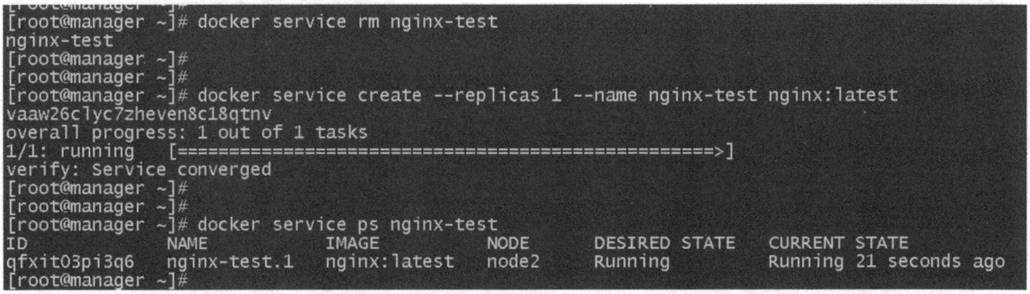

图 8-5　Docker Swarm 查看 Nginx 服务

（3）显示服务详细信息，操作指令如下，如图 8-6 所示。

```
docker service inspect --pretty nginx-test
```

```
[root@manager ~]# docker service inspect --pretty nginx-test
ID:             vaaw26c1yc7zheven8c18qtnv
Name:           nginx-test
Service Mode:   Replicated
 Replicas:      1
Placement:
UpdateConfig:
 Parallelism:   1
 On failure:    pause
 Monitoring Period: 5s
 Max failure ratio: 0
 Update order:      stop-first
RollbackConfig:
```

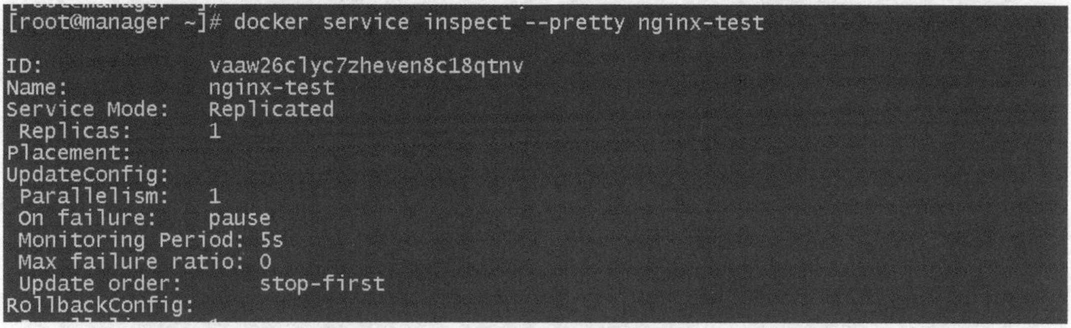

图 8-6　Docker Swarm 显示服务详细信息

（4）查看 JSON 格式，操作指令如下，如图 8-7 所示。

```
docker service inspect nginx-test
```

```
[root@manager ~]# docker service inspect nginx-test
[
    {
        "ID": "vaaw26c1yc7zheven8c18qtnv",
        "Version": {
            "Index": 145
        },
        "CreatedAt": "2021-04-27T02:59:04.677040141Z",
        "UpdatedAt": "2021-04-27T02:59:04.677040141Z",
        "Spec": {
            "Name": "nginx-test",
            "Labels": {},
            "TaskTemplate": {
                "ContainerSpec": {
                    "Image": "nginx:latest@sha256:75a55d33ecc73c2a242450a9f1cc85
```

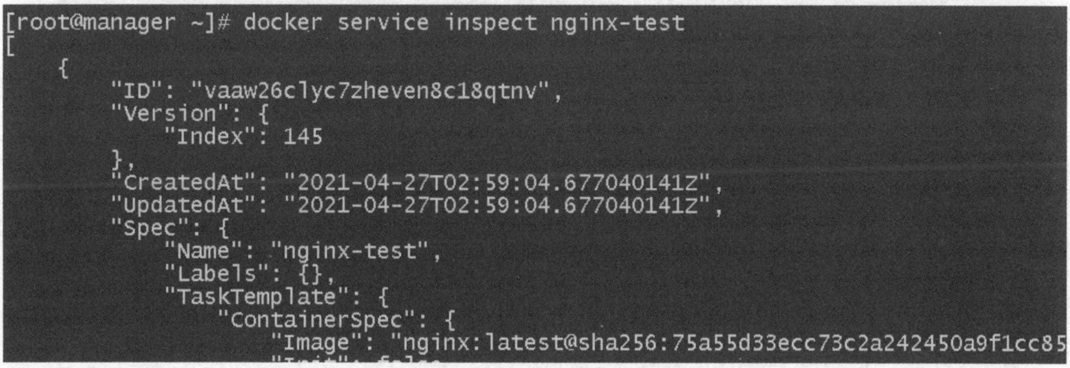

图 8-7　Docker Swarm 查看 JSON 格式

8.9　Swarm 服务扩容和升级

（1）Nginx 服务扩容和缩容，最终每个 Node 上分布了 1 个 Nginx 容器，操作指令如下，如图 8-8 所示。

```
docker service scale nginx-test=3
docker service ls
docker service ps nginx-test
```

（2）滚动更新服务，操作指令如下，如图 8-9 所示。

```
docker service update --image tomcat nginx-test
```

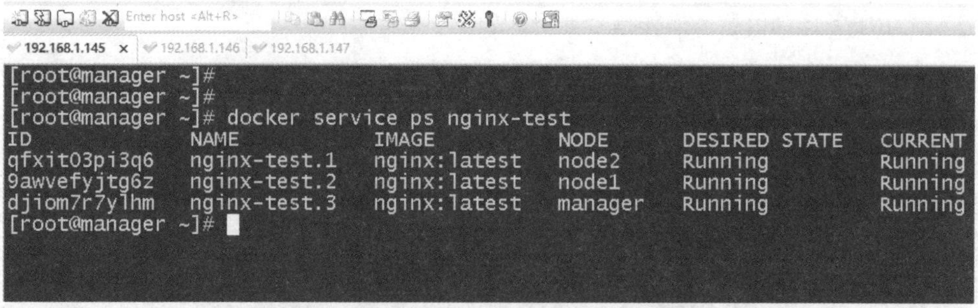

图 8-8　Docker Swarm 查看 Nginx 服务

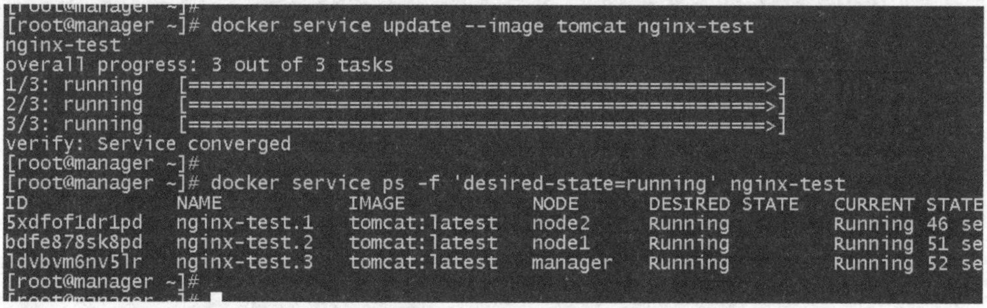

图 8-9　Docker Swarm 滚动更新服务

（3）创建服务时设定更新策略，操作指令如下：

```
docker service create \
--name nginx-test \
--replicas 10 \
--update-delay 10s \
--update-parallelism 2 \
--update-failure-action continue \
nginx:latest
```

（4）创建服务时设定回滚策略，操作指令如下：

```
docker service create \
--name nginx-test \
--replicas 10 \
--rollback-parallelism 2 \
--rollback-monitor 20s \
--rollback-max-failure-ratio .2 \
nginx:latest
```

（5）服务更新，操作指令如下：

```
docker service update --image nginx:latest nginx-test
```

（6）手动回滚，操作指令如下：

```
docker service update --rollback nginx-test
```

8.10 Manager 和 Node 角色切换

（1）Manager 和 Node 角色切换之前查看节点状态信息，如图 8-10 所示。

```
docker node ls
```

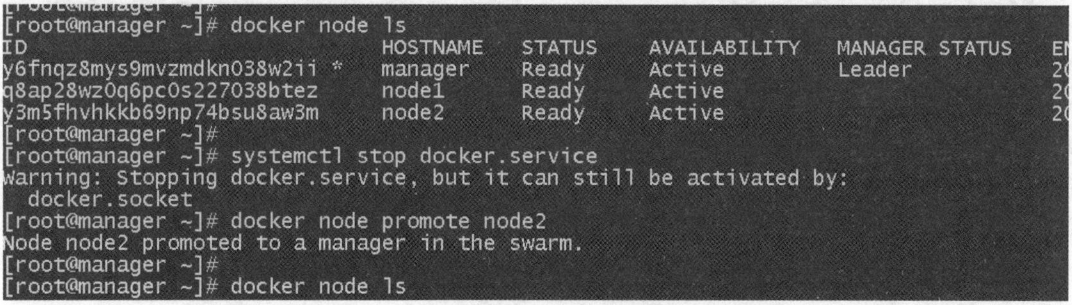

```
[root@manager ~]# docker node ls
ID                            HOSTNAME   STATUS   AVAILABILITY   MANAGER STATUS   E
y6fnqz8mys9mvzmdkn038w2ii *   manager    Ready    Active         Leader           2
q8ap28wz0q6pc0s227038btez     node1      Ready    Active                          2
y3m5fhvhkkb69np74bsu8aw3m     node2      Ready    Active                          2
[root@manager ~]#
[root@manager ~]# systemctl stop docker.service
Warning: Stopping docker.service, but it can still be activated by:
  docker.socket
[root@manager ~]# docker node promote node2
Node node2 promoted to a manager in the swarm.
[root@manager ~]# docker node ls
```

图 8-10　Docker Swarm 查看节点状态信息

（2）Manager 和 Node 角色切换，停止现有 Manager Docker 引擎服务，操作指令如下：

```
#停止 Docker 服务
systemctl stop docker.service
#将 node2 升级为 Manager
docker node promote node2
```

（3）Manager 和 Node 角色切换之后，查看节点状态信息，如图 8-11 所示。

```
docker node ls
```

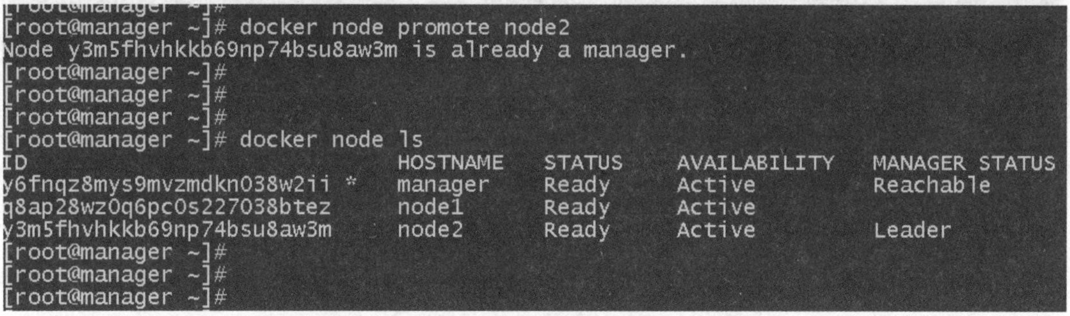

```
[root@manager ~]# docker node promote node2
Node y3m5fhvhkkb69np74bsu8aw3m is already a manager.
[root@manager ~]#
[root@manager ~]#
[root@manager ~]#
[root@manager ~]# docker node ls
ID                            HOSTNAME   STATUS   AVAILABILITY   MANAGER STATUS
y6fnqz8mys9mvzmdkn038w2ii *   manager    Ready    Active         Reachable
q8ap28wz0q6pc0s227038btez     node1      Ready    Active
y3m5fhvhkkb69np74bsu8aw3m     node2      Ready    Active         Leader
[root@manager ~]#
[root@manager ~]#
[root@manager ~]#
```

图 8-11　Docker Swarm 查看节点状态变化

8.11 Swarm 数据管理之 volume

Docker Swarm 数据管理方式有很多，其中 volume 方式管理数据比较常见，配置相对比较简单。其原理是在宿主机上创建一个 volume，默认目录为 /var/lib/docker/volume/your_custom_ volume/_data。

将容器的某个目录（例如容器网站数据目录）映射到宿主机的 volume 上，即使容器崩溃了，数据还会保留在宿主机的 volume 上。

```
#创建 Nginx 服务,Volume 映射
docker service create --replicas 1 --mount type=volume,src=nginx_data,
dst=/usr/share/nginx/html --name jfedu-nginx nginx:latest
#查看服务
docker service inspect jfedu-nginx
#查看数据卷
ls -l /var/lib/docker/volumes/
#查看数据信息
ls -l /var/lib/docker/volumes/nginx_data/_data/
```

可以看到，容器里面的 Nginx 数据目录已经挂载到宿主机的 nginx_data 了，如图 8-12 所示。

```
[root@node1 ~]# ll /var/lib/docker/volumes/
total 24
brw------- 1 root root  8, 2 Apr 27 10:39 backingFsBlockDev
-rw------- 1 root root 32768 Apr 27 11:55 metadata.db
drwx-----x 3 root root    19 Apr 27 11:55 nginx_data
[root@node1 ~]#
[root@node1 ~]#
[root@node1 ~]# docker volume inspect nginx_data
[
    {
        "CreatedAt": "2021-04-27T11:55:00+08:00",
        "Driver": "local",
        "Labels": null,
        "Mountpoint": "/var/lib/docker/volumes/nginx_data/_data",
        "Name": "nginx_data",
        "Options": null,
        "Scope": "local"
    }
]
```

图 8-12　volume 数据

8.12　Swarm 数据管理之 Bind

Bind mount 模式工作原理：将宿主机某个目录映射到 Docker 容器，很适合网站，同时把宿主机的这个目录作为 git 版本目录，每次更新代码的时候，容器就会更新。

（1）分别在 Manager、Node1、Node2 上创建 Web 网站目录。

```
mkdir -p /data/webapps/www/
```

（2）创建服务，操作指令如下：

```
docker service create --replicas 1 --mount type=bind,src=/data/webapps
/www/,dst=/usr/share/nginx/html --name nginx-v1 nginx:latest
```

（3）查看已创建的 nginx-v1 服务，操作指令如下：

```
docker service ps nginx-v1
docker service inspect nginx-v1
```

（4）测试宿主机的数据盘和容器是映射关系，如图 8-13 所示。

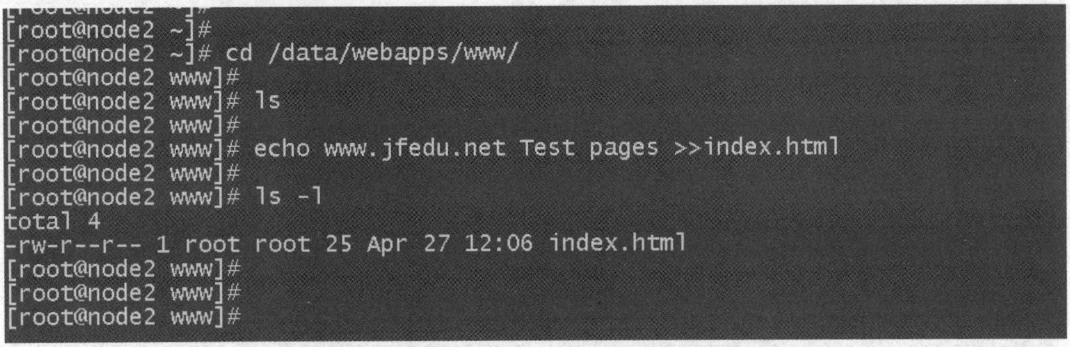

```
[root@node2 ~]#
[root@node2 ~]# cd /data/webapps/www/
[root@node2 www]#
[root@node2 www]# ls
[root@node2 www]#
[root@node2 www]# echo www.jfedu.net Test pages >>index.html
[root@node2 www]#
[root@node2 www]# ls -l
total 4
-rw-r--r-- 1 root root 25 Apr 27 12:06 index.html
[root@node2 www]#
[root@node2 www]#
[root@node2 www]#
```

图 8-13　测试宿主机的数据盘和容器是映射关系

（5）进入容器查看内容，如图 8-14 所示。

```
[root@node2 www]# docker ps
CONTAINER ID   IMAGE           COMMAND                  CREATED          STATUS
a7a6d87d774c   nginx:latest    "/docker-entrypoint.…"   3 minutes ago    Up 3 minutes
65cc9c7eba93   tomcat:latest   "catalina.sh run"        31 minutes ago   Up 31 minute
[root@node2 www]#
[root@node2 www]# docker exec -it a7a6d87d774c /bin/bash
root@a7a6d87d774c:/#
root@a7a6d87d774c:/# ls /usr/share/nginx/html/
index.html
root@a7a6d87d774c:/# cat /usr/share/nginx/html/index.html
www.jfedu.net Test pages
root@a7a6d87d774c:/#
root@a7a6d87d774c:/#
```

图 8-14　Docker 进入容器查看内容

可以看到，在宿主机上创建的 index.html 已经挂载到容器上。

8.13　Swarm 数据管理之 NFS

以上两种方式都是单机 Docker 上的数据共享方式，在集群中就不适用了，必须使用共享存储或网络存储。这里使用 NFS 测试。

（1）基于 Linux 平台构建 NFS 网络文件系统，配置指令如下：

```
#安装 NFS 文件服务
yum install nfs-utils -y
```

```
#配置共享目录及权限
vim /etc/exports
/data/ *(rw,sync,all_squash,anonuid=0,anongid=0)
#启动 NFS 服务
service nfs restart
#在 NFS 数据目录新建测试页面
echo 'www.jfedu.net Test Hello NFS' >/data/index.html
```

（2）创建 Nginx volume 名称为 jfeduv66，操作指令如下，如图 8-15 所示。

```
docker volume create --driver local --opt type=nfs --opt o=addr=192.
168.1.147,rw --opt device=:/data jfeduv66
```

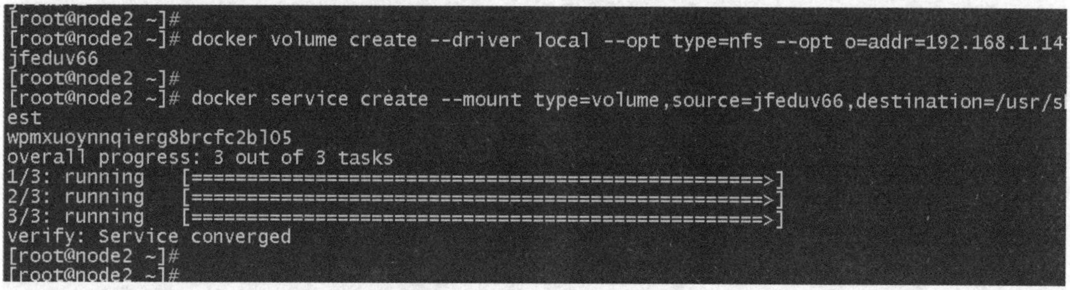

```
[root@node2 ~]#
[root@node2 ~]# docker volume create --driver local --opt type=nfs --opt o=addr=192.168.1.14
jfeduv66
[root@node2 ~]#
[root@node2 ~]# docker service create --mount type=volume,source=jfeduv66,destination=/usr/s
est
wpmxuoynnqierg8brcfc2bl05
overall progress: 3 out of 3 tasks
1/3: running   [==================================================>]
2/3: running   [==================================================>]
3/3: running   [==================================================>]
verify: Service converged
[root@node2 ~]#
[root@node2 ~]#
```

图 8-15　Docker Swarm 创建服务

（3）创建 Nginx 服务，绑定 jfeduv66 NFS 映射目录，操作指令如下：

```
docker service create --mount type=volume,source=jfeduv66,destination=
/usr/share/nginx/html/ --replicas 3 nginx:latest
```

（4）可以看到 NFS 已经挂载到 Manager 节点，进入容器查看并创建内容，如图 8-16 所示。

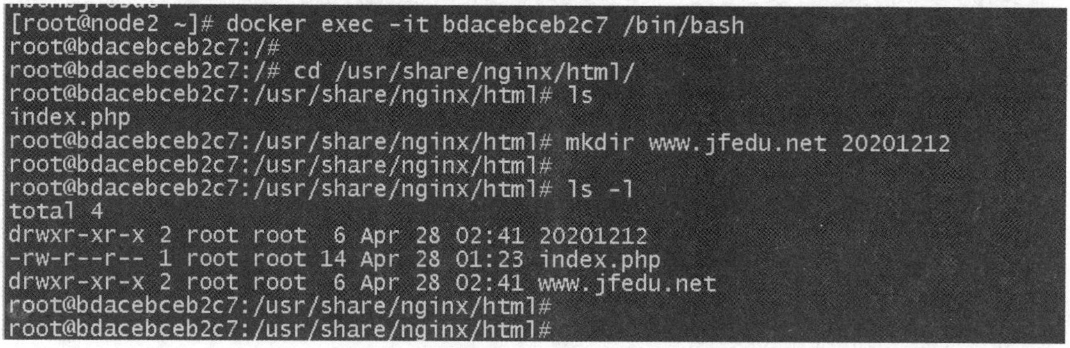

```
[root@node2 ~]# docker exec -it bdacebceb2c7 /bin/bash
root@bdacebceb2c7:/#
root@bdacebceb2c7:/# cd /usr/share/nginx/html/
root@bdacebceb2c7:/usr/share/nginx/html# ls
index.php
root@bdacebceb2c7:/usr/share/nginx/html# mkdir www.jfedu.net 20201212
root@bdacebceb2c7:/usr/share/nginx/html#
root@bdacebceb2c7:/usr/share/nginx/html# ls -l
total 4
drwxr-xr-x 2 root root  6 Apr 28 02:41 20201212
-rw-r--r-- 1 root root 14 Apr 28 01:23 index.php
drwxr-xr-x 2 root root  6 Apr 28 02:41 www.jfedu.net
root@bdacebceb2c7:/usr/share/nginx/html#
root@bdacebceb2c7:/usr/share/nginx/html#
```

图 8-16　进入容器查看并创建内容

（5）在 Nginx 容器中创建内容之后，退出容器，进入宿主机 NFS 服务器，进入/data/目录查看内容，如图 8-17 所示。

```
root@bdacebceb2c7:/usr/share/nginx/html#
root@bdacebceb2c7:/usr/share/nginx/html# exit
exit
[root@node2 ~]#
[root@node2 ~]# cd /data/
[root@node2 data]# ls
20201212   index.php   www.jfedu.net
[root@node2 data]# ls -l
total 4
drwxr-xr-x 2 root root  6 Apr 28 10:41 20201212
-rw-r--r-- 1 root root 14 Apr 28 09:23 index.php
drwxr-xr-x 2 root root  6 Apr 28 10:41 www.jfedu.net
[root@node2 data]#
[root@node2 data]#
```

图 8-17 进入/date/目录查看内容

8.14 Docker Swarm 新增节点

在生产环境正常运行中，随着企业业务的飞速增长，可能需要扩容 Swarm 节点，作为运维人员该如何操作呢？操作的方法和步骤有以下几点。

（1）在每台服务器上 hosts 文件中添加 Node3 和 IP 绑定记录，操作指令如下：

```
#添加 hosts 解析
cat >/etc/hosts<<EOF
127.0.0.1 localhost localhost.localdomain
192.168.1.145 manager
192.168.1.123 node1
192.168.1.147 node2
192.168.1.148 node3
EOF
#临时关闭 SELinux 和防火墙
sed -i '/SELINUX/s/enforcing/disabled/g'  /etc/sysconfig/selinux
setenforce  0
systemctl   stop     firewalld.service
systemctl   disable  firewalld.service
#同步节点时间
yum install ntpdate -y
ntpdate pool.ntp.org
#修改对应节点主机名
hostname 'cat /etc/hosts|grep $(ifconfig|grep broadcast|awk '{print $2}')|awk '{print $2}'';su
```

```
#关闭swapoff
swapoff -a
```

（2）在新节点Node3上，部署Docker引擎服务，操作指令如下：

```
#安装依赖软件包
yum install -y yum-utils device-mapper-persistent-data lvm2
#添加Docker repository,这里使用国内阿里云yum源
yum-config-manager --add-repo http://mirrors.aliyun.com/docker-ce/linux/centos/docker-ce.repo
#安装docker-ce,这里直接安装最新版本
yum install -y docker-ce
#修改Docker配置文件
mkdir /etc/docker
cat > /etc/docker/daemon.json <<EOF
{
  "exec-opts": ["native.cgroupdriver=systemd"],
  "log-driver": "json-file",
  "log-opts": {
    "max-size": "100m"
  },
  "storage-driver": "overlay2",
  "storage-opts": [
    "overlay2.override_kernel_check=true"
  ],
  "registry-mirrors": ["https://uyah70su.mirror.aliyuncs.com"]
}
EOF
#注意,由于国内拉取镜像较慢,配置文件最后增加了registry-mirrors
mkdir -p /etc/systemd/system/docker.service.d
sed -i '/^ExecStart/s/dockerd/dockerd -H tcp:\/\/0.0.0.0:2375/g' /usr/lib/systemd/system/docker.service
#重启Docker服务
systemctl daemon-reload
systemctl enable docker.service
systemctl restart docker.service
ps -ef|grep -aiE docker
```

（3）在已经初始化的机器（Manager）上执行如下指令，获取客户端加入集群的命令，如图8-18所示。

```
docker swarm join-token manager
docker swarm join --token SWMTKN-1-5lioovnms9y4zyctzj6g8npam11rntiotpkz
wlselo5kpk7h0t-6jtljl9wat92884f1z6j2hos0 192.168.1.147:2377
```

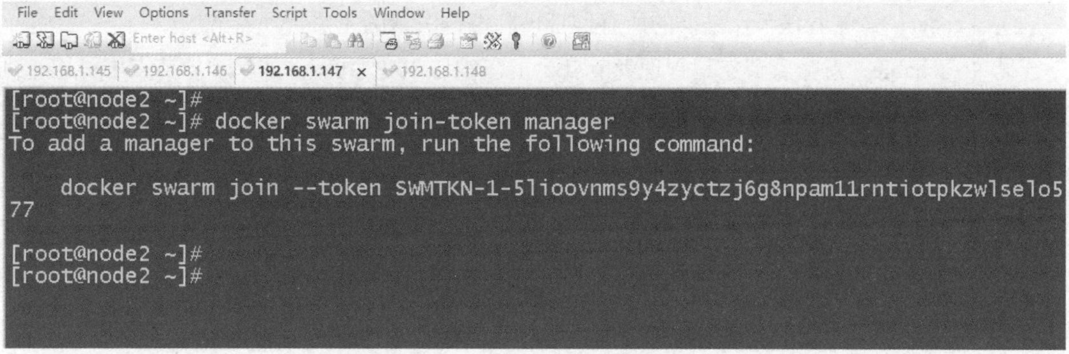

第 8 章 Docker Swarm 集群案例实战

图 8-18 获取客户端加入集群的命令

（4）将结果复制到新增 Node3 节点机器上执行即可，如图 8-19 所示。

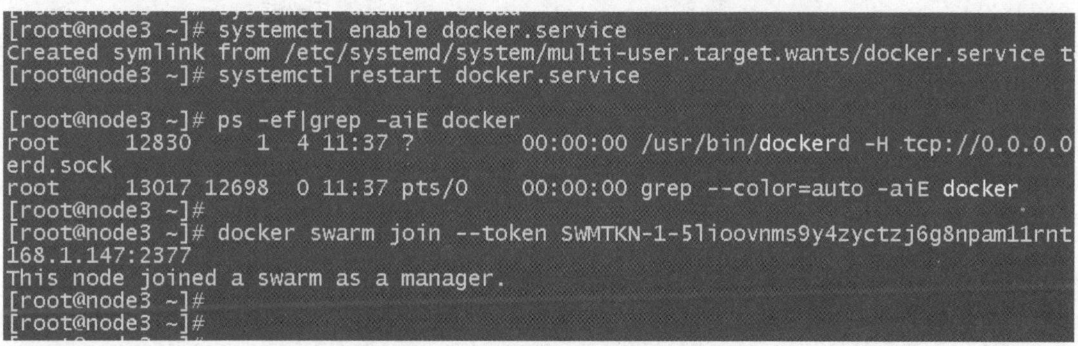

图 8-19 将 node3 加入 swarm 集群

查看集群节点，如图 8-20 所示。

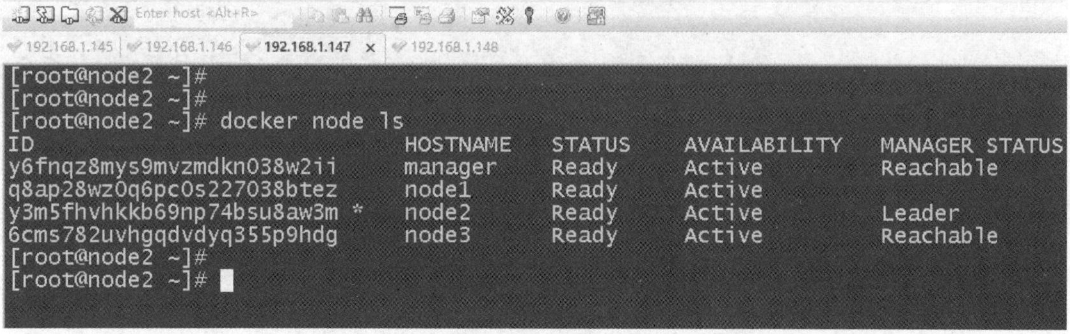

图 8-20 查看集群节点

MANAGER STATUS 列显示节点是属于 Manager 还是 Worker；没有值表示不参与群管理的工作节点，具体说明如下。

① Leader，意味着该节点是群的所有群管理和编排决策的主要管理器节点。

② Reachable，意味着节点是管理者节点，正在参与 Raft 共识。如果领导节点不可用，则该节点有资格被选为新领导者。

③ Unavailable（没有值），意味着节点是不能与其他管理器通信的管理器。如果管理器节点不可用，应该将新的管理器节点加入群集，或者将工作器节点升级为管理器。

节点 AVAILABILITY 列显示调度程序是否可以将任务分配给节点。

① Active，意味着调度程序可以将任务分配给节点。

② Pause，意味着调度程序不会将新任务分配给节点，但现有任务仍在运行。

③ Drain，意味着调度程序不会向节点分配新任务。调度程序关闭所有现有任务并在可用节点上调度它们。

8.15　Docker Swarm 删除节点

Docker Swarm 在生产环境正常运行中，随着服务器使用寿命长，难免有服务器过保、下架，此时需要删除 Swarm 节点，作为运维人员该如何操作呢？操作的方法和步骤有以下几点。

（1）将 node3 节点停用，该节点上的容器会迁移到其他节点，操作指令如下：

```
docker node update --availability drain node3
```

（2）检查容器迁移情况，当 node3 的容器都迁移完后，停止 Docker 服务即可，如图 8-21 所示。

```
[root@node3 ~]#
[root@node3 ~]# docker node  update --availability drain node3
node3
[root@node3 ~]#
[root@node3 ~]# docker ps
CONTAINER ID    IMAGE           COMMAND             CREATED       STATUS      PORTS      NAMES
[root@node3 ~]#
[root@node3 ~]# docker ps -a
CONTAINER ID    IMAGE           COMMAND                CREATED          STATUS
fd8ffa80c710   centos7-ssh:v1   "/bin/sh -c /usr/sbi…"  13 days ago     Exited (25
[root@node3 ~]#
[root@node3 ~]#
```

图 8-21　查看容器迁移情况

（3）停止 node3 Docker 服务，删除节点前，需先停止该节点的 Docker 服务。

```
systemctl stop docker.service
```

（4）登录到 Manager 上，将节点 node3 降级成 Worker，然后删除。只能删除 Worker 基本的节点。

```
docker node demote node3
docker node rm node3
```

8.16　Docker 自动化部署一

生产环境中，由于 Docker 指令过多，操作比较烦琐，可以编写 Shell 自动安装并配置 Docker 虚拟化及桥接网络，同时使用 pipework 脚本配置容器 IP，以实现容器的批量管理，此脚本适用于 CentOS 6.x 系统。

CentOS 6.x 系列 Docker 批量管理脚本代码如下：

```
#!/bin/bash
#auto install docker and Create VM
#by www.jfedu.net 2021
#Define PATH Variables
IPADDR=`ifconfig |grep "Bcast"|awk '{print $2}'|cut -d: -f2|grep 
"192.168"|head -1`
GATEWAY=`route -n|grep "UG"|awk '{print $2}'|grep "192.168"|head -1`
DOCKER_IPADDR=$1
IPADDR_NET=`ifconfig |grep "Bcast"|awk '{print $2}'|cut -d: -f2|grep 
"192.168"|head -1|awk -F. '{print $1"."$2"."$3".""xxx"}'`

NETWORK=(
    HWADDR=`ifconfig eth0 |egrep "HWaddr|Bcast" |tr "\n" " "|awk '{print 
$5,$7,$NF}'|sed -e 's/addr://g' -e 's/Mask://g'|awk '{print $1}'`
    IPADDR=`ifconfig eth0 |egrep "HWaddr|Bcast" |tr "\n" " "|awk '{print 
$5,$7,$NF}'|sed -e 's/addr://g' -e 's/Mask://g'|awk '{print $2}'`
    NETMASK=`ifconfig eth0 |egrep "HWaddr|Bcast" |tr "\n" " "|awk '{print 
$5,$7,$NF}'|sed -e 's/addr://g' -e 's/Mask://g'|awk '{print $3}'`
    GATEWAY=`route -n|grep "UG"|awk '{print $2}'`
)

if [ -z "$1" -o -z "$2" -o -z "$3" -o -z "$4" ];then

    echo -e "\033[32m------------------------------------\033[0m"
    echo -e "\033[32mPlease exec $0 IPADDR CPU(C) MEM(G) DISK(G),example $0 
$IPADDR_NET 16 32 50\033[0m"
    exit 0
fi


CPU=`expr $2 - 1`
if [ ! -e /usr/bin/bc ];then
    yum install bc -y >>/dev/null 2>&1
fi
```

```
MEM_F='echo $3 \* 1024|bc'
MEM=`printf "%.0f\n" $MEM_F`
DISK=$4
USER=$5
REMARK=$6

ping $DOCKER_IPADDR -c 1 >>/dev/null 2>&1

if [ $? -eq 0 ];then

    echo -e "\033[32m----------------------------------\033[0m"
    echo -e "\033[32mThe IP address to be used,Please change other IP,exit.\
033[0m"
    exit 0
fi

if [ ! -e /etc/init.d/docker ];then
    rpm -ivh http://dl.fedoraproject.org/pub/epel/6/x86_64/epel-release-6-8.noarch.rpm
    yum install docker-io -y
    yum install device-mapper* -y
    /etc/init.d/docker start
    if [ $? -ne 0 ];then
        echo "Docker install error ,please check."
        exit
    fi
fi


cd /etc/sysconfig/network-scripts/
    mkdir -p /data/backup/`date +%Y%m%d-%H%M`
    yes|cp ifcfg-eth* /data/backup/`date +%Y%m%d-%H%M`/
if
    [ -e /etc/sysconfig/network-scripts/ifcfg-br0 ];then
    echo
else
    cat >ifcfg-eth0 <<EOF
    DEVICE=eth0
    BOOTPROTO=none
    ${NETWORK[0]}
    NM_CONTROLLED=no
    ONBOOT=yes
    TYPE=Ethernet
    BRIDGE="br0"
    ${NETWORK[1]}
    ${NETWORK[2]}
```

```
        ${NETWORK[3]}
        USERCTL=no
EOF
        cat >ifcfg-br0 <<EOF
        DEVICE="br0"
        BOOTPROTO=none
        ${NETWORK[0]}
        IPV6INIT=no
        NM_CONTROLLED=no
        ONBOOT=yes
        TYPE="Bridge"
        ${NETWORK[1]}
        ${NETWORK[2]}
        ${NETWORK[3]}
        USERCTL=no
EOF

        /etc/init.d/network restart

fi
echo 'Your can restart Ethernet Service: /etc/init.d/network restart !'
echo '------------------------------------------------------------'
cd -
#######create docker container
service docker status >>/dev/null
if [ $? -ne 0 ];then
        /etc/init.d/docker restart
fi
NAME="Docker$$_'echo $DOCKER_IPADDR|awk -F"." '{print $(NF-1)"_"$NF}''"
IMAGES='docker images|grep -v "REPOSITORY"|grep -v "none"|head -1|awk '{print $1}''
CID=$(docker run -itd --cpuset-cpus=0-$CPU -m ${MEM}m --net=none --name=$NAME $IMAGES /bin/bash)

if [ -z $IMAGES ];then
        echo "Plesae Download Docker Centos Images,you can to be use docker search centos,and docker pull centos6.5-ssh,exit 0"
        exit 0
fi

if [ ! -f /usr/local/bin/pipework ];then
        yum install wget unzip zip -y
        wget https://github.com/jpetazzo/pipework/archive/master.zip
            unzip master
            cp pipework-master/pipework  /usr/local/bin/
            chmod +x /usr/local/bin/pipework
```

```
            rm -rf master
fi

ip netns >>/dev/null
if [ $? -ne 0 ];then
    rpm -e iproute --nodeps
        rpm -ivh https://repos.fedorapeople.org/openstack/EOL/openstack-grizzly/epel-6/iproute-2.6.32-130.el6ost.netns.2.x86_64.rpm
fi
pipework br0 $NAME  $DOCKER_IPADDR/24@$IPADDR

docker ps -a |grep "$NAME"

DEV=$(basename $(echo /dev/mapper/docker-*-$CID))
dmsetup table $DEV | sed "s/0 [0-9]* thin/0 $((${DISK}*1024*1024*1024/512)) thin/" | dmsetup load $DEV
dmsetup resume $DEV
resize2fs /dev/mapper/$DEV
docker start $CID
docker logs $CID
LIST="docker_vmlist.csv"
if [ ! -e $LIST ];then
    echo "编号,容器 ID,容器名称,CPU,内存,硬盘,容器 IP,宿主机 IP,使用人,备注" >$LIST
fi
###################
NUM=`cat docker_vmlist.csv |grep -v CPU|tail -1|awk -F, '{print $1}'`
if [[ $NUM -eq "" ]];then
        NUM="1"
else
        NUM=`expr $NUM + 1`
fi
##################
echo -e "\033[32mCreate virtual client Successfully.\n$NUM `echo $CID|cut -b 1-12` $NAME $2C ${MEM}M ${DISK}G $DOCKER_IPADDR $IPADDR $USER $REMARK\033[0m"
if [ -z $USER ];then
    USER="NULL"
    REMARK="NULL"
fi
echo $NUM, `echo $CID|cut -b 1-12`,$NAME,${2}C,${MEM}M,${DISK}G,$DOCKER_IPADDR,$IPADDR,$USER,$REMARK >>$LIST
rm -rf docker_vmlist_*
iconv -c -f utf-8 -t gb2312 docker_vmlist.csv -o docker_vmlist_`date +%H%M`.csv
```

8.17 Docker 自动化部署二

生产环境中 Docker 指令过多，操作比较烦琐，因此可以编写 Shell 自动安装并配置 Docker 虚拟化及桥接网络，同时使用 pipework 脚本来配置容器 IP，能够实现容器的批量管理，此脚本适用于 CentOS 7.x 系统。

CentOS 7.x 系列 Docker 批量管理脚本代码如下：

```bash
#!/bin/bash
#auto install docker and Create VM
#by www.jfedu.net 2021
#Define PATH Variables
IPADDR=`ifconfig|grep -E "\<inet\>"|awk '{print $2}'|grep "192.168"|head -1`
GATEWAY=`route -n|grep "UG"|awk '{print $2}'|grep "192.168"|head -1`
IPADDR_NET=`ifconfig|grep -E "\<inet\>"|awk '{print $2}'|grep "192.168"\
|head -1|awk -F. '{print $1"."$2"."$3"."}'`
LIST="/root/docker_vmlist.csv"
if [ ! -f /usr/sbin/ifconfig ];then
    yum install net-tools* -y
fi
for i in `seq 1 253`;do ping -c 1 ${IPADDR_NET}${i} ;[ $? -ne 0 ]&& DOCKER_
IPADDR="${IPADDR_NET}${i}" &&break;done >>/dev/null 2>&1
echo "###################"
echo -e "Dynamic get docker IP,The Docker IP address\n\n$DOCKER_IPADDR"
NETWORK=(
    HWADDR=`ifconfig eth0|grep ether|awk '{print $2}'`
    IPADDR=`ifconfig eth0|grep -E "\<inet\>"|awk '{print $2}'`
    NETMASK=`ifconfig eth0|grep -E "\<inet\>"|awk '{print $4}'`
    GATEWAY=`route -n|grep "UG"|awk '{print $2}'`
)
if [ -z "$1" -o -z "$2" ];then

    echo -e "\033[32m---------------------------------\033[0m"
    echo -e "\033[32mPlease exec $0 CPU(C) MEM(G),example $0 4 8\033[0m"
    exit 0
fi
#CPU=`expr $2 - 1`
if [ ! -e /usr/bin/bc ];then
    yum install bc -y >>/dev/null 2>&1
fi
CPU_ALL=`cat /proc/cpuinfo |grep processor|wc -l`
if [ ! -f $LIST ];then
    CPU_COUNT=$1
    CPU_1="0"
    CPU1=`expr $CPU_1 + 0`
```

```
        CPU2=`expr $CPU1 + $CPU_COUNT - 1`
        if [ $CPU2 -gt $CPU_ALL ];then
            echo -e "\033[32mThe System CPU count is $CPU_ALL,not more than it.\033[0m"
            exit
        fi
else
    CPU_COUNT=$1
    CPU_1=`cat $LIST|tail -1|awk -F"," '{print $4}'|awk -F"-" '{print $2}'`
    CPU1=`expr $CPU_1 + 1`
    CPU2=`expr $CPU1 + $CPU_COUNT - 1`
    if [ $CPU2 -gt $CPU_ALL ];then
        echo -e "\033[32mThe System CPU count is $CPU_ALL,not more than it.\033[0m"
        exit
    fi
fi
MEM_F=`echo $2 \* 1024|bc`
MEM=`printf "%.0f\n" $MEM_F`
DISK=20
USER=$3
REMARK=$4
ping $DOCKER_IPADDR -c 1 >>/dev/null 2>&1

if [ $? -eq 0 ];then

    echo -e "\033[32m------------------------------------\033[0m"
    echo -e "\033[32mThe IP address to be used,Please change other IP,exit.\033[0m"
    exit 0
fi

if [ ! -e /usr/bin/docker ];then
    yum install docker* device-mapper* lxc -y
    mkdir -p /export/docker/
    cd /var/lib/ ;rm -rf docker ;ln -s /export/docker/ .
    mkdir -p /var/lib/docker/devicemapper/devicemapper
    dd if=/dev/zero of=/var/lib/docker/devicemapper/devicemapper/data bs=1G count=0 seek=2000
    service docker start
    if [ $? -ne 0 ];then
        echo "Docker install error ,please check."
        exit
    fi
fi
```

```
    cd /etc/sysconfig/network-scripts/
    mkdir -p /data/backup/`date +%Y%m%d-%H%M`
    yes|cp ifcfg-eth* /data/backup/`date +%Y%m%d-%H%M`/
if
    [ -e /etc/sysconfig/network-scripts/ifcfg-br0 ];then
    echo
else
    cat >ifcfg-eth0<<EOF
    DEVICE=eth0
    BOOTPROTO=none
    ${NETWORK[0]}
    NM_CONTROLLED=no
    ONBOOT=yes
    TYPE=Ethernet
    BRIDGE="br0"
    ${NETWORK[1]}
    ${NETWORK[2]}
    ${NETWORK[3]}
    USERCTL=no
EOF
    cat >ifcfg-br0 <<EOF
    DEVICE="br0"
    BOOTPROTO=none
    ${NETWORK[0]}
    IPV6INIT=no
    NM_CONTROLLED=no
    ONBOOT=yes
    TYPE="Bridge"
    ${NETWORK[1]}
    ${NETWORK[2]}
    ${NETWORK[3]}
    USERCTL=no
EOF
    /etc/init.d/network restart

fi
echo 'Your can restart Ethernet Service: /etc/init.d/network restart !'
echo '--------------------------------------------------------------'

cd -
#######create docker container
service docker status >>/dev/null
if [ $? -ne 0 ];then
    service docker restart
fi
```

```bash
NAME="Docker_`echo $DOCKER_IPADDR|awk -F"." '{print $(NF-1)"_"$NF}'`"
IMAGES=`docker images|grep -v "REPOSITORY"|grep -v "none"|grep "centos"|head -1|awk '{print $1}'`
if [ -z $IMAGES ];then
    echo "Plesae Download Docker Centos Images,you can to be use docker search centos,and docker pull centos6.5-ssh,exit 0"
    if [ ! -f jfedu_centos68.tar ];then
        echo "Please upload jfedu_centos68.tar for docker server."
        exit
    fi
    cat jfedu_centos68.tar|docker import - jfedu_centos6.8
fi
IMAGES=`docker images|grep -v "REPOSITORY"|grep -v "none"|grep "centos"|head -1|awk '{print $1}'`
CID=$(docker run -itd --privileged --cpuset-cpus=${CPU1}-${CPU2} -m ${MEM}m --net=none --name=$NAME $IMAGES /bin/bash)
echo $CID
docker ps -a |grep "$NAME"
pipework br0 $NAME  $DOCKER_IPADDR/24@$IPADDR
docker exec $NAME /etc/init.d/sshd start
if [ ! -e $LIST ];then
    echo "编号,容器 ID,容器名称,CPU,内存,硬盘,容器 IP,宿主机 IP,使用人,备注" >$LIST
fi
####################
NUM=`cat $LIST |grep -v CPU|tail -1|awk -F, '{print $1}'`
if [[ $NUM -eq "" ]];then
      NUM="1"
else
      NUM=`expr $NUM + 1`
fi
#################
echo -e "\033[32mCreate virtual client Successfully.\n$NUM `echo $CID|cut -b1-12`,$NAME,$CPU1-$CPU2,${MEM}M,${DISK}G,$DOCKER_IPADDR,$IPADDR,$USER,$REMARK\033[0m"
if [ -z $USER ];then
    USER="NULL"
    REMARK="NULL"
fi
echo $NUM, `echo $CID|cut -b 1-12`,$NAME,$CPU1-$CPU2,${MEM}M,${DISK}G,$DOCKER_IPADDR,$IPADDR,$USER,$REMARK >>$LIST
rm -rf /root/docker_vmlist_*
iconv -c -f utf-8 -t gb2312 $LIST  -o /root/docker_vmlist_`date +%H%M`.csv
```

第 9 章 OpenStack+KVM 构建企业级私有云

9.1 OpenStack 入门简介

OpenStack 是一个由美国国家航空航天局（National Aeronautics and Space Administration，NASA）和 Rackspace 合作研发并发起的，以 Apache 许可证授权的自由软件和开放源代码项目。OpenStack 是一个开源的云计算管理平台项目，由几个主要的组件组合起来完成具体工作。

OpenStack 支持几乎所有类型的云环境，项目目标是提供实施简单、可大规模扩展、丰富、标准统一的云计算管理平台。OpenStack 通过各种互补的服务提供了基础设施即服务（Infrastructure as a Service，IaaS）的解决方案，每个服务提供 API 以进行集成，当然除了 IaaS 解决方案，还有主流的平台即服务（Platform as a Service，PaaS）和软件即服务（Software as a Service，SaaS），三者区别如图 9-1 所示。

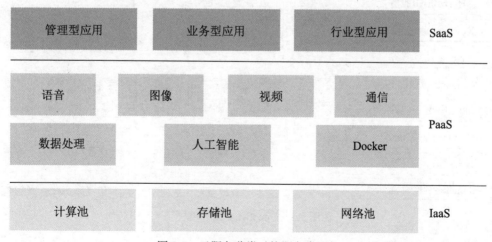

图 9-1 云服务分类（按服务类型）

OpenStack 是一个旨在为公共及私有云的建设与管理提供软件的开源项目。它的社区拥有超过 130 家企业及 1 350 位开发者，这些机构与个人都将 OpenStack 作为 IaaS 资源的通用前端。OpenStack 项目的首要任务是简化云的部署过程并为其带来良好的可扩展性。

OpenStack 云计算平台帮助服务商和企业内部实现类似于 Amazon EC2 和 S3 的云基础架构服务。OpenStack 包含两个主要模块：Nova 和 Swift，前者是 NASA 开发的虚拟服务器部署和业务计算模块，后者是 Rackspace 开发的分布式云存储模块，二者可以一起使用，也可以单独使用。OpenStack 除了有 Rackspace 和 NASA 的大力支持外，还有包括 Dell、Citrix、Cisco、Canonical 等重量级公司的贡献和支持，发展速度非常快，有取代另一个业界领先开源云平台 Eucalyptus 的态势。

9.2 OpenStack 核心组件

OpenStack 覆盖了网络、虚拟化、操作系统、服务器等各个方面。它是一个正在开发中的云计算平台项目，根据成熟及重要程度的不同，被分解成核心项目、孵化项目，以及支持项目和相关项目。每个项目都有自己的委员会和项目技术主管，且每个项目都不是一成不变的，孵化项目可以根据发展的成熟度和重要性，转变为核心项目。截至 Icehouse 版本，下面列出了 10 个 OpenStack 核心模块，如图 9-2 所示。

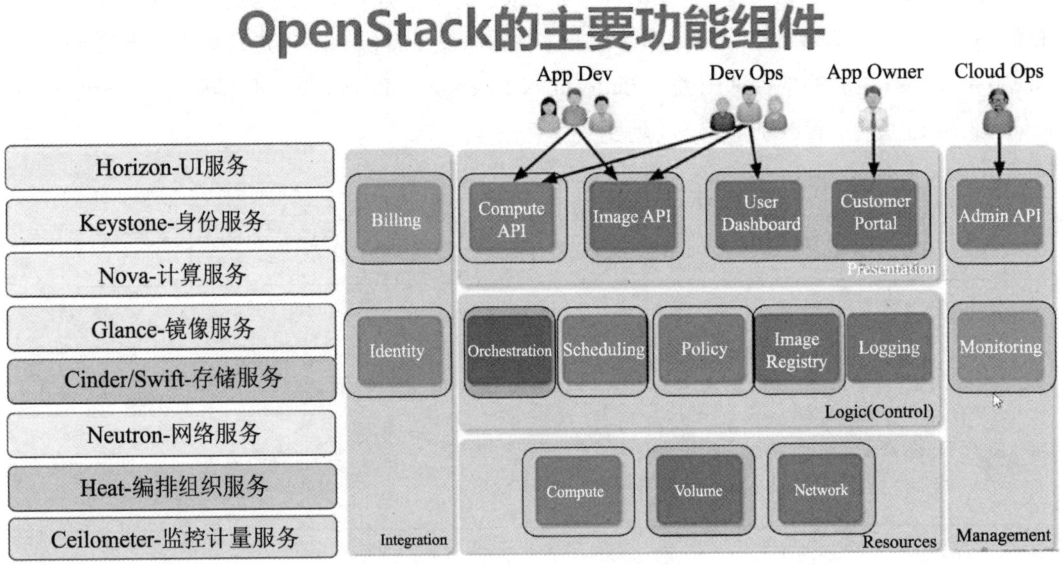

图 9-2 OpenStack 核心模块

（1）计算（Compute）：Nova，一套控制器，根据用户需求提供虚拟服务，负责虚拟机创建、开机、关机、挂起、暂停、调整、迁移、销毁等操作，配置 CPU、内存等信息规格。

（2）对象存储（Object Storage）：Swift，一套用于在大规模可扩展系统中通过内置冗余及高容错机制实现对象存储的系统，允许进行存储或者检索文件，可为 Glance 提供镜像存储，为 Cinder 提供卷备份服务。

（3）镜像服务（Image Service）：Glance，一套虚拟机镜像查找及检索系统，支持多种虚拟机镜像格式（AKI、AMI、ARI、ISO、QCOW2、Raw、VDI、VHD 和 VMDK），有创建上传镜像、删除镜像、编辑镜像基本信息的功能。

（4）身份服务（Identity Service）：Keystone，为 OpenStack 其他服务提供身份验证、服务规则和服务令牌的功能，管理 Domains、Projects、Users、Groups 和 Roles。

（5）网络和地址管理（Network）：Neutron，提供云计算的网络虚拟化技术，为 OpenStack 其他服务提供网络连接服务。为用户提供接口，可以定义 Network、Subnet、Router，配置 DHCP、DNS、负载均衡、L3 服务，网络支持 GRE、VLAN。插件架构支持许多主流的网络厂家和技术，如 OpenvSwitch。

（6）块存储（Block Storage）：Cinder，为运行实例提供稳定的数据块存储服务，它的插件驱动架构有利于块设备的创建和管理，如创建卷、删除卷，在实例上挂载和卸载卷。

（7）UI 界面（Dashboard）：Horizon，OpenStack 中各种服务的 Web 管理门户，用于简化用户对服务的操作，例如启动实例、分配 IP 地址、配置访问控制等。

（8）测量（Metering）：Ceilometer，像一个漏斗一样，能把 OpenStack 内部发生的几乎所有的事件都收集起来，然后为计费和监控以及其他服务提供数据支撑。

（9）部署编排（Orchestration）：Heat，提供了一种通过模板定义的协同部署方式，实现云基础设施软件运行环境（计算、存储和网络资源）的自动化部署。

（10）数据库服务（Database Service）：Trove，为用户在 OpenStack 的环境提供可扩展和可靠的关系和非关系数据库引擎服务。

9.3 OpenStack 准备环境

OpenStack 准备环境如下所示：

```
操作系统版本：CentOS Linux release 7.3.1611
192.168.1.120        node1 控制节点
192.168.1.121        node2 计算节点
192.168.1.122        node3 计算节点（可选）
控制节点主要用于操控计算节点,计算节点上创建虚拟机
```

9.4 Hosts 及防火墙设置

对 OpenStack Node1、Node2 节点进行如下配置：

```
cat >/etc/hosts<<EOF
127.0.0.1          localhost localhost.localdomain
#103.27.60.52      mirror.centos.org
66.241.106.180     mirror.centos.org
192.168.1.120      node1
192.168.1.121      node2
192.168.1.123      node3
EOF
sed -i '/SELINUX/s/enforcing/disabled/g'   /etc/sysconfig/selinux
setenforce  0
systemctl   stop      firewalld.service
systemctl   disable   firewalld.service
ntpdate  pool.ntp.org              #保持主节点,计算节点时间同步
hostname `cat /etc/hosts|grep $(ifconfig|grep broadcast|awk '{print $2}')|
awk '{print $2}'`;su
```

9.5 OpenStack 服务安装

（1）OpenStack Node1 主节点安装如下服务，操作方法和指令如下：

```
yum --enablerepo=centos-openstack-liberty clean metadata
#Base
yum install -y epel-release
yum install -y centos-release-openstack-liberty
yum install -y python-openstackclient
#MySQL
yum install -y mariadb mariadb-server MySQL-python mariadb-devel
#RabbitMQ
yum install -y rabbitmq-server
#Keystone
yum install -y openstack-keystone httpd httpd-devel mod_wsgi memcached python-memcached
#Glance
yum install -y openstack-glance python-glance python-glanceclient
#Nova
yum install -y openstack-nova-api openstack-nova-cert openstack-nova-conductor openstack-nova-console openstack-nova-novncproxy openstack-nova-scheduler python-novaclient
#Neutron
yum install -y openstack-neutron openstack-neutron-ml2 openstack-neutron-linuxbridge python-neutronclient ebtables ipset
#Dashboard
yum install -y openstack-dashboard
```

```
#Cinder
yum install -y openstack-cinder python-cinderclient
#Update Qemu
yum install -y centos-release-qemu-ev.noarch
yum -y install qemu-kvm qemu-img
sed -i '/\[mysqld\]/amax_connections=2000' /etc/my.cnf
systemctl enable mariadb.service
systemctl start  mariadb.service
```

（2）OpenStack Node1 节点创建数据库配置，操作方法和指令如下：

```
mysql
create database keystone;
grant all  on keystone.* to 'keystone'@'localhost' identified by 'keystone';
grant all  on keystone.* to 'keystone'@'%' identified by 'keystone';
grant all  on keystone.* to 'keystone'@'node1' identified by 'keystone';
create database glance;
grant all on glance.* to 'glance'@'localhost' identified by 'glance';
grant all on glance.* to 'glance'@'%' identified by 'glance';
grant all on glance.* to 'glance'@'node1' identified by 'glance';
create database nova;
grant all on nova.* to 'nova'@'localhost' identified by 'nova';
grant all on nova.* to 'nova'@'%' identified by 'nova';
grant all on nova.* to 'nova'@'node1' identified by 'nova';
create database neutron;
grant all  on neutron.* to 'neutron'@'localhost' identified by 'neutron';
grant all  on neutron.* to 'neutron'@'%' identified by 'neutron';
grant all  on neutron.* to 'neutron'@'node1' identified by 'neutron';
create database cinder;
grant all on cinder.* to 'cinder'@'localhost' identified by 'cinder';
grant all on cinder.* to 'cinder'@'%' identified by 'cinder';
grant all on cinder.* to 'cinder'@'node1' identified by 'cinder';
flush privileges;
exit;
```

9.6 MQ（消息队列）简介

消息队列（Message Queue，MQ）是一种应用程序对应用程序的通信方法，应用程序通过读写出入队列的消息（针对应用程序的数据）通信，无须两个业务系统直接相互连接。

消息队列中间件是分布式系统中重要的组件，主要解决应用耦合、异步消息、流量削锋等问题、实现高性能、高可用、可伸缩和最终一致性架构。主流消息队列包括 ActiveMQ、RabbitMQ、ZeroMQ、Kafka、MetaMQ、RocketMQ 等。

消息传递指的是程序之间通过在消息中发送数据进行通信，而不是通过直接调用彼此来通信，直接调用通常是用于诸如远程过程调用的技术。

排队指的是应用程序通过队列来通信，队列的使用除去了接收和发送应用程序同时执行的要求。RabbitMQ 是一个在 AMQP 基础上完整的、可复用的企业消息系统，遵循 GPL 开源协议。

OpenStack 的架构决定了需要使用消息队列机制实现不同模块间的通信，通过消息验证、消息转换、消息路由架构模式，带来的好处就是可以使模块之间最大程度解耦，客户端不需要关注服务端的位置和是否存在，只需通过消息队列进行信息的发送。

RabbitMQ 适合部署在一个拓扑灵活、易扩展的规模化系统环境中，有效保证不同模块、不同节点、不同进程之间消息通信的时效性，可有效支持 OpenStack 云平台系统的规模化部署、弹性扩展、灵活架构以及信息安全的需求。

9.7　MQ 应用场景

1）MQ 异步信息场景

MQ 应用的场合非常多，例如某个论坛网站，用户注册信息后，需要发注册邮件和注册短信，如图 9-3 所示。

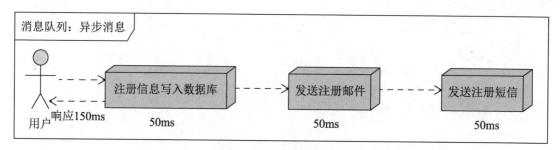

图 9-3　MQ 异步信息场景

将注册信息写入数据库成功后，发送注册邮件，再发送注册短信。以上三个任务全部完成后，返回给客户端，总共花费时间为 150ms。

引入消息队列后，将不是必需的业务逻辑异步处理。改造后的架构如图 9-4 所示。

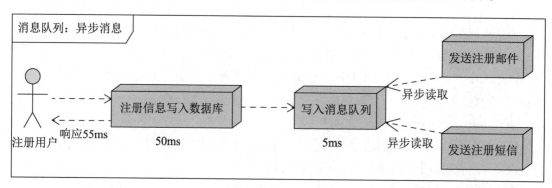

图 9-4　改造后的架构

用户的响应时间相当于注册信息写入数据库的时间，为 50ms。注册邮件，发送短信写入消息队列后，直接返回，这样写入消息队列的速度很快，耗时基本可以忽略，因此用户的响应时间可能是 55ms。

2）MQ 应用解耦场景

用户下单后，订单系统需要通知库存系统。传统做法是订单系统调用库存系统的接口，缺点是假如库存系统无法访问，则订单减库存将失败，从而导致订单失败，如图 9-5 所示。

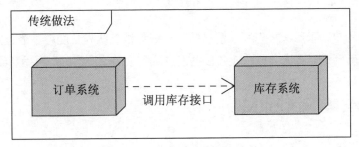

图 9-5 传统做法

引入应用消息队列后的方案如图 9-6 所示。

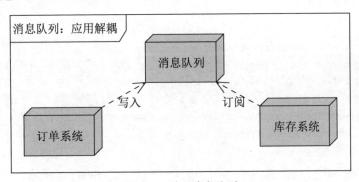

图 9-6 引入消息队列

（1）订单系统：用户下单后，订单系统完成持久化处理，将消息写入消息队列，返回用户订单下单成功。

（2）库存系统：订阅下单的消息，采用拉/推的方式，获取下单信息，库存系统根据下单信息，进行库存操作。下单时即使库存系统不能正常使用，也不影响正常下单，因为下单后，订单系统写入消息队列就不再关心其他后续操作，实现了订单系统与库存系统的应用解耦。

3）MQ 流量削峰场景

流量削峰也是消息队列中的常用场景，一般在秒杀或团抢活动中使用广泛。秒杀活动中，一般会因为流量过大，导致流量暴增，应用崩溃。为解决这个问题，一般需要在应用前端加入消息队列。其优点是可以控制活动的人数，缓解短时间内高流量压垮应用，如图 9-7 所示。

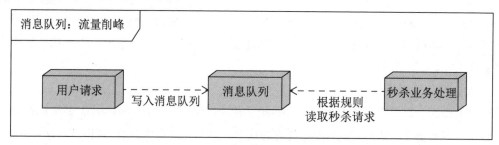

图 9-7 MQ 流量削峰场景

服务器接收用户请求后，首先写入消息队列，假如消息队列长度超过最大数量，则直接抛弃用户请求或跳转到错误页面，秒杀业务根据消息队列中的请求信息，再做后续处理。

9.8 安装配置 RabbitMQ

（1）启动 RabbitMQ 服务，默认监听端口为 TCP 15672，同时监听 5672、15672 两个端口。其中 5672 是 RabbitMQ 的主端口号，而 15672 是 RabbitMQ 的 Web 页面的端口号。添加 OpenStack 通信用户，操作指令如下：

```
systemctl enable rabbitmq-server.service
systemctl start rabbitmq-server.service
rabbitmqctl add_user openstack openstack
rabbitmqctl set_permissions openstack ".*" ".*" ".*"
rabbitmq-plugins list
rabbitmq-plugins enable rabbitmq_management
systemctl restart rabbitmq-server.service
lsof -i :15672
```

（2）访问 RabbitMQ，地址是 http://192.168.1.120:15672，如图 9-8 所示。

图 9-8 RabbitMQ 登录界面

(3)使用默认用户名 guest，密码 guest 登录，添加 OpenStack 用户组并登录测试，如图 9-9 所示。

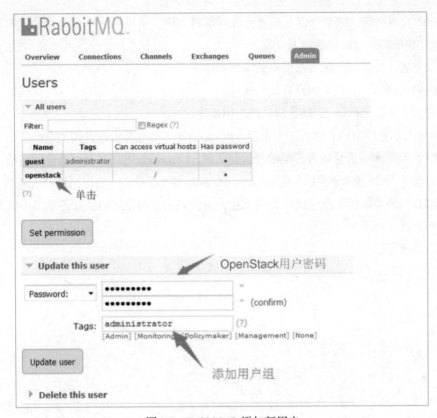

图 9-9　RabbitMQ 添加新用户

(4)创建完毕，使用 OpenStack 用户和密码登录，如图 9-10 所示。

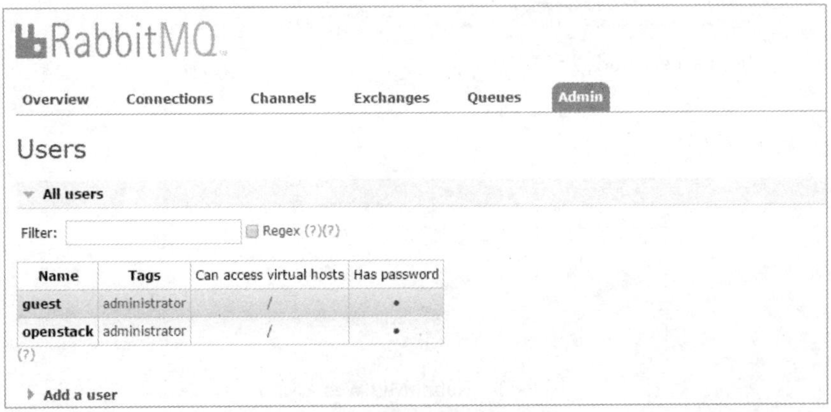

图 9-10　RabbitMQ 新用户登录

9.9 RabbitMQ 消息测试

RabbitMQ 消息测试如下。以上消息服务器部署完毕，可以进行简单消息的发布和订阅。RabbitMQ 完整的消息通信包括如下内容。

（1）发布者（Producer）：发布消息的应用程序。

（2）队列（Queue）：用于消息存储的缓冲。

（3）消费者（Consumer）：接收消息的应用程序。

RabbitMQ 消息模型核心理念：发布者（Producer）不会直接发送任何消息给队列，发布者（Producer）甚至不知道消息是否已经被投递到队列，（Producer）只需要把消息发送给一个交换器（Exchange）。交换器非常简单，它一边从发布者方接收消息，一边把消息推入队列。

交换器必须知道如何处理它接收到的消息，是应该推送到指定的队列还是多个队列，或者是直接忽略消息，如图 9-11 所示。

（a）

（b）

图 9-11 RabbitMQ Web 界面

（a）查看 RabbitMQ Overview 状态；（b）查看 RabbitMQ 队列状态

9.10 配置 Keystone 验证服务

Keystone 可以为 OpenStack 其他服务提供身份验证、服务规则和服务令牌的功能，管理 Domains、Projects、Users、Groups 和 Roles。

（1）配置 Keystone 服务，操作指令如下：

```
ID='openssl rand -hex 10';echo $ID
cat >/etc/keystone/keystone.conf<<EOF
[DEFAULT]
admin_token=$ID
verbose=true
[assignment]
[auth]
[cache]
[catalog]
[cors]
[cors.subdomain]
[credential]
[database]
connection=mysql://keystone:keystone@192.168.1.120/keystone
[domain_config]
[endpoint_filter]
[endpoint_policy]
[eventlet_server]
[eventlet_server_ssl]
[federation]
[fernet_tokens]
[identity]
[identity_mapping]
[kvs]
[ldap]
[matchmaker_redis]
[matchmaker_ring]
[memcache]
servers=192.168.1.120:11211
[oauth1]
[os_inherit]
[oslo_messaging_amqp]
[oslo_messaging_qpid]
[oslo_messaging_rabbit]
[oslo_middleware]
[oslo_policy]
[paste_deploy]
[policy]
```

```
[resource]
[revoke]
driver=sql
[role]
[saml]
[signing]
[ssl]
[token]
provider=uuid
driver=memcache
[tokenless_auth]
[trust]
EOF
#创建数据库表,使用命令同步
su -s /bin/sh -c "keystone-manage db_sync" keystone
tail -n 10 /var/log/keystone/keystone.log
```

(2)配置并启动 Memcached 和 Apache 服务,操作指令如下:

```
sed -i 's/OPTIONS=.*/OPTIONS=\"0.0.0.0\"/g' /etc/sysconfig/memcached
systemctl enable memcached
systemctl start memcached
cat>/etc/httpd/conf.d/wsgi-keystone.conf<<EOF
Listen 5000
Listen 35357
<VirtualHost *:5000>
WSGIDaemonProcess keystone-public processes=5 threads=1 user=keystone
group=keystone display-name=%{GROUP}
WSGIProcessGroup keystone-public
WSGIScriptAlias / /usr/bin/keystone-wsgi-public
WSGIApplicationGroup %{GLOBAL}
WSGIPassAuthorization On
<IfVersion >= 2.4>
ErrorLogFormat "%{cu}t %M"
</IfVersion>
ErrorLog /var/log/httpd/keystone-error.log
CustomLog /var/log/httpd/keystone-access.log combined
<Directory /usr/bin>
<IfVersion >= 2.4>
Require all granted
</IfVersion>
<IfVersion < 2.4>
Order allow,deny
Allow from all
</IfVersion>
</Directory>
</VirtualHost>
```

```
<VirtualHost *:35357>
WSGIDaemonProcess keystone-admin processes=5 threads=1 user=keystone
group=keystone display-name=%{GROUP}
WSGIProcessGroup keystone-admin
WSGIScriptAlias / /usr/bin/keystone-wsgi-admin
WSGIApplicationGroup %{GLOBAL}
WSGIPassAuthorization On
<IfVersion >= 2.4>
ErrorLogFormat "%{cu}t %M"
</IfVersion>
ErrorLog /var/log/httpd/keystone-error.log
CustomLog /var/log/httpd/keystone-access.log combined
<Directory /usr/bin>
<IfVersion >= 2.4>
Require all granted
</IfVersion>
<IfVersion < 2.4>
Order allow,deny
Allow from all
</IfVersion>
</Directory>
</VirtualHost>
EOF
systemctl enable httpd
systemctl restart httpd
netstat -lntup|grep httpd
```

（3）创建 Keystone 用户，临时设置 admin_token 用户的环境变量，用来创建用户。

```
export OS_TOKEN=$ID
export OS_URL=http://192.168.1.120:35357/v3
export OS_IDENTITY_API_VERSION=3
#创建 admin 项目,创建 admin 用户,密码均为 admin
openstack project create --domain default --description "Admin Project" admin
openstack user create --domain default --password-prompt admin
```

（4）创建一个普通用户 demo，密码为 demo，操作指令如下：

```
openstack role create admin
openstack role add --project admin --user admin admin
openstack project create --domain default --description "Demo Project" demo
openstack user create --domain default --password=demo demo
openstack role create user
openstack role add --project demo --user demo user
#创建 service 项目,用来管理其他服务
openstack project create --domain default --description "Service Project" service
```

（5）检查 OpenStack 用户是否正常，如图 9-12 所示。

```
openstack user list
openstack project list
```

```
[root@node1 ~]# openstack user list
openstack project list
+----------------------------------+---------+
| ID                               | Name    |
+----------------------------------+---------+
| 4db7d18e9f80470b9a06713374b565e9 | demo    |
| 8c7cd98fc2834fc7b0698405ea6b099b | neutron |
| cbda414d2ea0444a807f98c8f2a0990e | admin   |
| d8a827ad7097482dab6017f25902c41c | glance  |
| e71eb2fd9d7f4e3e850a8b9607662658 | nova    |
+----------------------------------+---------+
[root@node1 ~]# openstack project list
+----------------------------------+---------+
| ID                               | Name    |
+----------------------------------+---------+
| 27b67feaa0d14c0d8b310923404ed493 | admin   |
| 590c7de4a8e54810b2c10a190e68d192 | service |
| e435d0a25c2247f480daef375cfac04d | demo    |
+----------------------------------+---------+
```

图 9-12　Keystone 配置查看

（6）注册 Keystone 服务，分别为公共的、内部的、管理的类型，操作指令如下：

```
openstack service   create --name keystone --description "OpenStack Identity" identity
openstack endpoint  create --region RegionOne identity public http://192.168.1.120:5000/v2.0
openstack endpoint  create --region RegionOne identity internal http://192.168.1.120:5000/v2.0
openstack endpoint  create --region RegionOne identity admin http://192.168.1.120:35357/v2.0
#查看
openstack endpoint list
```

（7）验证，获取 token，只有获取到才能说明 Keystone 配置成功，操作指令如下：

```
#unset OS_TOKEN
#unset OS_URL
openstack --os-auth-url http://192.168.1.120:35357/v3 --os-project-domain-id default --os-user-domain-id default --os-project-name admin --os-username admin --os-auth-type password token issue
```

（8）使用环境变量获取 token，环境变量创建虚拟机时需要调用，操作指令如下：

```
cat>admin-openrc.sh<<EOF
export OS_PROJECT_DOMAIN_ID=default
export OS_USER_DOMAIN_ID=default
export OS_PROJECT_NAME=admin
export OS_TENANT_NAME=admin
export OS_USERNAME=admin
export OS_PASSWORD=admin
export OS_AUTH_URL=http://192.168.1.120:35357/v3
export OS_IDENTITY_API_VERSION=3
EOF
cat>demo-openrc.sh<<EOF
export OS_PROJECT_DOMAIN_ID=default
export OS_USER_DOMAIN_ID=default
export OS_PROJECT_NAME=demo
export OS_TENANT_NAME=demo
export OS_USERNAME=demo
export OS_PASSWORD=demo
export OS_AUTH_URL=http://192.168.1.120:5000/v3
export OS_IDENTITY_API_VERSION=3
EOF
unset OS_TOKEN
unset OS_URL
source admin-openrc.sh
openstack token issue
```

9.11 配置 Glance 镜像服务

Glance 是一套虚拟机镜像查找及检索系统，支持多种虚拟机镜像格式（AKI、AMI、ARI、ISO、QCOW2、Raw、VDI、VHD 和 VMDK），有创建上传镜像、删除镜像、编辑镜像基本信息的功能。

（1）修改配置文件/etc/glance/glance-api.conf，操作指令如下：

```
cat>/etc/glance/glance-api.conf<<EOF
[DEFAULT]
verbose=True
notification_driver=noop
#Galnce 不需要消息队列
[database]
connection=mysql://glance:glance@192.168.1.120/glance
[glance_store]
default_store=file
filesystem_store_datadir=/var/lib/glance/images/
[image_format]
[keystone_authtoken]
```

```
auth_uri=http://192.168.1.120:5000
auth_url=http://192.168.1.120:35357
auth_plugin=password
project_domain_id=default
user_domain_id=default
project_name=service
username=glance
password=glance
[matchmaker_redis]
[matchmaker_ring]
[oslo_concurrency]
[oslo_messaging_amqp]
[oslo_messaging_qpid]
[oslo_messaging_rabbit]
[oslo_policy]
[paste_deploy]
flavor=keystone
[store_type_location_strategy]
[task]
[taskflow_executor]
EOF
```

（2）修改配置文件/etc/glance/glance-registry.conf，操作指令如下：

```
cat>/etc/glance/glance-registry.conf<<EOF
[DEFAULT]
verbose=True
notification_driver=noop
[database]
connection=mysql://glance:glance@192.168.1.120/glance
[glance_store]
[keystone_authtoken]
auth_uri=http://192.168.1.120:5000
auth_url=http://192.168.1.120:35357
auth_plugin=password
project_domain_id=default
user_domain_id=default
project_name=service
username=glance
password=glance
[matchmaker_redis]
[matchmaker_ring]
[oslo_messaging_amqp]
[oslo_messaging_qpid]
[oslo_messaging_rabbit]
[oslo_policy]
[paste_deploy]
```

```
flavor=keystone
EOF
```

（3）创建数据库表，同步数据库 Glance 库，操作指令如下：

```
su -s /bin/sh -c "glance-manage db_sync" glance
su -s /bin/sh -c "glance-manage db_sync" glance
```

（4）创建 Glance 的 Keystone 用户，注册用户信息，操作指令如下：

```
source admin-openrc.sh
openstack user create --domain default --password=glance glance
openstack role add --project service --user glance admin
systemctl enable openstack-glance-api
systemctl enable openstack-glance-registry
systemctl start openstack-glance-api
systemctl start openstack-glance-registry
netstat -lnutp |grep 9191 #registry
netstat -lnutp |grep 9292 #api
```

（5）在 Keystone 上注册，操作指令如下：

```
source admin-openrc.sh
openstack service create --name glance --description "OpenStack Image service" image
openstack endpoint create --region RegionOne image public http://192.168.1.120:9292
openstack endpoint create --region RegionOne image internal http://192.168.1.120:9292
openstack endpoint create --region RegionOne image admin http://192.168.1.120:9292
```

（6）添加 Glance 环境变量并测试，操作指令如下：

```
echo "export OS_IMAGE_API_VERSION=2" | tee -a admin-openrc.sh demo-openrc.sh
glance image-list
```

（7）下载镜像并上传到 Glance，操作指令如下，最终结果如图 9-13 所示。

```
wget -q http://download.cirros-cloud.net/0.3.4/cirros-0.3.4-x86_64-disk.img
glance image-create --name "cirros" --file cirros-0.3.4-x86_64-disk.img --disk-format qcow2 --container-format bare --visibility public --progress
#wget http://cloud.centos.org/centos/7/images/CentOS-7-x86_64-GenericCloud.qcow2
#glance image-create --name "CentOS-7-x86_64" --file CentOS-7-x86_64-GenericCloud.qcow2 --disk-format qcow2 --container-format bare --visibility public --progress
glance image-list
ll /var/lib/glance/images/
```

```
[root@node1 ~]# netstat -lnutp |grep 9292 #api
tcp        0      0 0.0.0.0:9292            0.0.0.0:*               LISTEN
[root@node1 ~]# glance image-list
ll /var/lib/glance/images/
+--------------------------------------+-------------+
| ID                                   | Name        |
+--------------------------------------+-------------+
| a8e32bcb-0b05-4cf9-ba7a-f4d0caee291a | CentOS-7-x86_64 |
| a6f3ef4f-059c-462c-af35-99e507659921 | centos7     |
| 71da6aae-9609-4f79-ba55-c4af7e2428e1 | cirros      |
+--------------------------------------+-------------+
[root@node1 ~]# ll /var/lib/glance/images/
total 2061812
-rw-r------ 1 glance glance   13287936 Jul  7  2017 71da6aae-9609-4f79-ba55
-rw-r------ 1 glance glance  713031680 Jul  5 15:53 a6f3ef4f-059c-462c-af35
-rw-r------ 1 glance glance 1384972288 Jul  5 02:19 a8e32bcb-0b05-4cf9-ba7a
[root@node1 ~]#
```

图 9-13　Glance 镜像操作结果

9.12　Nova 控制节点配置

Nova 是一套控制器，用于为单个用户或使用群组管理虚拟机实例的整个生命周期，根据用户需求提供虚拟服务。负责虚拟机创建、开机、关机、挂起、暂停、调整、迁移、重启、销毁等操作，配置 CPU、内存等信息规格，如图 9-14 所示。

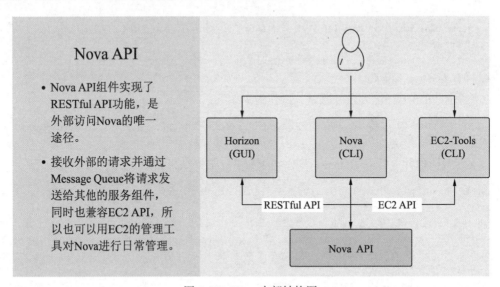

图 9-14　Nova 内部结构图

（1）配置 Nova 服务，操作指令如下：

```
cat>/etc/nova/nova.conf<<EOF
[DEFAULT]
```

```
my_ip=192.168.1.120
enabled_apis=osapi_compute,metadata
auth_strategy=keystone
network_api_class=nova.network.neutronv2.api.API
linuxnet_interface_driver=nova.network.linux_net.NeutronLinuxBridgeInterfaceDriver
security_group_api=neutron
firewall_driver=nova.virt.firewall.NoopFirewallDriver
debug=true
#vif_plugging_is_fatal=False
#vif_plugging_timeout=0
verbose=true
rpc_backend=rabbit
allow_resize_to_same_host=True
scheduler_default_filters=RetryFilter,AvailabilityZoneFilter,RamFilter,ComputeFilter,ComputeCapabilitiesFilter,ImagePropertiesFilter,ServerGroupAntiAffinityFilter,ServerGroupAffinityFilter
[api_database]
[barbican]
[cells]
[cinder]
[conductor]
[cors]
[cors.subdomain]
[database]
connection=mysql://nova:nova@192.168.1.120/nova
[ephemeral_storage_encryption]
[glance]
host=\$my_ip
[guestfs]
[hyperv]
[image_file_url]
[ironic]
[keymgr]
[keystone_authtoken]
auth_uri=http://192.168.1.120:5000
auth_url=http://192.168.1.120:35357
auth_plugin=password
project_domain_id=default
user_domain_id=default
project_name=service
username=nova
password=nova
[libvirt]
virt_type=qemu
#virt_type=kvm
```

```
[matchmaker_redis]
[matchmaker_ring]
[metrics]
[neutron]
url=http://192.168.1.120:9696
auth_url=http://192.168.1.120:35357
auth_plugin=password
project_domain_id=default
user_domain_id=default
region_name=RegionOne
project_name=service
username=neutron
password=neutron
service_metadata_proxy=True
metadata_proxy_shared_secret=neutron
lock_path=/var/lib/nova/tmp
[osapi_v21]
[oslo_concurrency]
[oslo_messaging_amqp]
[oslo_messaging_qpid]
[oslo_messaging_rabbit]
rabbit_host=192.168.1.120
rabbit_port=5672
rabbit_userid=openstack
rabbit_password=openstack
[oslo_middleware]
[rdp]
[serial_console]
[spice]
[ssl]
[trusted_computing]
[upgrade_levels]
[vmware]
[vnc]
novncproxy_base_url=http://192.168.1.120:6080/vnc_auto.html
vncserver_listen= \$my_ip
vncserver_proxyclient_address= \$my_ip
keymap=en-us
[workarounds]
[xenserver]
[zookeeper]
EOF
```

（2）同步 Nova 数据库并创建 Nova 用户及服务启动，操作指令如下：

```
su -s /bin/sh -c "nova-manage db sync" nova
su -s /bin/sh -c "nova-manage db sync" nova
#创建 Nova Keystone 用户
```

```
openstack user create --domain default --password=nova nova
openstack role add --project service --user nova admin
#启动 Nova 相关服务
systemctl enable openstack-nova-api.service openstack-nova-cert.service
openstack-nova-consoleauth.service openstack-nova-scheduler.service openstack-
nova-conductor.service openstack-nova-novncproxy.service
systemctl restart openstack-nova-api.service openstack-nova-cert.service
openstack-nova-consoleauth.service openstack-nova-scheduler.service openstack-
nova-conductor.service openstack-nova-novncproxy.service
#在 Keystone 上注册
source admin-openrc.sh
openstack service create --name nova --description "OpenStack Compute"
compute
openstack endpoint create --region RegionOne compute public http://192.168.
1.120:8774/v2/%\(tenant_id\)s
openstack endpoint create --region RegionOne compute internal http://192.
168.1.120:8774/v2/%\(tenant_id\)s
openstack endpoint create --region RegionOne compute admin http://192.
168.1.120:8774/v2/%\(tenant_id\)s
openstack host list
```

9.13 Nova 计算节点配置

根据以上 Nova 控制节点部署，接下来配置 Nova 计算节点。Nova Compute 介绍如图 9-15 所示。

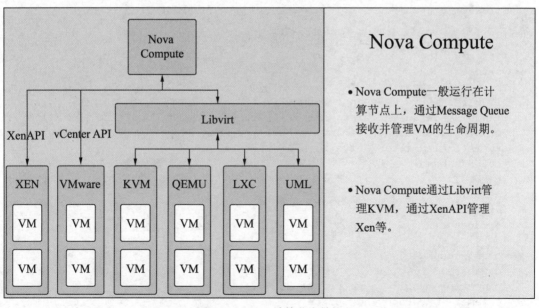

图 9-15 Nova 计算节点结构

(1) Node2 计算节点配置，操作指令如下：

```
#Base
yum install -y epel-release
yum install -y centos-release-openstack-liberty
yum install -y python-openstackclient
#Nova
yum install -y openstack-nova-compute sysfsutils
#Neutron
yum install -y openstack-neutron openstack-neutron-linuxbridge ebtables ipset
#Cinder
yum install -y openstack-cinder python-cinderclient targetcli python-oslo-policy
#Update Qemu
yum install -y centos-release-qemu-ev.noarch
yum install qemu-kvm qemu-img -y
```

(2) 修改 nova.conf 配置文件，操作指令如下：

```
cat> /etc/nova/nova.conf<<EOF
[DEFAULT]
my_ip=192.168.1.121
enabled_apis=osapi_compute,metadata
auth_strategy=keystone
network_api_class=nova.network.neutronv2.api.API
linuxnet_interface_driver=nova.network.linux_net.NeutronLinuxBridgeInterfaceDriver
security_group_api=neutron
firewall_driver=nova.virt.firewall.NoopFirewallDriver
debug=true
#vif_plugging_is_fatal=False
#vif_plugging_timeout=0
verbose=true
rpc_backend=rabbit
allow_resize_to_same_host=True
scheduler_default_filters=RetryFilter,AvailabilityZoneFilter,RamFilter,ComputeFilter,ComputeCapabilitiesFilter,ImagePropertiesFilter,ServerGroupAntiAffinityFilter,ServerGroupAffinityFilter
[api_database]
[barbican]
[cells]
[cinder]
[conductor]
[cors]
[cors.subdomain]
[database]
connection=mysql://nova:nova@192.168.1.120/nova
[ephemeral_storage_encryption]
[glance]
```

```
host=192.168.1.120
[guestfs]
[hyperv]
[image_file_url]
[ironic]
[keymgr]
[keystone_authtoken]
auth_uri=http://192.168.1.120:5000
auth_url=http://192.168.1.120:35357
auth_plugin=password
project_domain_id=default
user_domain_id=default
project_name=service
username=nova
password=nova
[libvirt]
virt_type=qemu
#virt_type=kvm
[matchmaker_redis]
[matchmaker_ring]
[metrics]
[neutron]
url=http://192.168.1.120:9696
auth_url=http://192.168.1.120:35357
auth_plugin=password
project_domain_id=default
user_domain_id=default
region_name=RegionOne
project_name=service
username=neutron
password=neutron
service_metadata_proxy=True
metadata_proxy_shared_secret=neutron
lock_path=/var/lib/nova/tmp
[osapi_v21]
[oslo_concurrency]
[oslo_messaging_amqp]
[oslo_messaging_qpid]
[oslo_messaging_rabbit]
rabbit_host=192.168.1.120
rabbit_port=5672
rabbit_userid=openstack
rabbit_password=openstack
[oslo_middleware]
[rdp]
[serial_console]
[spice]
[ssl]
```

```
[trusted_computing]
[upgrade_levels]
[vmware]
[vnc]
novncproxy_base_url=http://192.168.1.120:6080/vnc_auto.html
vncserver_listen=0.0.0.0
vncserver_proxyclient_address=\$my_ip
keymap=en-us
[workarounds]
[xenserver]
[zookeeper]
EOF
```

(3）启动 Nova 节点相关服务，操作指令如下：

```
systemctl enable libvirtd openstack-nova-compute
systemctl restart libvirtd openstack-nova-compute
```

9.14　OpenStack 节点测试

根据如上步骤和指令的配置，OpenStack 控制节点与计算节点配置完毕，以下为在计算节点进行测试，如图 9-16 所示，操作指令如下：

```
openstack host list
glance image-list
nova image-list
```

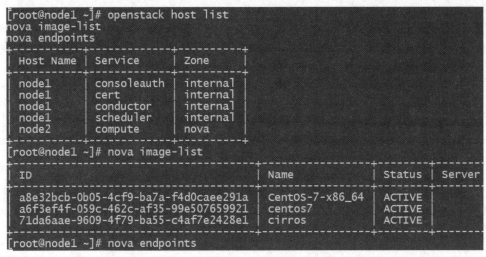

(a)

图 9-16　OpenStack 节点测试和验证

(a）查看 OpenStack host 列表

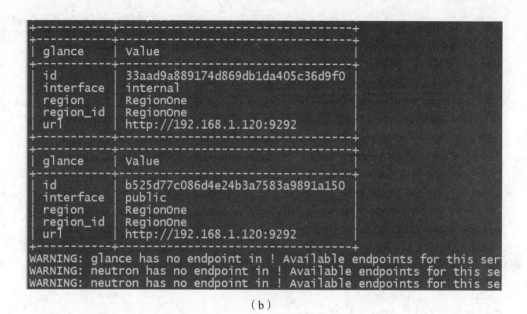

(b)

图 9-16 OpenStack 节点测试和验证（续）

（b）查看 OpenStack glance 列表

9.15 Neutron 控制节点配置

Neutron 是提供云计算的网络虚拟化技术，为 OpenStack 其他服务提供网络连接服务。为用户提供接口，可以定义 Network、Subnet、Router，配置 DHCP、DNS、负载均衡、L3 服务，网络支持 GRE、VLAN。插件架构支持许多主流的网络厂家和技术，如 OpenSwitch，如图 9-17 所示。

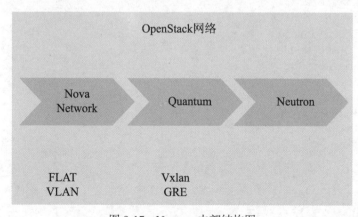

图 9-17 Neutron 内部结构图

（1）Neutron 控制节点配置，修改/etc/neutron/neutron.conf 文件，操作指令如下：

```
cat>/etc/neutron/neutron.conf<<EOF
[DEFAULT]
state_path=/var/lib/neutron
core_plugin=ml2
service_plugins=router
auth_strategy=keystone
notify_nova_on_port_status_changes=True
notify_nova_on_port_data_changes=True
nova_url=http://192.168.1.120:8774/v2
rpc_backend=rabbit
[matchmaker_redis]
[matchmaker_ring]
[quotas]
[agent]
[keystone_authtoken]
auth_uri=http://192.168.1.120:5000
auth_url=http://192.168.1.120:35357
auth_plugin=password
project_domain_id=default
user_domain_id=default
project_name=service
username=neutron
password=neutron
admin_tenant_name=%SERVICE_TENANT_NAME%
admin_user=%SERVICE_USER%
admin_password=%SERVICE_PASSWORD%
[database]
connection=mysql://neutron:neutron@192.168.1.120:3306/neutron
[nova]
auth_url=http://192.168.1.120:35357
auth_plugin=password
project_domain_id=default
user_domain_id=default
region_name=RegionOne
project_name=service
username=nova
password=nova
[oslo_concurrency]
lock_path=$state_path/lock
[oslo_policy]
[oslo_messaging_amqp]
[oslo_messaging_qpid]
[oslo_messaging_rabbit]
rabbit_host=192.168.1.120
rabbit_port=5672
```

```
rabbit_userid=openstack
rabbit_password=openstack
[qos]
EOF
```

（2）按照以上步骤配置完成之后，需要创建配置文件 ml2_conf.ini、linuxbridge_agent.ini、dhcp_agent.ini、metadata_agent.ini，操作指令如下：

```
cat>/etc/neutron/plugins/ml2/ml2_conf.ini<<EOF
[ml2]
type_drivers=flat,vlan,gre,vxlan,geneve
tenant_network_types=vlan,gre,vxlan,geneve
mechanism_drivers=openvswitch,linuxbridge
extension_drivers=port_security
[ml2_type_flat]
flat_networks=physnet1
[ml2_type_vlan]
[ml2_type_gre]
[ml2_type_vxlan]
[ml2_type_geneve]
[securitygroup]
enable_ipset=True
EOF
cat>/etc/neutron/plugins/ml2/linuxbridge_agent.ini<<EOF
[linux_bridge]
physical_interface_mappings=physnet1:eth0
[vxlan]
enable_vxlan=false
[agent]
prevent_arp_spoofing=True
[securitygroup]
firewall_driver=neutron.agent.linux.iptables_firewall.IptablesFirewallDriver
enable_security_group=True
EOF
cat>/etc/neutron/dhcp_agent.ini<<EOF
[DEFAULT]
interface_driver=neutron.agent.linux.interface.BridgeInterfaceDriver
dhcp_driver=neutron.agent.linux.dhcp.Dnsmasq
enable_isolated_metadata=true
[AGENT]
EOF
cat>/etc/neutron/metadata_agent.ini<<EOF
[DEFAULT]
auth_uri=http://192.168.1.120:5000
```

```
auth_url=http://192.168.1.120:35357
auth_region=RegionOne
auth_plugin=password
project_domain_id=default
user_domain_id=default
project_name=service
username=neutron
password=neutron
nova_metadata_ip=192.168.1.120
metadata_proxy_shared_secret=neutron
admin_tenant_name=%SERVICE_TENANT_NAME%
admin_user=%SERVICE_USER%
admin_password=%SERVICE_PASSWORD%
[AGENT]
EOF
```

（3）创建连接并创建 Keystone 的用户 Neutron，更新数据库信息及注册至 Keystone 中，操作指令如下：

```
#创建连接并创建Keystone的用户
ln -s /etc/neutron/plugins/ml2/ml2_conf.ini /etc/neutron/plugin.ini
openstack user create --domain default --password=neutron neutron
openstack role add --project service --user neutron admin
#更新数据库
su -s /bin/sh -c "neutron-db-manage --config-file /etc/neutron/neutron.conf --config-file /etc/neutron/plugins/ml2/ml2_conf.ini upgrade head" neutron
su -s /bin/sh -c "neutron-db-manage --config-file /etc/neutron/neutron.conf --config-file /etc/neutron/plugins/ml2/ml2_conf.ini upgrade head" neutron
#注册Keystone
source admin-openrc.sh
openstack service create --name neutron --description "OpenStack Networking" network
openstack endpoint create --region RegionOne network public http://192.168.1.120:9696
openstack endpoint create --region RegionOne network internal http://192.168.1.120:9696
openstack endpoint create --region RegionOne network admin http://192.168.1.120:9696
```

（4）因为 Neutron 和 Nova 有联系，操作 Neutron 时修改 Nova 的配置文件，Nova.conf 与 Neutron 关联配置，需要重启 OpenStack-Nova-API 服务，操作指令如下：

```
systemctl restart openstack-nova-api.service openstack-nova-cert.service openstack-nova-consoleauth.service openstack-nova-scheduler.service openstack-nova-conductor.service openstack-nova-novncproxy.service
systemctl enable neutron-server.service neutron-linuxbridge-agent.service
```

```
neutron-dhcp-agent.service neutron-metadata-agent.service
systemctl restart neutron-server.service neutron-linuxbridge-agent.service
neutron-dhcp-agent.service neutron-metadata-agent.service
mkdir -p /lock;chmod 777 -R /lock
```

（5）Neutron 配置检查操作指令如下，如图 9-18 所示。

```
neutron agent-list
openstack endpoint list
```

```
[root@node1 ~]# cd
[root@node1 ~]# neutron agent-list
+--------------------------------------+--------------------+-------+
| id                                   | agent_type         | host  |
+--------------------------------------+--------------------+-------+
| 191c5ff0-4720-4804-bf88-60e29db4832c | Linux bridge agent | node1 |
| 6ba3694d-c590-4fe3-a75f-122c5407c2b0 | Metadata agent     | node1 |
| 97c99b0b-2dba-48c4-b67e-5465577f228f | Linux bridge agent | node2 |
| c74452f5-7b8e-4b7b-961d-2657f91ed8ef | DHCP agent         | node1 |
+--------------------------------------+--------------------+-------+
```

(a)

```
[root@node1 ~]# openstack endpoint list
+----------------------------------+-----------+--------------+--------------+
| ID                               | Region    | Service Name | Service Type |
+----------------------------------+-----------+--------------+--------------+
| 10703c763f474bd1bbec86450bd48979 | RegionOne | keystone     | identity     |
:5000/v2.0                         |
| 22f01b832edc441a8a3eb1ce6c4ce435 | RegionOne | nova         | compute      |
:8774/v2/%(tenant_id)s             |
| 339b9ec5b6454841ad322189f788de97 | RegionOne | glance       | image        |
:9292                              |
| 33aad9a889174d869db1da405c36d9f0 | RegionOne | glance       | image        |
:9292                              |
| 67aa2437cbb44814a8d2fffe75eb0897 | RegionOne | nova         | compute      |
:8774/v2/%(tenant_id)s             |
| 6e69e7bc50c24751b5e2c582f62b9d33 | RegionOne | neutron      | network      |
:9696                              |
| 7f08b9bd301b4565a2e3b1157001a8aa | RegionOne | keystone     | identity     |
```

(b)

图 9-18　Neutron 控制节点配置验证

（a）查看 Neutron agent 列表信息；（b）查看 OpenStack endpoint 列表信息

9.16　Neutron 计算节点配置

（1）修改 Neutron 计算节点配置文件/etc/neutron/neutron.conf，操作指令如下：

```
cat>/etc/neutron/neutron.conf<<EOF
[DEFAULT]
state_path=/var/lib/neutron
core_plugin=ml2
service_plugins=router
auth_strategy=keystone
notify_nova_on_port_status_changes=True
notify_nova_on_port_data_changes=True
nova_url=http://192.168.1.120:8774/v2
rpc_backend=rabbit
[matchmaker_redis]
[matchmaker_ring]
[quotas]
[agent]
[keystone_authtoken]
auth_uri=http://192.168.1.120:5000
auth_url=http://192.168.1.120:35357
auth_plugin=password
project_domain_id=default
user_domain_id=default
project_name=service
username=neutron
password=neutron
admin_tenant_name=%SERVICE_TENANT_NAME%
admin_user=%SERVICE_USER%
admin_password=%SERVICE_PASSWORD%
[database]
connection=mysql://neutron:neutron@192.168.1.120:3306/neutron
[nova]
auth_url=http://192.168.1.120:35357
auth_plugin=password
project_domain_id=default
user_domain_id=default
region_name=RegionOne
project_name=service
username=nova
password=nova
[oslo_concurrency]
lock_path=$state_path/lock
[oslo_policy]
[oslo_messaging_amqp]
```

```
[oslo_messaging_qpid]
[oslo_messaging_rabbit]
rabbit_host=192.168.1.120
rabbit_port=5672
rabbit_userid=openstack
rabbit_password=openstack
[qos]
EOF
```

（2）修改相关配置文件 linuxbridge_agent.ini、ml2_conf.ini，操作指令如下：

```
cat>/etc/neutron/plugins/ml2/linuxbridge_agent.ini<<EOF
[linux_bridge]
physical_interface_mappings=physnet1:eth0
[vxlan]
enable_vxlan=false
[agent]
prevent_arp_spoofing=True
[securitygroup]
firewall_driver=neutron.agent.linux.iptables_firewall.IptablesFirewallDriver
enable_security_group=True
EOF
cat>/etc/neutron/plugins/ml2/ml2_conf.ini<<EOF
[ml2]
type_drivers=flat,vlan,gre,vxlan,geneve
tenant_network_types=vlan,gre,vxlan,geneve
mechanism_drivers=openvswitch,linuxbridge
extension_drivers=port_security
[ml2_type_flat]
flat_networks=physnet1
[ml2_type_vlan]
[ml2_type_gre]
[ml2_type_vxlan]
[ml2_type_geneve]
[securitygroup]
enable_ipset=True
EOF
mkdir -p /lock;chmod 777 -R /lock
ln -s /etc/neutron/plugins/ml2/ml2_conf.ini /etc/neutron/plugin.ini
systemctl enable neutron-linuxbridge-agent.service
systemctl restart neutron-linuxbridge-agent.service
```

9.17　OpenStack 控制节点网桥

（1）检查控制节点及计算节点信息，操作指令如下，如图 9-19 所示。

```
neutron agent-list
```

```
[root@node1 ~]# neutron agent-list
+--------------------------------------+--------------------+-------+-------+
| id                                   | agent_type         | host  | alive |
+--------------------------------------+--------------------+-------+-------+
| 191c5ff0-4720-4804-bf88-60e29db4832c | Linux bridge agent | node1 | :-)   |
| 6ba3694d-c590-4fe3-a75f-122c5407c2b0 | Metadata agent     | node1 | :-)   |
| 97c99b0b-2dba-48c4-b67e-5465577f228f | Linux bridge agent | node2 | :-)   |
| c74452f5-7b8e-4b7b-961d-2657f91ed8ef | DHCP agent         | node1 | :-)   |
+--------------------------------------+--------------------+-------+-------+
```

图 9-19 OpenStack 控制节点及计算节点信息

（2）在 OpenStack 控制节点创建新的网桥信息，操作指令如下，如图 9-20 所示。

```
source admin-openrc.sh
neutron net-create flat --shared --provider:physical_network physnet1
--provider:network_type flat
neutron subnet-create flat 192.168.1.0/24 --name flat-subnet --allocation
-pool start=192.168.1.140,end=192.168.1.200 --dns-nameserver 192.168.1.1
--gateway 192.168.1.1
neutron net-list
neutron subnet-list
```

```
[root@node1 ~]# neutron net-list
neutron subnet-list
+--------------------------------------+------+--------------------------------------+
| id                                   | name | subnets                              |
+--------------------------------------+------+--------------------------------------+
| 2d642a8c-d0c6-444e-828c-6a988539ddad | flat | a22d1bde-2ff1-4073-9703-6f7eca626bb8 19
+--------------------------------------+------+--------------------------------------+
[root@node1 ~]# neutron subnet-list
+--------------------------------------+-------------+----------------+-----------------+
| id                                   | name        | cidr           | allocation_pool |
+--------------------------------------+-------------+----------------+-----------------+
| a22d1bde-2ff1-4073-9703-6f7eca626bb8 | flat-subnet | 192.168.1.0/24 | {"start": "192.
+--------------------------------------+-------------+----------------+-----------------+
```

图 9-20 OpenStack 控制节点网桥信息

（3）创建虚拟机，为 vm 分配内网 IP，创建 key，操作指令如下：

```
#创建 key
ssh-keygen -t rsa -P '' -f ~/.ssh/id_rsa
```

```
nova keypair-add --pub-key /root/.ssh/id_rsa.pub mykey
nova keypair-list
#创建安全组
nova secgroup-add-rule default icmp -1 -1 0.0.0.0/0
nova secgroup-add-rule default tcp 22 22 0.0.0.0/0
nova secgroup-add-rule default tcp 80 80 0.0.0.0/0
#查看支持的虚拟机类型
nova flavor-list
#查看镜像
nova image-list
#查看网络
neutron net-list
```

（4）网桥配置完毕，接下来创建 OpenStack 虚拟机，操作指令如下，如图 9-21 所示。

```
nova boot --flavor m1.tiny --image cirros --nic net-id=6277d20f-d033-
42b8-96bc-6565ff07e8a3 --security-group default --key-name mykey hello-
instance
```

（5）创建虚拟机时，OpenStack 在 Neutron 组网内采用 dhcp-agent 自动分配 IP，可以在创建虚拟机时指定固定 IP。

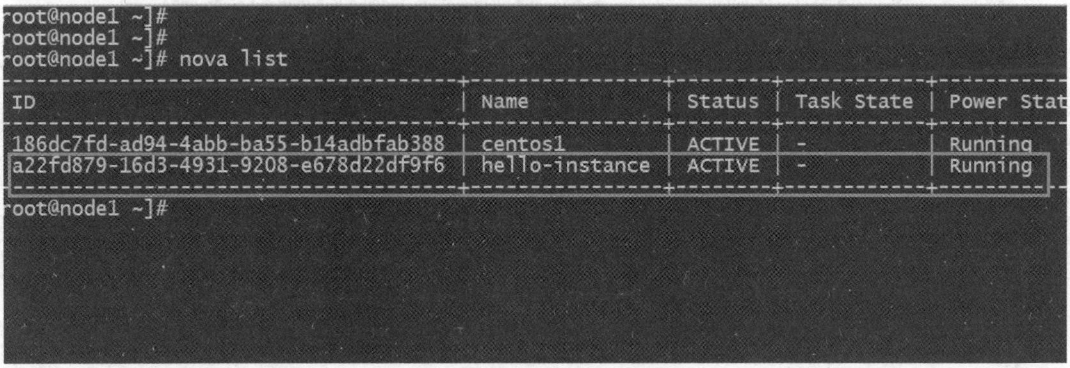

图 9-21　OpenStack 虚拟机创建

9.18　Dashboard 控制节点配置

OpenStack Web 平台主要用于简化用户对服务的操作，例如启动实例、分配 IP 地址、配置访问控制等。通过 Dashboard Web 界面可以管理 OpenStack，操作指令如下：

```
yum install openstack-dashboard -y
cat> /etc/openstack-dashboard/local_settings<<EOF
import os
from django.utils.translation import import ugettext_lazy as _
```

```
from openstack_dashboard import exceptions
from openstack_dashboard.settings import HORIZON_CONFIG
DEBUG=False
TEMPLATE_DEBUG=DEBUG
WEBROOT is the location relative to Webserver root
should end with a slash.
WEBROOT='/dashboard/'
LOGIN_URL=WEBROOT + 'auth/login/'
LOGOUT_URL=WEBROOT + 'auth/logout/'
LOGIN_REDIRECT_URL can be used as an alternative for
HORIZON_CONFIG.user_home, if user_home is not set.
Do not set it to '/home/', as this will cause circular redirect loop
LOGIN_REDIRECT_URL=WEBROOT
Required for Django 1.5.
If horizon is running in production (DEBUG is False), set this
with the list of host/domain names that the application can serve.
For more information see:
https://docs.djangoproject.com/en/dev/ref/settings/#allowed-hosts
ALLOWED_HOSTS=['*']
Set SSL proxy settings:
For Django 1.4+ pass this header from the proxy after terminating the SSL,
and don't forget to strip it from the client's request.
For more information see:
https://docs.djangoproject.com/en/1.4/ref/settings/#secure-proxy-ssl-header
SECURE_PROXY_SSL_HEADER=('HTTP_X_FORWARDED_PROTOCOL', 'https')
https://docs.djangoproject.com/en/1.5/ref/settings/#secure-proxy-ssl-header
SECURE_PROXY_SSL_HEADER=('HTTP_X_FORWARDED_PROTO', 'https')
If Horizon is being served through SSL, then uncomment the following two
settings to better secure the cookies from security exploits
CSRF_COOKIE_SECURE=True
SESSION_COOKIE_SECURE=True
Overrides for OpenStack API versions. Use this setting to force the
OpenStack dashboard to use a specific API version for a given service API.
Versions specified here should be integers or floats, not strings.
NOTE: The version should be formatted as it appears in the URL for the
service API. For example, The identity service APIs have inconsistent
use of the decimal point, so valid options would be 2.0 or 3.
OPENSTACK_API_VERSIONS={
    "data-processing": 1.1,
    "identity": 3,
    "volume": 2,
}

Set this to True if running on multi-domain model. When this is enabled,
```

it
will require user to enter the Domain name in addition to username for login.
OPENSTACK_KEYSTONE_MULTIDOMAIN_SUPPORT=False
Overrides the default domain used when running on single-domain model
with Keystone V3. All entities will be created in the default domain.
OPENSTACK_KEYSTONE_DEFAULT_DOMAIN='Default'
Set Console type:
valid options are "AUTO"(default), "VNC", "SPICE", "RDP", "SERIAL" or None
Set to None explicitly if you want to deactivate the console.
ONSOLE_TYPE="AUTO"
Show backdrop element outside the modal, do not close the modal
after clicking on backdrop.
HORIZON_CONFIG["modal_backdrop"]="static"
Specify a regular expression to validate user passwords.
HORIZON_CONFIG["password_validator"]={
    "regex": '.*',
    "help_text": _("Your password does not meet the requirements."),
}
Disable simplified floating IP address management for deployments with
multiple floating IP pools or complex network requirements.
HORIZON_CONFIG["simple_ip_management"]=False
Turn off browser autocompletion for forms including the login form and
the database creation workflow if so desired.
HORIZON_CONFIG["password_autocomplete"]="off"
Setting this to True will disable the reveal button for password fields,
including on the login form.
HORIZON_CONFIG["disable_password_reveal"]=False
LOCAL_PATH='/tmp'
Set custom secret key:
You can either set it to a specific value or you can let horizon generate a
default secret key that is unique on this machine, e.i. regardless of the
amount of Python WSGI workers (if used behind Apache+mod_wsgi): However,
there may be situations where you would want to set this explicitly, e.g.
when multiple dashboard instances are distributed on different machines
(usually behind a load-balancer). Either you have to make sure that a session
gets all requests routed to the same dashboard instance or you set the same
SECRET_KEY for all of them.
SECRET_KEY='36c739e5c252f9e014d9'
We recommend you use memcached for development; otherwise after every reload
of the django development server, you will have to login again. To use
memcached set CACHES to something like
CACHES={
    'default': {
        'BACKEND': 'django.core.cache.backends.memcached.MemcachedCache',
        'LOCATION': '192.168.1.120:11211',
    }

```
}
CACHES={
    'default': {
        'BACKEND': 'django.core.cache.backends.locmem.LocMemCache',
    }
}
Send email to the console by default
EMAIL_BACKEND='django.core.mail.backends.console.EmailBackend'
Or send them to /dev/null
EMAIL_BACKEND='django.core.mail.backends.dummy.EmailBackend'
Configure these for your outgoing email host
EMAIL_HOST='smtp.my-company.com'
EMAIL_PORT=25
EMAIL_HOST_USER='djangomail'
EMAIL_HOST_PASSWORD='top-secret!'
For multiple regions uncomment this configuration, and add (endpoint, title).
AVAILABLE_REGIONS=[
    ('http://cluster1.example.com:5000/v2.0', 'cluster1'),
    ('http://cluster2.example.com:5000/v2.0', 'cluster2'),
]
OPENSTACK_HOST="192.168.1.120"
OPENSTACK_KEYSTONE_URL="http://%s:5000/v2.0" % OPENSTACK_HOST
OPENSTACK_KEYSTONE_DEFAULT_ROLE="user"

Enables keystone web single-sign-on if set to True.
WEBSSO_ENABLED=False
Determines which authentication choice to show as default.
WEBSSO_INITIAL_CHOICE="credentials"
The list of authentication mechanisms
which include keystone federation protocols.
Current supported protocol IDs are 'saml2' and 'oidc'
which represent SAML 2.0, OpenID Connect respectively.
Do not remove the mandatory credentials mechanism.
WEBSSO_CHOICES=(
    ("credentials", _("Keystone Credentials")),
    ("oidc", _("OpenID Connect")),
    ("saml2", _("Security Assertion Markup Language")))
Disable SSL certificate checks (useful for self-signed certificates):
OPENSTACK_SSL_NO_VERIFY=True
The CA certificate to use to verify SSL connections
OPENSTACK_SSL_CACERT='/path/to/cacert.pem'
The OPENSTACK_KEYSTONE_BACKEND settings can be used to identify the
capabilities of the auth backend for Keystone.
If Keystone has been configured to use LDAP as the auth backend then set
can_edit_user to False and name to 'ldap'.
```

```
TODO(tres): Remove these once Keystone has an API to identify auth backend.
OPENSTACK_KEYSTONE_BACKEND={
    'name': 'native',
    'can_edit_user': True,
    'can_edit_group': True,
    'can_edit_project': True,
    'can_edit_domain': True,
    'can_edit_role': True,
}
```

Setting this to True, will add a new "Retrieve Password" action on instance, allowing Admin session password retrieval/decryption.
OPENSTACK_ENABLE_PASSWORD_RETRIEVE=False

The Launch Instance user experience has been significantly enhanced. You can choose whether to enable the new launch instance experience, the legacy experience, or both. The legacy experience will be removed in a future release, but is available as a temporary backup setting to ensure compatibility with existing deployments. Further development will not be done on the legacy experience. Please report any problems with the new experience via the Launchpad tracking system.

Toggle LAUNCH_INSTANCE_LEGACY_ENABLED and LAUNCH_INSTANCE_NG_ENABLED to determine the experience to enable. Set them both to true to enable both.
LAUNCH_INSTANCE_LEGACY_ENABLED=True
LAUNCH_INSTANCE_NG_ENABLED=False

The Xen Hypervisor has the ability to set the mount point for volumes attached to instances (other Hypervisors currently do not). Setting can_set_mount_point to True will add the option to set the mount point from the UI.
```
OPENSTACK_HYPERVISOR_FEATURES={
    'can_set_mount_point': False,
    'can_set_password': False,
    'requires_keypair': False,
}
```

The OPENSTACK_CINDER_FEATURES settings can be used to enable optional services provided by cinder that is not exposed by its extension API.
```
OPENSTACK_CINDER_FEATURES={
    'enable_backup': False,
}
```

The OPENSTACK_NEUTRON_NETWORK settings can be used to enable optional

services provided by neutron. Options currently available are load balancer service, security groups, quotas, VPN service.
```
OPENSTACK_NEUTRON_NETWORK={
    'enable_router': True,
    'enable_quotas': True,
    'enable_ipv6': True,
    'enable_distributed_router': False,
    'enable_ha_router': False,
    'enable_lb': True,
    'enable_firewall': True,
    'enable_vpn': True,
    'enable_fip_topology_check': True,
```

Neutron can be configured with a default Subnet Pool to be used for IPv4 subnet-allocation. Specify the label you wish to display in the Address pool selector on the create subnet step if you want to use this feature.
`'default_ipv4_subnet_pool_label': None,`

Neutron can be configured with a default Subnet Pool to be used for IPv6 subnet-allocation. Specify the label you wish to display in the Address pool selector on the create subnet step if you want to use this feature. You must set this to enable IPv6 Prefix Delegation in a PD-capable environment.
`'default_ipv6_subnet_pool_label': None,`

The profile_support option is used to detect if an external router can be
configured via the dashboard. When using specific plugins the profile_support can be turned on if needed.
`'profile_support': None,`
`'profile_support': 'cisco',`

Set which provider network types are supported. Only the network types in this list will be available to choose from when creating a network. Network types include local, flat, vlan, gre, and vxlan.
`'supported_provider_types': ['*'],`

Set which VNIC types are supported for port binding. Only the VNIC types in this list will be available to choose from when creating a port.
VNIC types include 'normal', 'macvtap' and 'direct'.
Set to empty list or None to disable VNIC type selection.
`'supported_vnic_types': ['*']`
}

The OPENSTACK_IMAGE_BACKEND settings can be used to customize features

in the OpenStack Dashboard related to the Image service, such as the list
of supported image formats.
```
OPENSTACK_IMAGE_BACKEND={
    'image_formats': [
        ('', _('Select format')),
        ('aki', _('AKI - Amazon Kernel Image')),
        ('ami', _('AMI - Amazon Machine Image')),
        ('ari', _('ARI - Amazon Ramdisk Image')),
        ('docker', _('Docker')),
        ('iso', _('ISO - Optical Disk Image')),
        ('ova', _('OVA - Open Virtual Appliance')),
        ('qcow2', _('QCOW2 - QEMU Emulator')),
        ('raw', _('Raw')),
        ('vdi', _('VDI - Virtual Disk Image')),
        ('vhd', ('VHD - Virtual Hard Disk')),
        ('vmdk', _('VMDK - Virtual Machine Disk')),
    ]
}
```

The IMAGE_CUSTOM_PROPERTY_TITLES settings is used to customize the titles for
image custom property attributes that appear on image detail pages.
```
IMAGE_CUSTOM_PROPERTY_TITLES={
    "architecture": _("Architecture"),
    "kernel_id": _("Kernel ID"),
    "ramdisk_id": _("Ramdisk ID"),
    "image_state": _("Euca2ools state"),
    "project_id": _("Project ID"),
    "image_type": _("Image Type"),
}
```

The IMAGE_RESERVED_CUSTOM_PROPERTIES setting is used to specify which image
custom properties should not be displayed in the Image Custom Properties
table.
```
IMAGE_RESERVED_CUSTOM_PROPERTIES=[]
```

OPENSTACK_ENDPOINT_TYPE specifies the endpoint type to use for the endpoints
in the Keystone service catalog. Use this setting when Horizon is running
external to the OpenStack environment. The default is 'publicURL'.
```
OPENSTACK_ENDPOINT_TYPE="publicURL"
```

SECONDARY_ENDPOINT_TYPE specifies the fallback endpoint type to use in the
case that OPENSTACK_ENDPOINT_TYPE is not present in the endpoints
in the Keystone service catalog. Use this setting when Horizon is running
external to the OpenStack environment. The default is None. This
value should differ from OPENSTACK_ENDPOINT_TYPE if used.

```
SECONDARY_ENDPOINT_TYPE="publicURL"

The number of objects (Swift containers/objects or images) to display
on a single page before providing a paging element (a "more" link)
to paginate results.
API_RESULT_LIMIT=1000
API_RESULT_PAGE_SIZE=20

The size of chunk in bytes for downloading objects from Swift
SWIFT_FILE_TRANSFER_CHUNK_SIZE=512 * 1024

Specify a maximum number of items to display in a dropdown.
DROPDOWN_MAX_ITEMS=30

The timezone of the server. This should correspond with the timezone
of your entire OpenStack installation, and hopefully be in UTC.
TIME_ZONE="Asia/Shanghai"

When launching an instance, the menu of available flavors is
sorted by RAM usage, ascending. If you would like a different sort order,
you can provide another flavor attribute as sorting key. Alternatively,
you
can provide a custom callback method to use for sorting. You can also provide
a flag for reverse sort. For more info, see
http://docs.python.org/2/library/functions.html#sorted
CREATE_INSTANCE_FLAVOR_SORT={
    'key': 'name',
    #or
    'key': my_awesome_callback_method,
    'reverse': False,
}

Set this to True to display an 'Admin Password' field on the Change Password
form to verify that it is indeed the admin logged-in who wants to change
the password.
ENFORCE_PASSWORD_CHECK=False

Modules that provide /auth routes that can be used to handle different types
of user authentication. Add auth plugins that require extra route handling
to
this list.
AUTHENTICATION_URLS=[
    'openstack_auth.urls',
]

The Horizon Policy Enforcement engine uses these values to load per service
```

```
policy rule files. The content of these files should match the files the
OpenStack services are using to determine role based access control in the
target installation.


Map of local copy of service policy files.
Please insure that your identity policy file matches the one being used on
your keystone servers. There is an alternate policy file that may be used
in the Keystone v3 multi-domain case, policy.v3cloudsample.json.
This file is not included in the Horizon repository by default but can be
found at
http://git.openstack.org/cgit/openstack/keystone/tree/etc/ \
policy.v3cloudsample.json
Having matching policy files on the Horizon and Keystone servers is essential
for normal operation. This holds true for all services and their policy
files.
POLICY_FILES_PATH='/etc/openstack-dashboard'
POLICY_FILES_PATH='/etc/openstack-dashboard'
Map of local copy of service policy files
POLICY_FILES={
    'identity': 'keystone_policy.json',
    'compute': 'nova_policy.json',
    'volume': 'cinder_policy.json',
    'image': 'glance_policy.json',
    'orchestration': 'heat_policy.json',
    'network': 'neutron_policy.json',
    'telemetry': 'ceilometer_policy.json',
}

Trove user and database extension support. By default support for
creating users and databases on database instances is turned on.
To disable these extensions set the permission here to something
unusable such as ["!"].
TROVE_ADD_USER_PERMS=[]
TROVE_ADD_DATABASE_PERMS=[]

Change this patch to the appropriate static directory containing
two files: _variables.scss and _styles.scss
CUSTOM_THEME_PATH='themes/default'

LOGGING={
    'version': 1,
    When set to True this will disable all logging except
    for loggers specified in this configuration dictionary. Note that
    if nothing is specified here and disable_existing_loggers is True,
```

```python
        django.db.backends will still log unless it is disabled explicitly.
    'disable_existing_loggers': False,
    'handlers': {
        'null': {
            'level': 'DEBUG',
            'class': 'django.utils.log.NullHandler',
        },
        'console': {
            Set the level to "DEBUG" for verbose output logging.
            'level': 'INFO',
            'class': 'logging.StreamHandler',
        },
    },
    'loggers': {
        Logging from django.db.backends is VERY verbose, send to null
        by default.
        'django.db.backends': {
            'handlers': ['null'],
            'propagate': False,
        },
        'requests': {
            'handlers': ['null'],
            'propagate': False,
        },
        'horizon': {
            'handlers': ['console'],
            'level': 'DEBUG',
            'propagate': False,
        },
        'openstack_dashboard': {
            'handlers': ['console'],
            'level': 'DEBUG',
            'propagate': False,
        },
        'novaclient': {
            'handlers': ['console'],
            'level': 'DEBUG',
            'propagate': False,
        },
        'cinderclient': {
            'handlers': ['console'],
            'level': 'DEBUG',
            'propagate': False,
        },
        'keystoneclient': {
            'handlers': ['console'],
```

```python
        'level': 'DEBUG',
        'propagate': False,
    },
    'glanceclient': {
        'handlers': ['console'],
        'level': 'DEBUG',
        'propagate': False,
    },
    'neutronclient': {
        'handlers': ['console'],
        'level': 'DEBUG',
        'propagate': False,
    },
    'heatclient': {
        'handlers': ['console'],
        'level': 'DEBUG',
        'propagate': False,
    },
    'ceilometerclient': {
        'handlers': ['console'],
        'level': 'DEBUG',
        'propagate': False,
    },
    'troveclient': {
        'handlers': ['console'],
        'level': 'DEBUG',
        'propagate': False,
    },
    'swiftclient': {
        'handlers': ['console'],
        'level': 'DEBUG',
        'propagate': False,
    },
    'openstack_auth': {
        'handlers': ['console'],
        'level': 'DEBUG',
        'propagate': False,
    },
    'nose.plugins.manager': {
        'handlers': ['console'],
        'level': 'DEBUG',
        'propagate': False,
    },
    'django': {
        'handlers': ['console'],
        'level': 'DEBUG',
```

```
                'propagate': False,
            },
            'iso8601': {
                'handlers': ['null'],
                'propagate': False,
            },
            'scss': {
                'handlers': ['null'],
                'propagate': False,
            },
        }
    }
```

'direction' should not be specified for all_tcp/udp/icmp.
It is specified in the form.
```
SECURITY_GROUP_RULES={
    'all_tcp': {
        'name': _('All TCP'),
        'ip_protocol': 'tcp',
        'from_port': '1',
        'to_port': '65535',
    },
    'all_udp': {
        'name': _('All UDP'),
        'ip_protocol': 'udp',
        'from_port': '1',
        'to_port': '65535',
    },
    'all_icmp': {
        'name': _('All ICMP'),
        'ip_protocol': 'icmp',
        'from_port': '-1',
        'to_port': '-1',
    },
    'ssh': {
        'name': 'SSH',
        'ip_protocol': 'tcp',
        'from_port': '22',
        'to_port': '22',
    },
    'smtp': {
        'name': 'SMTP',
        'ip_protocol': 'tcp',
        'from_port': '25',
        'to_port': '25',
    },
```

```
'dns': {
    'name': 'DNS',
    'ip_protocol': 'tcp',
    'from_port': '53',
    'to_port': '53',
},
'http': {
    'name': 'HTTP',
    'ip_protocol': 'tcp',
    'from_port': '80',
    'to_port': '80',
},
'pop3': {
    'name': 'POP3',
    'ip_protocol': 'tcp',
    'from_port': '110',
    'to_port': '110',
},
'imap': {
    'name': 'IMAP',
    'ip_protocol': 'tcp',
    'from_port': '143',
    'to_port': '143',
},
'ldap': {
    'name': 'LDAP',
    'ip_protocol': 'tcp',
    'from_port': '389',
    'to_port': '389',
},
'https': {
    'name': 'HTTPS',
    'ip_protocol': 'tcp',
    'from_port': '443',
    'to_port': '443',
},
'smtps': {
    'name': 'SMTPS',
    'ip_protocol': 'tcp',
    'from_port': '465',
    'to_port': '465',
},
'imaps': {
    'name': 'IMAPS',
    'ip_protocol': 'tcp',
    'from_port': '993',
```

```
            'to_port': '993',
        },
        'pop3s': {
            'name': 'POP3S',
            'ip_protocol': 'tcp',
            'from_port': '995',
            'to_port': '995',
        },
        'ms_sql': {
            'name': 'MS SQL',
            'ip_protocol': 'tcp',
            'from_port': '1433',
            'to_port': '1433',
        },
        'mysql': {
            'name': 'MYSQL',
            'ip_protocol': 'tcp',
            'from_port': '3306',
            'to_port': '3306',
        },
        'rdp': {
            'name': 'RDP',
            'ip_protocol': 'tcp',
            'from_port': '3389',
            'to_port': '3389',
        },
}

Deprecation Notice:

The setting FLAVOR_EXTRA_KEYS has been deprecated.
Please load extra spec metadata into the Glance Metadata Definition Catalog.

The sample quota definitions can be found in:
<glance_source>/etc/metadefs/compute-quota.json

The metadata definition catalog supports CLI and API:
$glance --os-image-api-version 2 help md-namespace-import
$glance-manage db_load_metadefs <directory_with_definition_files>

See Metadata Definitions on: http://docs.openstack.org/developer/glance/

Indicate to the Sahara data processing service whether or not
automatic floating IP allocation is in effect. If it is not
in effect, the user will be prompted to choose a floating IP
pool for use in their cluster. False by default. You would want
```

```
to set this to True if you were running Nova Networking with
auto_assign_floating_ip=True.
SAHARA_AUTO_IP_ALLOCATION_ENABLED=False

The hash algorithm to use for authentication tokens. This must
match the hash algorithm that the identity server and the
auth_token middleware are using. Allowed values are the
algorithms supported by Python's hashlib library.
OPENSTACK_TOKEN_HASH_ALGORITHM='md5'

Hashing tokens from Keystone keeps the Horizon session data smaller, but it
doesn't work in some cases when using PKI tokens. Uncomment this value and
set it to False if using PKI tokens and there are 401 errors due to token hashing.
OPENSTACK_TOKEN_HASH_ENABLED=True

AngularJS requires some settings to be made available to
the client side. Some settings are required by in-tree / built-in horizon
features. These settings must be added to REST_API_REQUIRED_SETTINGS in the
form of ['SETTING_1','SETTING_2'], etc.

You may remove settings from this list for security purposes, but do so at
the risk of breaking a built-in horizon feature. These settings are required
for horizon to function properly. Only remove them if you know what you
are doing. These settings may in the future be moved to be defined within
the enabled panel configuration.
You should not add settings to this list for out of tree extensions.
See: https://wiki.openstack.org/wiki/Horizon/RESTAPI
REST_API_REQUIRED_SETTINGS=['OPENSTACK_HYPERVISOR_FEATURES']

Additional settings can be made available to the client side for
extensibility by specifying them in REST_API_ADDITIONAL_SETTINGS
!! Please use extreme caution as the settings are transferred via HTTP/S
and are not encrypted on the browser. This is an experimental API and
may be deprecated in the future without notice.
REST_API_ADDITIONAL_SETTINGS=[]
DISALLOW_IFRAME_EMBED can be used to prevent Horizon from being embedded
within an iframe. Legacy browsers are still vulnerable to a Cross-Frame
Scripting (XFS) vulnerability, so this option allows extra security hardening
where iframes are not used in deployment. Default setting is True.
For more information see:
```

```
http://tinyurl.com/anticlickjack
DISALLOW_IFRAME_EMBED=True
EOF
systemctl restart httpd
```

9.19 OpenStack GUI 配置

根据以上所有步骤和指令操作，OpenStack Web 平台部署成功，通过浏览器可以直接访问其 URL 地址 http://192.168.1.120/dashboard/，输入用户名和密码，如图 9-22 所示。

图 9-22 OpenStack 登录界面

（1）查看 OpenStack 实例列表，如图 9-23 所示。

图 9-23 OpenStack 实例列表

（2）查看 OpenStack 具体实例 centos1 列表信息，如图 9-24 所示。
（3）进入 OpenStack 实例 centos1 控制台，如图 9-25 所示。

图 9-24 OpenStack 具体实例 centos1 列表信息

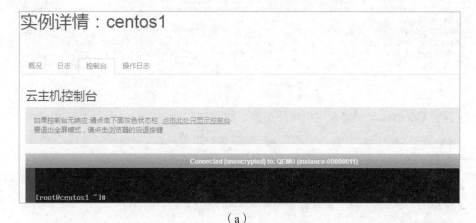

（a）

（b）

图 9-25 OpenStack 实例 centos1 控制台

（a）打开 OpenStack centos1 控制台；（b）进入 OpenStack centos1 控制台，查看磁盘分区和进程状态

（4）查看 OpenStack 具体实例 centos1 日志信息，如图 9-26 所示。

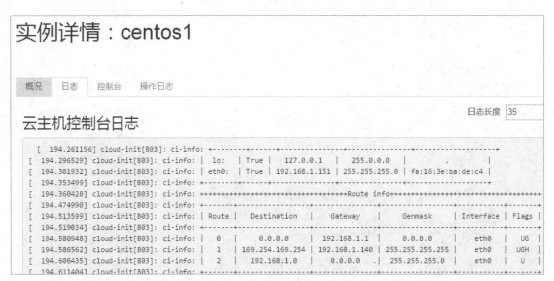

图 9-26 OpenStack 具体实例 centos1 控制日志

（5）查看 OpenStack 创建云主机可选类型，如图 9-27 所示。

	云主机类型名称	虚拟内核	内存	根磁盘	临时磁盘	交换盘空间	ID	公有	元数据
	m1.tiny	1	512MB	1GB	0GB	0 MB	1	True	{}
	m1.small	1	512MB	10GB	0GB	0 MB	c13fce6e-1210-4395-b8a9-a4b0e415c7a3	True	{}
	m1.medium	2	4GB	40GB	0GB	0 MB	3	True	{}
	m1.large	4	8GB	80GB	0GB	0 MB	4	True	{}
	m1.xlarge	8	16GB	160GB	0GB	0 MB	5	True	{}

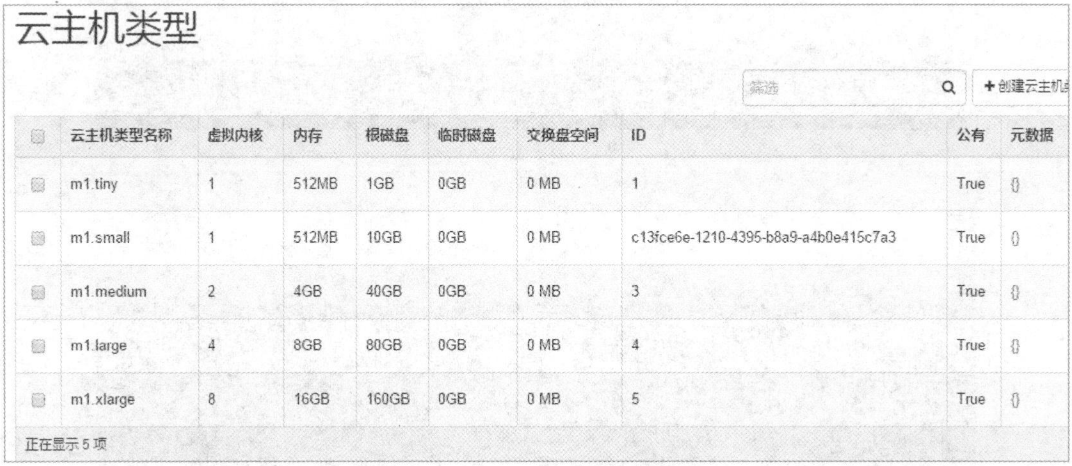

图 9-27 OpenStack 云主机可选类型

（6）查看 OpenStack 虚拟机管理器，可以看到所有的虚拟机及资源占用情况，如图 9-28 所示。

（7）查看 OpenStack 平台创建的所有用户列表信息，如图 9-29 所示。

（8）查看 OpenStack 平台创建的所有项目列表信息，如图 9-30 所示。

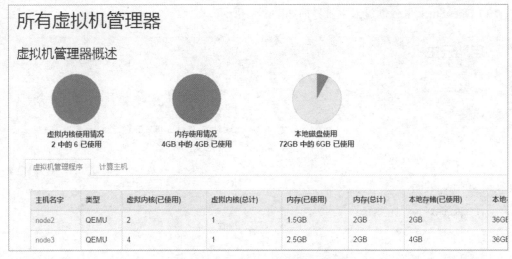

图 9-28　OpenStack 虚拟机管理器

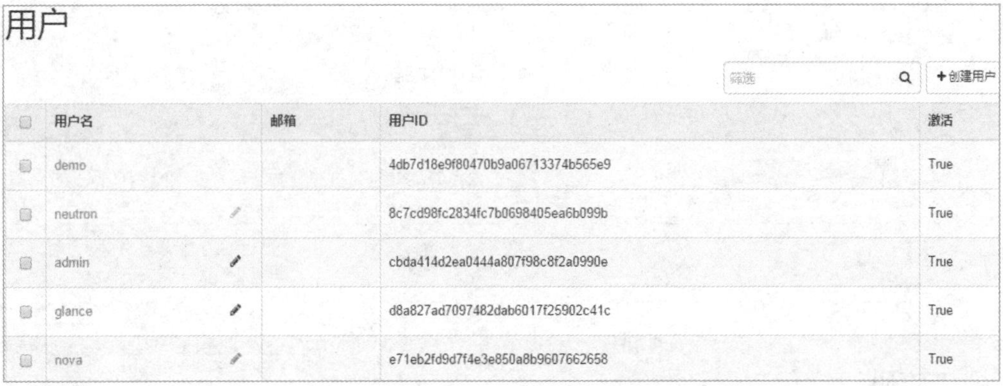

图 9-29　OpenStack 所有用户列表信息

图 9-30　OpenStack 项目列表信息

（9）OpenStack 平台创建云主机界面如图 9-31 所示。

（a）

（b）

图 9-31　OpenStack 创建云主机界面

（a）创建云主机并选择云主机的资源配置；（b）查看 OpenStack 云主机运行状态

（10）SSH 登录 OpenStack 创建的云主机控制台，如图 9-32 所示。

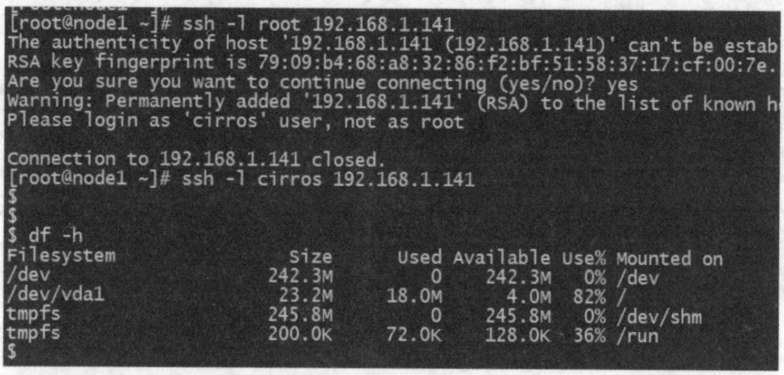

（a）

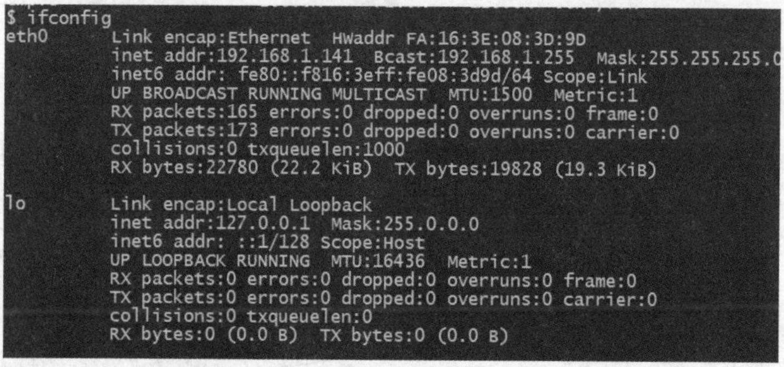

（b）

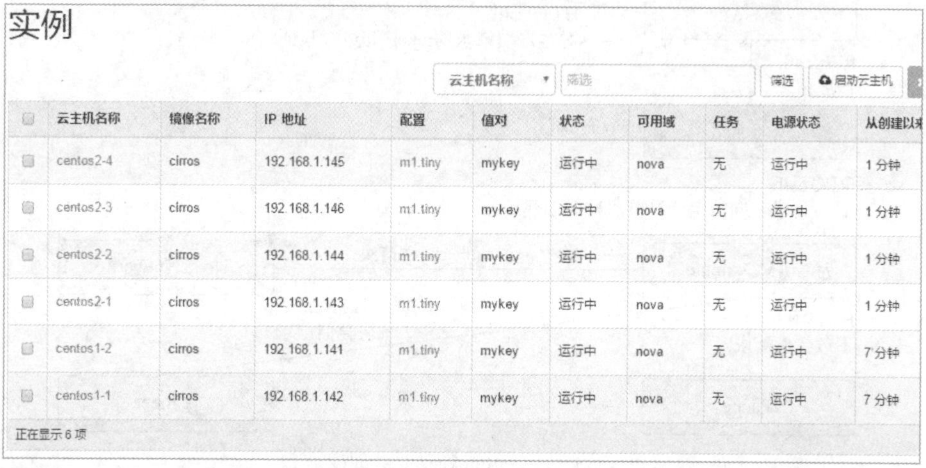

（c）

图 9-32　SSH 登录 OpenStack 创建的云主机控制台

（a）远程登录 192.168.1.141 虚拟机；（b）查看 192.168.1.141 虚拟机的 IP 地址；（c）查看云主机的详细信息

（11）在 OpenStack 创建的云主机上部署 Nginx 网站，通过浏览器访问，如图 9-33 所示。

图 9-33 OpenStack 云主机应用部署

9.20 OpenStack 核心流程

OpenStack 由不同的组件组成，学习 OpenStack 最难之处在于理解各个组件的含义，以及整个 OpenStack 创建流程。OpenStack 云主机创建流程如图 9-34 所示。

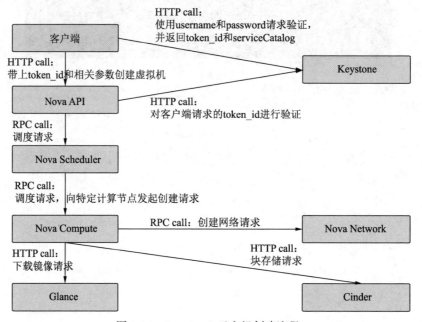

图 9-34 OpenStack 云主机创建流程

（1）OpenStack 创建云主机流程中，涉及的每个组件和服务详解如下。

客户端：Web 页面或者 Horizon、命令行的 Nova 客户端。

Nova API：用于接收和处理客户端发送的 HTTP 请求。

Nova Scheduler：Nova 的调度宿主机的服务，决定虚拟机创建的各个节点。

Nova Compute：Nova 核心的服务，负责虚拟机的生命周期的管理。

Nova Conductor：数据访问权限的控制操作，可以理解为数据库代理服务。

其他服务：Nova cert 管理证书，为了兼容 aws；Nova vncproxy 和 consoleauth 控制台服务。

不同的模块之间通过 HTTP 请求 REST API 服务。

同一个模块不同组件之间（如 Nova-scheduler 请求 Nova-compute）是 RPC 远程调用，通过 RabbitMQ 来实现。

（2）OpenStack 组件调用步骤如图 9-35 所示。

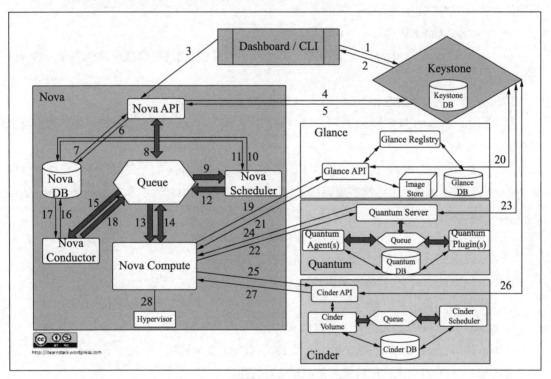

图 9-35　OpenStack 组件调用步骤

① 客户端使用自己的用户名、密码请求认证。

② Keystone 通过查询在 Keystone 数据库 user 表中保存的 user 相关信息，包括 password 加密后的 hash 值，并返回一个 token_id(令牌)和 serviceCatalog(一些服务的 endpoint 地址，cinder、glance-api 后面下载镜像和创建块存储时会用到)。

③ 客户端带上 KeyStone 返回的 token_id 和创建虚拟机的相关参数，Post 请求 Nova API 创建虚拟机。

④ Nova API 接收到请求后，首先使用请求携带的 token_id 来访问该 API，以验证请求是否有效。

⑤ Keystone 验证通过后返回更新后的认证信息。

⑥ Nova API 检查创建虚拟机参数是否有效与合法。检查虚拟机 name 是否符合命名规范，flavor_id 是否在数据库中存在，image_uuid 是否是正确的 uuid 格式。检查 instance、vcpu、ram 的数量是否超过配额。

⑦ 当且仅当所有传参都有效合法时，更新 Nova 数据库，新建一条 instance 记录，vm_states 设为 BUILDING，task_state 设为 SCHEDULING。

⑧ Nova API 远程调用传递请求、参数给 Nova Scheduler，把消息"请给我创建一台虚拟机"丢到消息队列，然后定期查询虚拟机的状态。

⑨ Nova Scheduler 从 Queue 中获取到这条消息。

⑩ Nova Scheduler 访问 Nova 数据库，通过调度算法过滤出一些合适的计算节点，然后进行排序。

⑪ 更新虚拟机节点信息，返回一个最优节点 ID 给 Nova Scheduler。

⑫ Nova Scheduler 选定 Host 之后，通过 rpc 调用 Nova Compute 服务，把"创建虚拟机请求"消息丢给 MQ。

⑬ Nova Compute 收到创建虚拟机请求的消息。Nova Compute 有个定时任务，定期从数据库中查找到运行在该节点上的所有虚拟机信息，统计得到空闲内存大小和空闲磁盘大小。然后更新数据库 compute_node 信息，以保证调度的准确性。

⑭ Nova Compute 通过 rpc 查询 Nova 数据库中虚拟机的信息，例如主机模板和 ID。

⑮ Nova Conductor 从消息队列中拿到请求查询数据库。

⑯ Nova Conductor 查询 Nova 数据库。

⑰ 数据库返回虚拟机信息。

⑱ Nova Compute 从消息队列中获取信息。

⑲ Nova Compute 请求 Glance 的 rest API，下载所需要的镜像，一般是 qcow2。

⑳ Glance API 也会去验证请求的 Token 的有效性。

㉑ Glance API 返回镜像信息给 Nova Compute。

㉒ Nova Compute 请求 Neutron API 配置网络，例如获取虚拟机 IP 地址。

㉓ 验证 Token 的有效性。

㉔ Neutron 返回网络信息。

㉕ 该步骤同 Glance、Neutron 验证 Token 返回块设备信息。

㉖ 据上面配置的虚拟机信息生成 xml，写入 libvirt、xml 文件，然后调用 libvirt driver 使用 libvirt.xml 文件启动虚拟机。

第 10 章 Kubernetes 组件概念

10.1 云计算概念

云计算技术其实是将硬件设备、操作系统、软件服务、网络带宽、流量、计费系统等资源组成一个大的资源池（动态扩容、弹性伸缩），然后可以再将所有的资源池分配给租户使用，租户可以根据自身的需求，按需购买资源。

云计算技术强调的是资源池，是租户的概念。虚拟化技术是云计算技术框架中的一个小模块、组件技术。云计算技术最终的产物包括硬件设备、操作系统、软件服务、网络带宽等。每个产物都可以租给用户使用，用户可以自行购买。

对于云计算技术的资源池，租户不需要了解云计算底层框架、架构，只要清楚自身对资源池的需求，需要多少台服务器、多少云主机、多大网络带宽，最终按需付费即可。

10.2 云计算技术的分类

云计算技术按照实现方式分为 3 种类型：基础设施即服务（Infrastructure as a Service，IaaS）、平台即服务（Platform as a Service，PaaS）、软件即服务（Software as a Service，SaaS），如图 10-1 所示。每种云的特点如下。

（1）基础设施即服务。
① 租户无须管理底层硬件设备、网络、服务器、存储、虚拟化技术。
② 租户只需对操作系统、中间件、数据、应用做维护即可。
（2）平台即服务。
① 租户无须管理底层硬件设备、网络、服务器、存储、虚拟化技术、操作系统、中间件。
② 租户只需对应用服务、软件程序做维护，无须操作系统和底层设施。

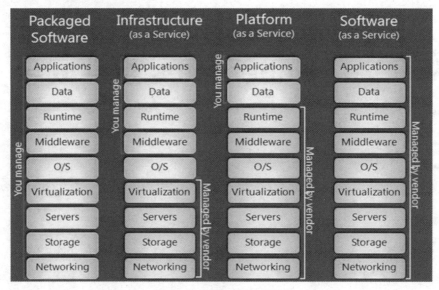

图 10-1　云计算技术分类

（3）软件即服务。

① 租户无须管理底层硬件设备、网络、服务器、存储、虚拟化技术、操作系统、中间件、应用服务、软件程序等。

② 租户只需花钱、付费，提交业务需求，运营商将满足租户所有需求。

10.3　Kubernetes 入门及概念介绍

Kubernetes 又称为 K8S（首字母为 K，首字母与尾字母之间有 8 个字符。尾字母为 S，所以简称 K8S）或者简称为 kube，是一种可以自动实施 Linux 容器操作的开源平台。

Kubernetes 可以帮助用户省去应用容器化过程中的许多手动部署和扩展操作。我们可以将运行 Linux 容器的多组主机聚集在一起，由 Kubernetes 帮助用户轻松高效地管理这些集群，而且这些集群可跨公共云、私有云或混合云部署主机。因此，对于要求快速扩展的云原生应用而言（例如，借助 Apache Kafka 进行的实时数据流处理），Kubernetes 是理想的托管平台。

Kubernetes 最初由 Google 公司的工程师开发和设计。Google 是最早研发 Linux 容器技术的企业之一（组建了 cgroups），曾公开分享介绍 Google 如何将一切都运行于容器之中（这是 Google 云服务背后的技术）。Google 每周会启用超过 20 亿个容器——全都由内部平台 Borg 支撑。Borg 是 Kubernetes 的前身，多年来开发 Borg 的经验教训成了影响 Kubernetes 中许多技术的主要因素。

在企业生产环境中，应用会涉及多个容器（主机），这些容器必须跨多个服务器主机进行部署。容器安全性需要多层部署，因此可能比较复杂。但 Kubernetes 有助于解决这一问题。Kubernetes 可以提供所需的编排和管理功能，以便针对这些工作负载大规模部署容器。借助 Kubernetes 编排功能，可以构建跨多个容器的应用服务、跨集群调度、扩展这些容器，并长期持续管理这些容器的健康状况。

Kubernetes（K8S）是自动化容器操作的开源平台，这些操作包括部署、调度和节点集群间扩展。如果用户曾经用过 Docker 容器技术部署容器，可以将 Docker 看成 Kubernetes 内部使用的低级别组件。Kubernetes 不仅支持 Docker，还支持 Rocket，这是另一种容器技术。使用 Kubernetes 可以实现如下功能：

（1）自动化容器的部署和复制。
（2）跨多台主机进行容器编排和管理。
（3）有效管控应用部署和更新，并实现自动化操作。
（4）挂载和增加存储，用于运行有状态的应用。
（5）能够快速、按需扩展容器化应用及其资源。
（6）对服务进行声明式管理，保证部署的应用始终按照部署的方式运行。
（7）更加充分地利用硬件，最大程度获取运行企业应用所需的资源。
（8）利用自动布局、自动重启、自动复制及自动扩展功能，对应用实施状况进行检查和自我修复。

10.4 Kubernetes 平台组件概念

Kubernetes 集群中主要存在两种类型的节点：master 节点和 node 节点。

1）master 节点

master 节点主要负责对外提供一系列管理集群的 API 接口，并且通过与 node 节点交互实现对集群的操作管理。以下为 master 节点的组件和用途。

（1）Apiserver：用户和 Kubernetes 集群交互的入口，封装了核心对象的增、删、改、查操作，提供了 RESTful 风格的 API 接口，通过 etcd 实现持久化并维护对象的一致性。

（2）Scheduler：负责集群资源的调度和管理。例如，当有 Pod 异常退出需要重新分配机器时，Scheduler 通过一定的调度算法找到最合适的节点。

（3）controller-manager：主要用于保证 replication controller 定义的复制数量和实际运行的 Pod（容器）数量一致，并保证从 service 到 Pod 的映射关系总是最新的。

2）node 节点

node 节点主要负责接收 master 发送的指令，同时与本地 Docker 引擎交互等操作。以下为 node 节点的组件和用途。

（1）kubelet：运行在 node 节点，负责与节点上的 Docker 交互。例如，启停容器、监控运行状态等。

（2）kube-proxy：运行在 node 节点上，负责为 Pod 提供代理功能，会定期从 etcd 获取 service 信息，并根据 service 信息通过修改 iptables 实现流量转发（最初的版本是直接通过程序提供转发功能，效率较低），将流量转发到要访问的 Pod 所在的节点上。

Kubernetes 云计算平台除了 master 和 node 节点上相应的组件模块之外，还需要 etcd、Flannel、Docker 等插件。以下为相关插件的用途。

（1）etcd：etcd 是一个分布式一致性 key-value 存储系统数据库，可用于注册与发现配置共享和服务，用来存储 Kubernetes 的信息。etcd 组件作为一个高可用、强一致性的服务发现存储仓库，渐渐为开发人员所关注。在云计算时代，如何让服务快速透明地接入计算集群，如何让共享配置信息快速被集群中的所有机器发现，更为重要的是，如何构建这样一套高可用、安全、易于部署以及响应快速的服务集群？etcd 的诞生就是为了解决该问题。

（2）Flannel：Flannel 是 CoreOS 团队针对 Kubernetes 设计的一个覆盖网络（Overlay Network）工具，目的是为集群中的所有节点重新制定 IP 地址的使用规则，从而使不同节点上的容器能够获得同属一个内网且不重复的 IP 地址，并让属于不同节点上的容器能够直接通过内网 IP 通信。

（3）Docker：Docker 是一款轻量级的虚拟化软件，主要是为了解决企业轻量级服务器操作系统和应用容器而诞生的，在 Kubernetes 集群中，Docker 被看成 Kubernetes 集群中的低级别组件，负责接收 node 节点上 kubelet 组件进行交互操作，如启动、停止、删除 Docker 容器，监控 Docker 容器运行状态等。

10.5 Kubernetes 工作原理剖析

Kubernetes 集群是一组节点，这些节点可以是物理服务器或者虚拟机，在其上安装 Kubernetes 平台。Kubernetes 云计算架构如图 10-2 所示，图中为了强调核心概念有所简化。

图 10-2（a）中相关模块和组件的名称如下：

（1）Pod（容器组）。

（2）container（容器）。

（3）labels(🏷)（标签）。

（4）service（🔲）（服务）。

（5）node（Kubernetes 计算节点）。

（6）Replication Controller（副本控制器）。

（7）Kubernetes master（Kubernetes 主节点）。

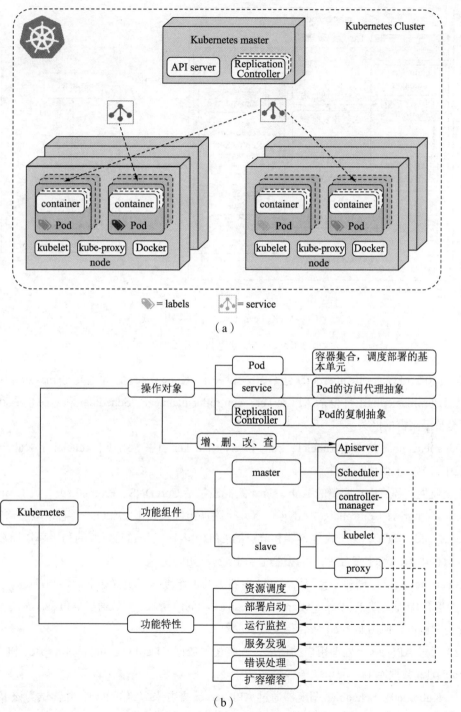

图 10-2　Kubernetes 云计算架构

（a）Kubernetes master 节点和 node 节点结构；（b）Kubernetes 操作对象、功能组件和特性

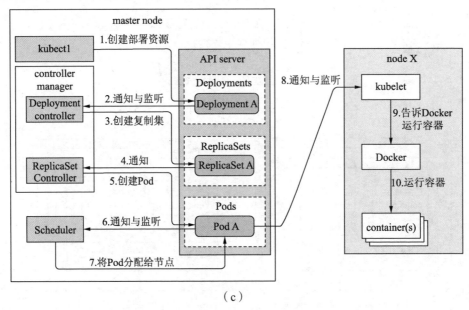

图 10-2　Kubernetes 云计算架构（续）

（c）Kubernetes 创建 Pod 容器的过程

从图 10-2 可以看到 Kubernetes 组件和逻辑关系：Kubernetes 集群主要由 master 和 node 两类节点组成；master 的组件包括 API server、controller-manager、Scheduler 和 etcd 等，其中 API server 是整个集群的网关。

Kubernetes node 主要由 kubelet、kube-proxy、Docker 引擎等组成。kubelet 是 Kubernetes 集群的工作与节点上的代理组件。

在企业生产环境中，完整的 Kubernetes 集群还包括 CoreDNS、Prometheus（或 HeapSter）、Dashboard、Ingress Controller、cAdvisor 等几个附加组件。其中 cAdvisor 组件作用于各个节点（master 和 node 节点），用于收集容器和节点的 CPU、内存以及磁盘资源的利用率指标数据，这些统计数据由 Heapster 聚合后，可以通过 API server 访问。

Kubernetes 集群中创建一个资源（Pod 容器）的工作流程和步骤如下。

（1）客户端提交创建（Deployment、Namespace、Pod）请求，可以通过 API server 的 RESTful API 提交，也可以使用 kubectl 命令行工具提交。

（2）通过 API server 处理用户的请求，并将相关数据（Deployment、Namespace 和 Pod）存储到 etcd 配置数据库中。

（3）Kubernetes Scheduler 调度器通过 API server 查看未绑定的 Pod，并尝试为该 Pod 分配 node 主机资源。

（4）过滤主机（调度预选）：调度器用一组规则过滤掉不符合要求的主机。例如，Pod 指定了所需要的资源量，那么可用资源比 Pod 需要的资源量少的主机就会被过滤掉。

（5）主机打分（调度优选）：对第一步筛选出的符合要求的主机进行打分。在主机打分阶段，调度器会考虑一些整体优化策略，例如，把 Replication Controller 的副本分布到不同的主机上，使用最低负载的主机等。

（6）选择主机：选择打分最高的主机，进行 binding 操作，将结果存储到 etcd 中。

（7）node 节点上的 kubelet 根据调度结果，调用主机上的 Docker 引擎执行 Pod 创建操作，绑定成功后，Scheduler 会调用 API server 的 API 在 etcd 中创建一个 boundpod 对象，描述在一个工作节点上绑定运行的所有 Pod 信息。

（8）同时运行在每个工作节点上的 kubelet 也会定期与 etcd 同步 boundpod 信息，一旦发现应该在该工作节点上运行的 boundpod 对象没有更新，将调用 Docker API 创建并启动 Pod 内的容器。

10.6 Pod 概念剖析

Pod（容器组）是 Kubernetes 中最小的可部署单元，一般部署在 node 节点上，包含一组容器和卷。同一个 Pod 中的容器共享同一个网络命名空间，可以使用 localhost 互相通信，Pod 是短暂的，不是持续性实体。关于 Pod 的常见问题和解答如下：

（1）Pod 一般是短暂的，如何才能持久化容器数据，使其能够跨重启而存在呢？Kubernetes 支持卷的概念，因此可以使用持久化的卷类型。

（2）如果要创建同一个容器的多份副本，需要逐个创建出来吗？可以手动创建单个 Pod，也可以使用 Replication Controller 中的 Pod 模板创建出多份副本。

（3）Pod 是短暂的，重启 Pod 时 IP 地址可能会改变，那么怎样才能从前端容器正确可靠地指向后台容器呢？答案是可以使用 service。

10.7 label 概念剖析

Kubernetes label 是标记到 Pod 的一个键/值对，用来传递用户定义的属性。例如，用户可能创建了一个 tier 和 app 标签，可通过 label（tier=frontend, app=jfedu-app）标记前端 Pod 容器，使用 label（tier=backend, app=jfedu-app）标记后台 Pod。

可以使用 Selectors 选择带有特定 label（标签）的 Pod，并且将 service 或者 Replication Controller 应用于这些 Pod。

10.8　Replication Controller 概念剖析

Replication Controller 确保任意时间都有指定数量的 Pod"副本"在运行。如果为某个 Pod 创建了 Replication Controller 并指定 3 个副本，它会创建 3 个 Pod，并持续监控它们。如果某个 Pod 不响应，那么 Replication Controller 会替换它，并保持总数为 3，如图 10-3 所示。

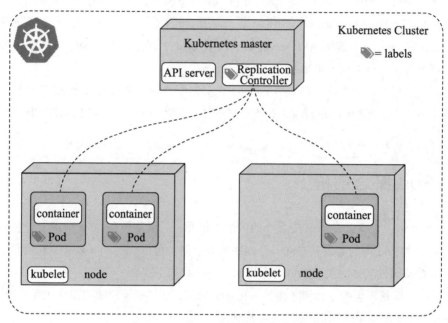

图 10-3　Replication Controller 副本结构

如果之前不响应的 Pod 恢复，就有了 4 个 Pod，那么 Replication Controller 会将其中一个 Pod 终止，以保持总数为 3。如果在运行中将副本总数改为 5，Replication Controller 会立刻启动 2 个新 Pod，保证总数为 5。还可以按照这样的方式减少 Pod 副本，这个特性在执行滚动升级时很有用。当创建 Replication Controller 时，需要指定以下两个内容。

（1）Pod 模板：用来创建 Pod 副本的模板。
（2）label：Replication Controller 需要监控的 Pod 的标签。

10.9　service 概念剖析

service 是定义一系列 Pod 以及访问这些 Pod 的策略的一层抽象。service 通过 label 找到 Pod 组。因为 service 是抽象的，所以通常看不到它们的存在，这就让这一概念更难以理解。Kubernetes service 内部结构如图 10-4 所示。

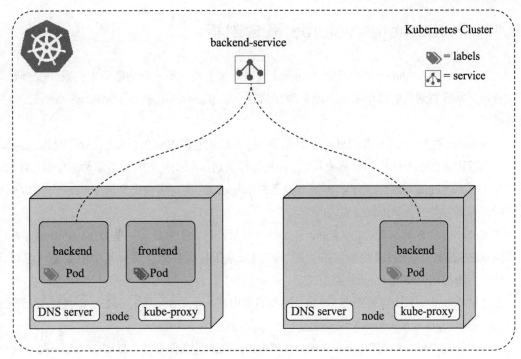

图 10-4　Kubernetes service 内部结构

假设创建了 2 个 Pod 容器，并定义后台 service 的名称为 backend-service，label 选择器为（tier=backend, app=jfedu-app），backend-service 的 service 会完成如下操作：

（1）会为 service 创建一个本地集群的 DNS 入口，因此前端 Pod 只需要 DNS 查找主机名为 backend-service，就能够解析出前端应用程序可用的 IP 地址。

（2）service 为创建的 2 个 Pod 容器提供透明的负载均衡，并将请求分发给其中的任意一个，通过每个 node 上运行的代理（kube-proxy）完成。

10.10　node 概念剖析

node（节点）是物理或虚拟机器，作为 Kubernetes worker，通常称为计算节点。每个节点都运行以下 Kubernetes 关键组件。

（1）kubelet：是 node 节点上的主程序，负责接收 master 发送的指令。

（2）kube-proxy：service 使用 kube-proxy 将请求转发至 Pod。

（3）Docker 或 Rocket：Kubernetes 使用 Docker 或 Rocket 容器技术创建容器。

10.11　Kubernetes volume 概念剖析

在 Docker 中有 volume 这个概念，volume 只是磁盘上或其他容器中的简单目录，生命周期不受管理，并且直到最近都是基于本地后端存储的。Docker 现在也提供了 volume driver，但是功能较弱。

Kubernetes 的 volume 有着明显的生命周期——和使用它的 Pod 生命周期一致。因此，volume 生命周期就比运行在 Pod 中的容器要长久，即使容器重启，volume 上的数据依然保存着。当 Pod 不再存在时，volume 也就消失了。更重要的是，Kubernetes 支持多种类型的 volume，且 Pod 可以同时使用多种类型的 volume。

在 Kubernetes 内部实现中，volume 只是一个目录，目录中可能有一些数据，Pod 的容器可以访问这些数据。这个目录是如何产生的？它后端基于什么存储介质？其中的数据内容是什么？这些都由使用的特定 volume 类型决定。

要使用 volume，Pod 需要指定 volume 的类型和内容（spec.volumes 字段），以及映射到容器的位置（spec.containers.volumeMounts 字段）。

容器中的进程可以看到 Docker image 和 volumes 组成的文件系统。Docker image 处于文件系统架构的 root，任何 volume 都映射在镜像的特定路径上。volume 不能映射到其他 volume 上，也不能硬链接到其他 volume。容器中的每个容器必须指定它们要映射的 volume。

Kubernetes 支持多种类型的 volume，包括 emptyDir、hostPath、gcePersistentDisk、awsElasticBlockStore、nfs、iscsi、flocker、glusterfs、rbd、cephfs、gitRepo、secret、persistentVolumeClaim 等。

10.12　Deployment 概念剖析

Deployment 为 Pod 和 ReplicaSet 提供声明式更新和部署，只需要在 Deployment 中描述想要的目标状态，Deployment controller 就会将 Pod 和 ReplicaSet 的实际状态改变到目标状态。

可以定义一个全新的 Deployment 创建 ReplicaSet，或者删除已有的 Deployment 并创建一个新的替换。

注意：无须手动管理由 Deployment 创建的 ReplicaSet，否则就越俎代庖了。

10.13　DaemonSet 概念剖析

DaemonSet 对象能确保其创建的 Pod 在集群中的每一台（或指定）node 上都运行一个副本。如果集群中动态加入了新的 node，DaemonSet 中的 Pod 也会被添加到新加入的 node 上。删除一个 DaemonSet 也会级联删除所有其创建的 Pod。DaemonSet 和 Deployment 的区别如下。

（1）Deployment 部署 Pod 会分布在各个 node 上，node 可能运行好几个副本。

（2）DaemonSet 部署 Pod 会分布在各个 node 上，每个 node 只能运行一个 Pod。

以下为 DaemonSet 的使用场景：

（1）在每台节点上运行一个集群存储服务，如 glusterd、ceph。

（2）在每台节点上运行一个日志收集服务，如 fluentd、logstash。

（3）在每台节点上运行一个节点监控服务，如 Prometheus node exporter、collectd、Datadog agent、New Relic agent 或 Ganglia gmond。

10.14　StatefulSet 概念剖析

Kubernetes RC、Deployment 和 DaemonSet 都是面向无状态的服务，它们所管理的 Pod 的 IP、名字、启停顺序等都是随机的，而 StatefulSet 是什么？顾名思义，StatefulSet 是有状态的集合，管理所有有状态的服务，如 MySQL、MongoDB 集群等。

StatefulSet 本质上是 Deployment 的一种变体，在 v1.9 版本中已成为 GA 版本。为了解决有状态服务的问题，它所管理的 Pod 拥有固定的 Pod 名称、启停顺序，在 StatefulSet 中，Pod 名字称为网络标识（hostname），还必须用到共享存储。

Deployment 中的服务是 service，而 StatefulSet 中的服务是 headless service，即无头服务，与 service 的区别就是后者没有 Cluster IP，解析其名称时将返回该 headless service 对应的全部 Pod 的 Endpoint 列表。

10.15　ConfigMap 概念剖析

在企业生产环境中，经常会遇到需要修改配置文件的情况。例如，Web 网站连接数据库的配置、业务系统之间相互调用配置、Kubernetes Pod 配置信息等，如果使用传统手动方式修改，则不仅会影响到服务的正常运行，操作步骤也很烦琐。

对于传统的应用服务而言，每个服务都有自己的配置文件，各自配置文件存储在服务所在的节点。对于单体应用，这种存储没有任何问题，但是随着用户数量的激增，一个节点不能满足线上用户的使用，故服务可能从一个节点扩展到十个节点，这就导致，如果有一个配置出现变更，就需要对应修改十次配置文件。

这种人工处理方式显然不能满足线上部署要求，故引入了各种类似于 ZooKeeper 中间件实现的配置中心，但配置中心属于"侵入式"设计，需要修改引入第三方类库，它要求每个业务都调用特定的配置接口，破坏了系统本身的完整性。

Kubernetes 利用 volume 功能完整设计了一套配置中心，其核心对象就是 ConfigMap，使用过程中不用修改任何原有设计，即可无缝对接 ConfigMap。

Kubernetes 项目从 1.2.0 版本开始引入了 ConfigMap 功能，主要用于将应用的配置信息与程序分离。这种方式不仅可以实现应用程序被复用，还可以通过不同的配置实现更灵活的功能。

在创建容器时，用户可以将应用程序打包为容器镜像，通过环境变量或外接挂载文件的方式进行配置注入。ConfigMap 非常灵活，相当于把配置文件信息单独存储在某处，需要时直接引用、挂载即可。

ConfigMap 以 key:value（K-V）的形式保存配置项，既可以表示一个变量的值（例如，config=info），也可以表示一个完整配置文件的内容（例如，server.xml=<?xml…>…）。ConfigMap 在容器中使用的典型用法如下：

（1）将配置项设置为容器内的环境变量。

（2）将启动参数设置为环境变量。

（3）以 volume 的形式挂载到容器内部的文件或目录。

可以基于 kubectl create 指令创建 ConfigMap。例如，使用 ConfigMap 存储 MySQL 数据库的 IP 地址和端口信息，如图 10-5 所示。

```
kubectl create configmap mysql-config --from-literal=db.host=192.168.1.111 --from-literal=db.port=3306
kubectl get configmap mysql-config -o yaml
```

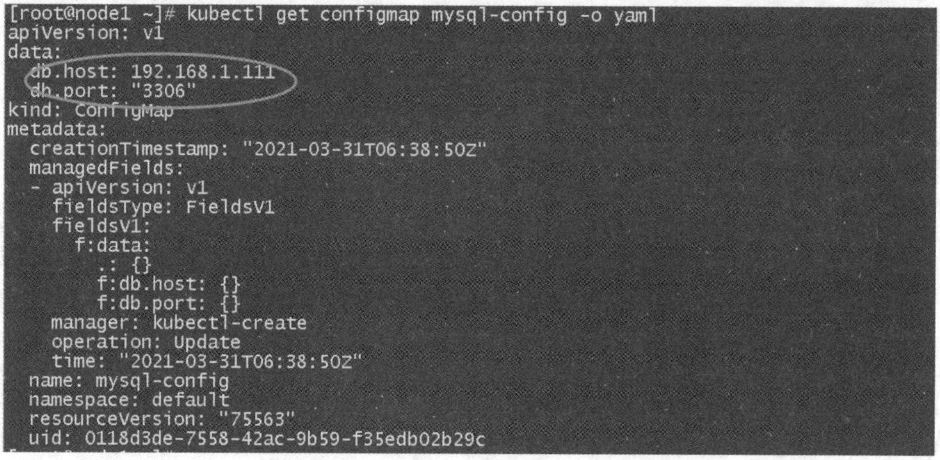

```
[root@node1 ~]# kubectl get configmap mysql-config -o yaml
apiVersion: v1
data:
  db.host: 192.168.1.111
  db.port: "3306"
kind: ConfigMap
metadata:
  creationTimestamp: "2021-03-31T06:38:50Z"
  managedFields:
  - apiVersion: v1
    fieldsType: FieldsV1
    fieldsV1:
      f:data:
        .: {}
        f:db.host: {}
        f:db.port: {}
    manager: kubectl-create
    operation: Update
    time: "2021-03-31T06:38:50Z"
  name: mysql-config
  namespace: default
  resourceVersion: "75563"
  uid: 0118d3de-7558-42ac-9b59-f35edb02b29c
```

图 10-5　Kubernetes ConfigMap 配置界面

10.16　Secrets 概念剖析

在 Kubernetes 集群中，Secrets 通常用于存储和管理一些敏感数据，如密码、token、密钥等敏感信息。它把 Pod 想要访问的加密数据存放到 etcd 中。

用户可以通过在 Pod 容器里挂载 volume 或环境变量的方式访问这些 Secret 里保存的信息。Secret 有以下 3 种类型。

（1）Opaque：用来存储密码、密钥等，采用 base64 编码格式。但数据也可以通过 base64_decode 解码得到原始数据，加密性很弱。

（2）Service Account：用来访问 Kubernetes API，由 Kubernetes 自动创建，会自动挂载到 Pod 的 /run/secrets/kubernetes.io/serviceaccount 目录中。

（3）kubernetes.io/dockerconfigjson：用来存储私有 Docker registry 的认证信息。

10.17 CronJob 概念剖析

CronJob 用于创建基于时间调度的任务（Jobs）。一个 CronJob 对象就像 Crontab (Cron table) 文件中的一行。它用 Cron 格式进行编写，并周期性地在给定的调度时间执行。所有 CronJob 的 schedule 都是基于 kube-controller-manager 的时区。

如果 Kubernetes 在 Pod 或裸容器中运行了 kube-controller-manager，那么为该容器所设置的时区将会决定 CronJob 的控制器所使用的时区。

为 CronJob 资源创建清单时，请确保所提供的名称是一个合法的 DNS 子域名，名称不能超过 52 个字符。这是因为 CronJob 控制器将自动在提供的任务名称后附加 11 个字符，且存在一个限制，即任务名称的最大长度不能超过 63 个字符。

CronJob 对于创建周期性的、多次重复的任务很有用。例如，执行数据备份或者发送邮件。CronJob 也可以计划在指定时间执行的独立任务。例如，计划当集群看起来很空闲时执行某个任务。

（1）CronJob 案例会在每分钟打印出当前时间和问候消息，代码如下，如图 10-6 所示。

```
apiVersion: batch/v1beta1
kind: CronJob
metadata:
  name: hello
spec:
  schedule: "*/1 * * * *"
  jobTemplate:
    spec:
      template:
        spec:
          containers:
          - name: hello
            image: busybox
```

```
      imagePullPolicy: IfNotPresent
      command:
      - /bin/sh
      - -c
      - date; echo Hello from the Kubernetes cluster
    restartPolicy: OnFailure
```

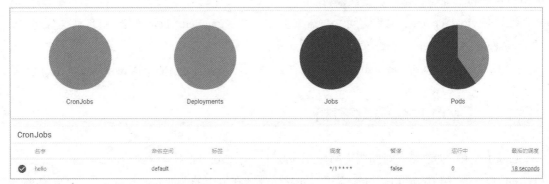

图 10-6　Kubernetes CronJob 任务创建

（2）CronJob 时间表语法如下所示。

```
# ┌───────────── 分 (0 ~ 59)
# │ ┌───────────── 时 (0 ~23)
# │ │ ┌───────────── 日 (1 ~31)
# │ │ │ ┌───────────── 月 (1 ~12)
# │ │ │ │ ┌───────────── 星期 (0 ~6)（星期日到星期一;在某些系统上,
# 7 也是星期日）
# │ │ │ │ │
# │ │ │ │ │
# │ │ │ │ │
# * * * * *
```

10.18　Kubernetes 证书剖析和制作实战

Kubernetes 需要公钥基础设施（Public Key Infrastructure，PKI）证书才能进行基于安全传输层协议（Transport Layer Security，TLS，主要用于在两个通信应用程序之间提供保密性和数据完整性）的身份验证。如果用户是通过 kubeadm 安装的 Kubernetes，则会自动生成集群所需的证书。用户还可以生成自己的证书。例如，不将私钥存储在 API 服务器上，可以让私钥更加安全。如果是通过 kubeadm 安装的 Kubernetes，则所有证书都存放在 /etc/kubernetes/pki 目录下。本文所有相关的路径都是基于该路径的相对路径。

PKI 采用证书进行公钥管理，通过第三方的可信任机构（认证中心，即 CA）把用户的公钥和其他标识信息捆绑在一起，其中包括用户名和电子邮件地址等信息，以在互联网上验证用户的身份。PKI 把公钥密码和对称密码结合起来，在互联网上实现密钥的自动管理，保证网上数据的安全传输。

Kubernetes 集群中需要的证书环节如下：

（1）etcd 对外提供服务，需要一套 etcd server 证书。

（2）etcd 各节点之间进行通信，需要一套 etcd peer 证书。

（3）kube-apiserver 访问 etcd，需要一套 etcd client 证书。

（4）kube-apiserver 对外提供服务，需要一套 kube-apiserver server 证书。

（5）kube-scheduler、kube-controller-manager、kube-proxy、kubelet 和其他可能用到的组件，需要访问 Kube-apiserver，需要一套 kube-apiserver client 证书。

（6）kube-controller-manager 要生成服务的 service account，需要一套用来签署 service account 的证书（CA 证书）。

（7）kubelet 对外提供服务，需要一套 kubelet server 证书。

（8）kube-apiserver 需要访问 kubelet，需要一套 kubelet client 证书。

同一个套里的证书必须是用同一个 CA 签署的，签署不同套里的证书的 CA 可以相同，也可以不同。例如，所有 etcd server 证书需要是同一个 CA 签署的，所有的 etcd peer 证书也是同一个 CA 签署的，而一个 etcd server 证书和一个 etcd peer 证书完全可以是两个 CA 签署的，彼此没有任何关系，这里就算是两套证书。

虽然可以用多套证书，但是维护多套 CA 实在过于繁杂，这里还是用一个 CA 签署所有证书。

（1）需要准备的证书如下：

```
admin.pem
ca.-key.pem
ca.pem
admin-key.pem
kube-scheduler-key.pem
kube-scheduler.pem
kube-controller-manager-key.pem
kube-controller-manager.pem
kube-proxy-key.pem
kube-proxy.pem
kubernetes-key.pem
kubernetes.pem
```

（2）使用证书的组件如下：

① etcd：ca.pem、kubernetes-key.pem、kubernetes.pem。

② Kube-apiserver：ca.pem、ca-key.pem、kubernetes-key.pem、kubernetes.pem。

③ Kubelet：ca.pem。

④ kube-proxy：ca.pem、kube-proxy-key.pem、kube-proxy.pem。

⑤ kubectl：ca.pem、admin-key.pem、admin.pem。

⑥ kube-controller-manager：ca-key.pem、ca.pem、kube-controller-manager-key.pem、kube-controller-manager.pem。

⑦ kube-scheduler：kube-scheduler-key.pem、kube-scheduler.pem。

此处使用 CFSSL 制作证书，CFSSL 是 Cloudflare 开发的一个开源的 PKI 工具，是一个完备的 CA 服务系统，可以签署、撤销证书等，覆盖了一个证书的整个生命周期。后面只用到了它的命令行工具。

注：一般情况下，Kubernetes 中的证书只需要创建一次，以后在向集群中添加新节点时只要将/etc/kubernetes/ssl 目录下的证书复制到新节点上即可。

```
wget https://pkg.cfssl.org/R1.2/cfssl_linux-amd64
wget https://pkg.cfssl.org/R1.2/cfssljson_linux-amd64
wget https://pkg.cfssl.org/R1.2/cfssl-certinfo_linux-amd64
chmod +x cfssl_linux-amd64 cfssljson_linux-amd64 cfssl-certinfo_linux-amd64
mv cfssl_linux-amd64 /usr/local/bin/cfssl
mv cfssljson_linux-amd64 /usr/local/bin/cfssljson
mv cfssl-certinfo_linux-amd64 /usr/bin/cfssl-certinfo
```

（3）创建 CA 证书，配置文件操作指令如下：

```
cat>ca-config.json<<EOF
{
  "signing": {
    "default": {
      "expiry": "87600h"
    },
    "profiles": {
      "kubernetes": {
        "usages": [
            "signing",
            "key encipherment",
            "server auth",
            "client auth"
        ],
        "expiry": "87600h"
      }
    }
  }
}
EOF
```

（4）Kubernetes 证书配置文件参数剖析。

① ca-config.json：可以定义多个 profiles，指定不同的过期时间、使用场景等参数。后续在签名证书时使用某个 profile。

② signing：可以签名其他证书，生成的 ca.pem 证书中 CA=TRUE。

③ server auth：client 可以用该 CA 对 server 提供的证书进行验证。

④ client auth：server 可以用该 CA 对 client 提供的证书进行验证。

⑤ expiry：设置证书过期时间。

（5）创建 CA 证书签名请求文件，操作指令如下：

```
cat>ca-csr.json<<EOF
{
  "CN": "kubernetes",
  "key": {
    "algo": "rsa",
    "size": 2048
  },
  "names": [
    {
      "C": "CN",
      "ST": "BeiJing",
      "L": "BeiJing",
      "O": "Kubernetes",
      "OU": "System"
    }
  ],
    "ca": {
      "expiry": "87600h"
    }
}
EOF
```

（6）CA 证书签名请求文件配置参数剖析。

① CN：Common Name，Kube-apiserver 从证书中提取该字段作为请求的用户名（User Name）。浏览器使用该字段验证网站是否合法。

② O：Organization，Kube-apiserver 从证书中提取该字段作为请求用户所属的组（Group）。

（7）生成 CA 证书和私钥，操作指令如下：

```
cfssl gencert -initca ca-csr.json | cfssljson -bare ca
ls | grep ca
ca-config.json
ca.csr
ca-csr.json
ca-key.pem
```

```
ca.pem
```

其中,ca-key.pem 是 CA 的私钥;ca.csr 是一个签署请求;ca.pem 是 CA 证书,是后面 Kubernetes 组件会用到的 RootCA。

(8)创建 Kubernetes 证书,并创建 Kubernetes 证书签名请求文件,操作指令如下:

```
cat>kubernetes-csr.json<<EOF
{
    "CN": "kubernetes",
    "hosts": [
      "127.0.0.1",
      "192.168.1.145",
      "192.168.1.146",
      "etcd01",
      "kubernetes",
      "kube-api.jd.com",
      "kubernetes.default",
      "kubernetes.default.svc",
      "kubernetes.default.svc.cluster",
      "kubernetes.default.svc.cluster.local"
    ],
    "key": {
        "algo": "rsa",
        "size": 2048
    },
    "names": [
        {
            "C": "CN",
            "ST": "BeiJing",
            "L": "BeiJing",
            "O": "Kubernetes",
            "OU": "System"
        }
    ]
}
EOF
```

如果 hosts 字段不为空,则需要指定授权使用该证书的 IP 地址或域名列表。由于该证书后续被 etcd 集群和 Kubernetes master 集群使用,将 etcd、master 节点的 IP 地址都填上,同时还有 service 网络的首 IP 地址(一般是 Kube-apiserver 指定的 service-cluster-ip-range 网段的第一个 IP 地址,如 10.0.0.1)。以上物理节点的 IP 地址也可以更换为主机名。

(9)生成 Kubernetes 证书和私钥,操作指令如下:

```
cfssl gencert -ca=ca.pem -ca-key=ca-key.pem -config=ca-config.json -profile=kubernetes kubernetes-csr.json | cfssljson -bare kubernetes
ls |grep kubernetes
```

```
kubernetes.csr
kubernetes-csr.json
kubernetes-key.pem
kubernetes.pem
```

(10) 创建 admin 证书,并创建 admin 证书签名请求文件,操作指令如下:

```
cat>admin-csr.json<<EOF
{
  "CN": "admin",
  "hosts": [],
  "key": {
    "algo": "rsa",
    "size": 2048
  },
  "names": [
    {
      "C": "CN",
      "ST": "BeiJing",
      "L": "BeiJing",
      "O": "system:masters",
      "OU": "System"
    }
  ]
}
EOF
```

Kube-apiserver 使用 RBAC 对客户端(如 kubelet、kube-proxy、Pod)请求进行授权。Kube-apiserver 预定义了一些 RBAC 使用的 RoleBindings。例如,cluster-admin 将 Group system: masters 与 Role cluster-admin 绑定,该 Role 授予了调用 Kube-apiserver 的所有 API 的权限。

O 指定该证书的 Group 为 system:masters,Kubelet 使用该证书访问 Kube-apiserver 时,由于证书被 CA 签名,所以认证通过,同时由于证书用户组为经过预授权的 system:masters,所以被授予访问所有 API 的权限。

注:这个 admin 证书用于生成管理员用的 kube config 配置文件,一般建议使用 RBAC 对 Kubernetes 进行角色权限控制,Kubernetes 将证书中的 CN 字段作为 User,O 字段作为 Group。

(11) 生成 admin 证书和私钥,操作指令如下:

```
cfssl gencert -ca=ca.pem -ca-key=ca-key.pem -config=ca-config.json -profile=kubernetes admin-csr.json | cfssljson -bare admin
ls | grep admin
admin.csr
admin-csr.json
admin-key.pem
admin.pem
```

（12）创建 kube-proxy 证书及 kube-proxy 证书签名请求文件，操作指令如下：

```
cat>kube-proxy-csr.json<<EOF
{
  "CN": "system:kube-proxy",
  "hosts": [],
  "key": {
    "algo": "rsa",
    "size": 2048
  },
  "names": [
    {
      "C": "CN",
      "ST": "BeiJing",
      "L": "BeiJing",
      "O": "Kubernetes",
      "OU": "System"
    }
  ]
}
EOF
```

Kube-apiserver 预定义的 RoleBinding system:node-proxier 将 User system:kube-proxy 与 Role system:node-proxier 绑定，该 Role 授予了调用 Kube-apiserver Proxy 相关 API 的权限。CN 指定证书的 User 为 system:kube-proxy。

因为该证书只会被 kubectl 当作 client 证书使用，所以 hosts 字段为空，生成 kube-proxy 证书和私钥。

```
cfssl gencert -ca=ca.pem -ca-key=ca-key.pem -config=ca-config.json -profile=kubernetes kube-proxy-csr.json | cfssljson -bare kube-proxy
ls |grep kube-proxy
kube-proxy.csr
kube-proxy-csr.json
kube-proxy-key.pem
kube-proxy.pem
```

（13）创建 kube-controller-manager 证书，并创建其证书签名请求文件，操作指令如下：

```
cat>kube-controller-manager-csr.json<<EOF
{
    "CN": "system:kube-controller-manager",
    "key": {
        "algo": "rsa",
        "size": 2048
    },
    "hosts": [
      "127.0.0.1",
```

```
            "192.168.1.145",
            "192.168.1.146",
            "Kubernetes-master1",
    ],
    "names": [
        {
            "C": "CN",
            "ST": "BeiJing",
            "L": "BeiJing",
            "O": "system:kube-controller-manager",
            "OU": "system"
        }
    ]
}
EOF
```

其中，hosts 列表包含所有 kube-controller-manager 节点 IP；CN 为 system:kube-controller-manager；O 为 system:kube-controller-manager。

（14）生成 kube-controller-manager 证书和私钥，操作指令如下：

```
cfssl gencert -ca=ca.pem -ca-key=ca-key.pem -config=ca-config.json -profile=kubernetes kube-controller-manager-csr.json | cfssljson -bare kube-controller-manager
```

（15）创建 kube-scheduler 证书签名请求文件，操作指令如下：

```
cat>kube-scheduler-csr.json<<EOF
{
    "CN": "system:kube-scheduler",
    "hosts": [
        "127.0.0.1",
        "192.168.1.145",
        "192.168.1.146",
        "Kubernetes-master1",
    ],
    "key": {
        "algo": "rsa",
        "size": 2048
    },
    "names": [
        {
            "C": "CN",
            "ST": "BeiJing",
            "L": "BeiJing",
            "O": "system:kube-scheduler",
            "OU": "4Paradigm"
```

```
            }
        ]
}
EOF
```

（16）根据以上创建的 CA 证书和 JSON 文件，接下来创建 Kubernetes 各个服务所需证书，操作指令如下：

```
cfssl gencert -ca=ca.pem -ca-key=ca-key.pem -config=ca-config.json -profile=kubernetes kube-scheduler-csr.json| cfssljson -bare kube-scheduler
ls | grep pem
admin-key.pem
admin.pem
ca-key.pem
ca.pem
kube-proxy-key.pem
kube-proxy.pem
kubernetes-key.pem
kubernetes.pem
kube-controller-manager-key.pem
kube-controller-manager.pem
kube-scheduler-key.pem
kube-scheduler.pem
```

（17）查看证书信息，操作指令如下：

```
cfssl-certinfo -cert kubernetes.pem
```

部署 Kubernetes 云计算平台时，将这些证书文件分发至此集群中其他节点机器对应目录即可。至此，TLS 证书创建完毕。

第 11 章 Kubernetes 云计算平台配置实战

部署 Kubernetes 云计算平台，至少准备 2 台服务器，此处为 4 台，包括 1 台 Docker 仓库。

```
Kubernetes master1 节点：192.168.1.145
Kubernetes node1 节点：192.168.1.146
Kubernetes node2 节点：192.168.1.147
Docker 私有库节点：192.168.1.148
```

11.1 Kubernetes 节点 hosts 及防火墙设置

Kubernetes master1、node1、node2 节点均执行如下代码，设置 hosts 和防火墙配置，操作指令如下：

```
#添加 hosts 解析
cat >/etc/hosts<<EOF
127.0.0.1 localhost localhost.localdomain
192.168.1.145 master1
192.168.1.146 node1
192.168.1.147 node2
EOF
#临时关闭 SELinux 和防火墙
sed -i '/SELINUX/s/enforcing/disabled/g'  /etc/sysconfig/selinux
setenforce  0
systemctl   stop     firewalld.service
systemctl   disable    firewalld.service
#同步节点时间
yum install ntpdate -y
ntpdate  pool.ntp.org
#修改对应节点主机名
hostname `cat /etc/hosts|grep $(ifconfig|grep broadcast|awk '{print $2}')|awk '{print $2}'`;su
```

```
#关闭swapoff
swapoff -a
```

11.2 Linux 内核参数设置和优化

Kubernetes master1、node1、node2 节点均执行如下代码，设置 IPVS 模式并调整内核参数，操作指令如下：

```
cat > /etc/modules-load.d/ipvs.conf <<EOF
#Load IPVS at boot
ip_vs
ip_vs_rr
ip_vs_wrr
ip_vs_sh
nf_conntrack_ipv4
EOF
systemctl enable --now systemd-modules-load.service
#确认内核模块加载成功
lsmod | grep -e ip_vs -e nf_conntrack_ipv4
#安装ipset、ipvsadm
yum install -y ipset ipvsadm
#配置内核参数
cat <<EOF > /etc/sysctl.d/Kubernetes.conf
net.bridge.bridge-nf-call-ip6tables=1
net.bridge.bridge-nf-call-iptables=1
EOF
sysctl --system
```

11.3 Docker 虚拟化案例实战

Kubernetes master1、node1、node2 节点均执行如下代码，部署 Docker 虚拟化软件，操作指令如下：

```
#安装依赖软件包
yum install -y yum-utils device-mapper-persistent-data lvm2
#添加Docker repository,这里使用国内阿里云yum源
yum-config-manager   --add-repo http://mirrors.aliyun.com/docker-ce/linux/centos/docker-ce.repo
#安装docker-ce,这里直接安装最新版本
yum install -y docker-ce
#修改Docker配置文件
mkdir /etc/docker
cat > /etc/docker/daemon.json <<EOF
```

```
{
  "exec-opts": ["native.cgroupdriver=systemd"],
  "log-driver": "json-file",
  "log-opts": {
    "max-size": "100m"
  },
  "storage-driver": "overlay2",
  "storage-opts": [
    "overlay2.override_kernel_check=true"
  ],
  "registry-mirrors": ["https://uyah70su.mirror.aliyuncs.com"]
}
EOF
#注意,由于国内拉取镜像较慢,配置文件最后增加了代码 registry-mirrors
#mkdir -p /etc/systemd/system/docker.service.d
#重启 Docker 服务
systemctl daemon-reload
systemctl enable docker.service
systemctl start docker.service
ps -ef|grep -aiE docker
```

11.4 Kubernetes 添加部署源

Kubernetes master1、node1、node2 节点均执行如下代码,添加 Kubernetes 网络安装源,操作指令如下:

```
cat>>/etc/yum.repos.d/kubernetes.repo<<EOF
[kubernetes]
name=Kubernetes
baseurl=https://mirrors.aliyun.com/kubernetes/yum/repos/kubernetes-el7-x86_64
enabled=1
gpgcheck=0
repo_gpgcheck=0
gpgkey=https://mirrors.aliyun.com/kubernetes/yum/doc/yum-key.gpg
EOF
```

11.5 Kubernetes Kubeadm 案例实战

Kubeadm 是一个 Kubernetes 部署工具,它提供了 kubeadm init 及 kubeadm join 这两个命令快速创建 Kubernetes 集群。

Kubeadm 通过执行必要的操作启动和运行一个最小可用的集群。它被设计为只关心启动集群,而不是之前的节点准备工作。诸如安装各种各样值得拥有的插件,如 Kubernetes Dashboard、监控解决方案以及特定云提供商的插件，这些都不在 Kubeadm 负责的范围内。

相反，我们期望由一个基于 Kubeadm、从更高层设计的、更加合适的工具来做这些事情；并且在理想情况下，使用 Kubeadm 作为所有部署的基础将会更容易创建一个符合期望的集群。

Kubeadm 是用于初始化 Kubernetes Cluster 的工具。kubelet 运行在 Cluster 所有节点上，负责调用 Docker 指令，启动 Pod 和容器。kubectl:kubectl 是 Kubenetes 命令行工具，通过 kubectl 可以部署和管理应用，查看各种资源，创建、删除和更新组件。

在计算机上手动安装 Docker、Kubeadm、kubelet、kubectl 几个二进制文件，然后才能再容器化部署其他 Kubernetes 组件。主要通过 Kubeadm init 初始化，如图 11-1 所示。初始化 Kubernetes 集群的流程如下。

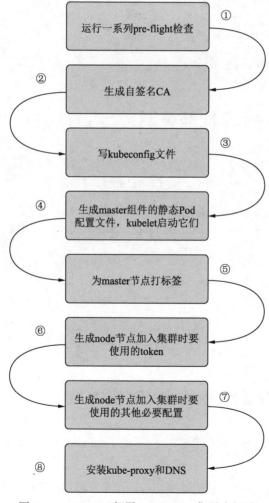

图 11-1　Kubeadm 部署 Kubernetes 集群流程图

（1）检查工作（Preflight Checks）：检查 Linux 内核版本、Cgroups 模块可用性、组件版本、端口占用情况、Docker 等的依赖情况。

（2）生成对外提供服务的 CA 证书及对应的目录。

（3）生成其他组件访问 kube-apiserver 所需的配置文件。

（4）为 master 组件生成 Pod 配置文件，利用这些配置文件，通过 Kubernetes 中特殊的容器启动方法 Static Pod（kubelet 启动时自动加载固定目录的 Pod YAML 文件并启动）能够以 Pod 方式部署 kube-apiserver、kube-controller-manager、kube-scheduler 三个 master 组件，同时还会生成 etcd 的 Pod YAML 文件。

（5）为 master 节点打标签。

（6）为集群生成一个 Bootstrap token，其他节点加入集群的计算机和 API server 打交道，需要获取相应的证书文件，所以 Bootstrap token 需要扮演安全验证的角色。

（7）生成 node 节点加入集群时要使用的其他必要配置。

（8）安装默认插件，如 kube-proxy 和 Core DNS，分别提供集群的服务发现和 DNS 功能。

安装 Kubeadm 工具，操作指令如下：

```
#安装 Kubeadm
yum install -y kubeadm-1.20.4 kubelet-1.20.4 kubectl-1.20.4
#启动 kubelet 服务
systemctl enable kubelet.service
systemctl start kubelet.service
```

Kubeadm 常见指令操作详解如下：

```
kubeadm init        #用于启动一个 Kubernetes 主节点
kubeadm join        #用于启动一个 Kubernetes 工作节点并将其加入集群
kubeadm upgrade     #用于更新一个 Kubernetes 集群到新版本
kubeadm config      #如果使用 v1.7.x 或更低版本的 kubeadm 初始化集群,则需要对集
                    #群做一些配置以便使用 kubeadm upgrade 命令
kubeadm token       #用于管理 kubeadm join 使用的令牌
kubeadm reset       #用于还原 kubeadm init 或者 kubeadm join 对主机所做的任何更改
kubeadm version     #用于打印 kubeadm 版本
kubeadm alpha       #用于预览一组可用的新功能,以便从社区搜集反馈
```

11.6 Kubernetes master 节点实战

（1）执行 kubeadm init 命令初始化安装 Master 相关软件。

```
kubeadm init    --control-plane-endpoint=192.168.1.145:6443 --image-
repository registry.aliyuncs.com/google_containers   --kubernetes-version
```

```
v1.20.4   --service-cidr=10.10.0.0/16   --pod-network-cidr=10.244.0.0/16
--upload-certs
```

（2）根据以上指令操作，执行成功，如图 11-2 所示。

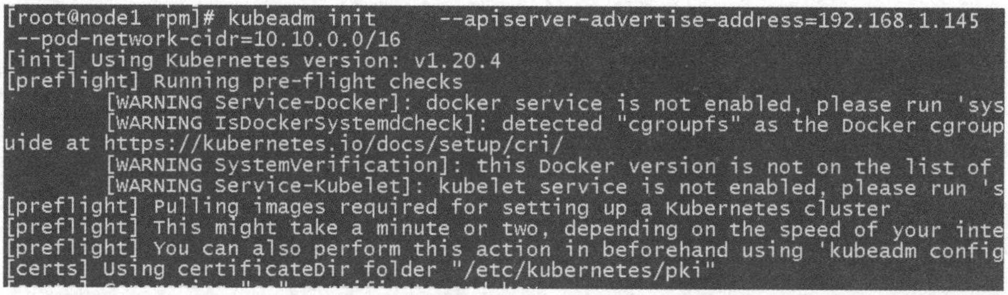

```
[root@node1 rpm]# kubeadm init     --apiserver-advertise-address=192.168.1.145
 --pod-network-cidr=10.10.0.0/16
[init] Using Kubernetes version: v1.20.4
[preflight] Running pre-flight checks
        [WARNING Service-Docker]: docker service is not enabled, please run 'sys
        [WARNING IsDockerSystemdCheck]: detected "cgroupfs" as the Docker cgroup
uide at https://kubernetes.io/docs/setup/cri/
        [WARNING SystemVerification]: this Docker version is not on the list of
        [WARNING Service-Kubelet]: kubelet service is not enabled, please run 's
[preflight] Pulling images required for setting up a Kubernetes cluster
[preflight] This might take a minute or two, depending on the speed of your inte
[preflight] You can also perform this action in beforehand using 'kubeadm config
[certs] Using certificateDir folder "/etc/kubernetes/pki"
```

（a）

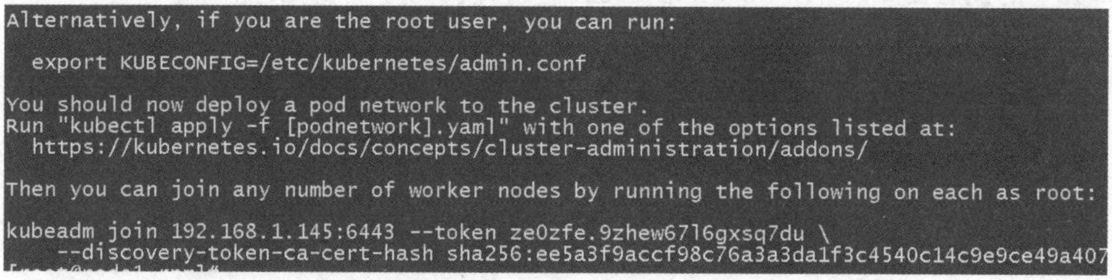

```
Alternatively, if you are the root user, you can run:

  export KUBECONFIG=/etc/kubernetes/admin.conf

You should now deploy a pod network to the cluster.
Run "kubectl apply -f [podnetwork].yaml" with one of the options listed at:
  https://kubernetes.io/docs/concepts/cluster-administration/addons/

Then you can join any number of worker nodes by running the following on each as root:

kubeadm join 192.168.1.145:6443 --token ze0zfe.9zhew67l6gxsq7du \
    --discovery-token-ca-cert-hash sha256:ee5a3f9accf98c76a3a3da1f3c4540c14c9e9ce49a407
```

（b）

图 11-2　Kubeadm 初始化 Kubernetes 集群

（a）执行 init 指令初始化 Kubernetes 集群；（b）查看 Kubernetes 客户端加入 master 集群信息

（3）根据图 11-2 提示，接下来需要手动执行如下指令，复制 admin 配置文件，操作指令如下：

```
mkdir -p $HOME/.kube
sudo cp -i /etc/kubernetes/admin.conf $HOME/.kube/config
sudo chown $(id -u):$(id -g) $HOME/.kube/config
```

（4）将 master 节点初始化之后，要将 node 节点也加入 Kubernetes 集群，操作指令如下：

```
kubeadm join 192.168.1.145:6443 --token ze0zfe.9zhew67l6gxsq7du \
    --discovery-token-ca-cert-hash
sha256:ee5a3f9accf98c76a3a3da1f3c4540c14c9e9ce49a4070de4b832aa8cb3a8f31
```

11.7　Kubernetes 集群节点和删除

根据 11.6 节中的所有操作和步骤，Kubernetes 集群部署成功，如果要将 node2 节点也加入 Kubernetes 集群，需要执行的操作指令如下：

```
#启动node1节点上的docker引擎服务
systemctl start docker.service
#将node1节点加入Kubernetes集群
kubeadm join 192.168.1.145:6443 --token ze0zfe.9zhew67l6gxsq7du \
--discovery-token-ca-cert-hash sha256:ee5a3f9accf98c76a3a3da1f3c4540c14
c9e9ce49a4070de4b832aa8cb3a8f31
#执行命令kubeadm init时没有记录下加入集群的指令,可以通过以下命令重新创建
kubeadm token create --print-join-command
#登录Kubernetes master节点验证节点信息
kubectl get nodes
#删除一个节点前,先驱赶掉上面的Pod容器
kubectl drain node2 --delete-local-data --force --ignore-daemonsets
#在master端执行如下指令
kubectl delete node/node2
```

11.8 Kubernetes 节点网络配置

Kubernetes 整个集群所有服务器(master、Minion)配置 Flannel,必须安装 Pod 网络插件,以便 Pod 之间相互通信。必须在全部应用程序之前部署网络,CoreDNS 不会在安装网络插件之前启动。

(1) 安装 Flannel 网络插件。Flannel 定义 Pod 的网段为 10.244.0.0/16,Pod 容器的 IP 地址会自动分配 10.244 开头的网段 IP。部署 Flannel 网络插件操作指令如下:

```
#下载Flannel插件YML文件
yum install wget -y
wget https://raw.githubusercontent.com/coreos/flannel/master/Documentation/
kube-flannel.yml
#提前下载Flanneld组建所需镜像
for i in $(cat kube-flannel.yml |grep image|awk -F: '{print $2":"$3}'
|uniq );do docker pull $i ;done
#应用YML文件
kubectl apply -f kube-flannel.yml
```

(2) 如果 Kube-flannel.yml 配置文件无法下载,也可以将以下代码直接复制并且写入 Kube-flannel.yml。文件代码如下:

```
---
apiVersion: policy/v1beta1
kind: PodSecurityPolicy
metadata:
  name: psp.flannel.unprivileged
  annotations:
    seccomp.security.alpha.kubernetes.io/allowedProfileNames: docker/
```

```
    default
        seccomp.security.alpha.kubernetes.io/defaultProfileName: docker/
default
        apparmor.security.beta.kubernetes.io/allowedProfileNames: runtime/
default
        apparmor.security.beta.kubernetes.io/defaultProfileName: runtime/
default
spec:
  privileged: false
  volumes:
  - configMap
  - secret
  - emptyDir
  - hostPath
  allowedHostPaths:
  - pathPrefix: "/etc/cni/net.d"
  - pathPrefix: "/etc/kube-flannel"
  - pathPrefix: "/run/flannel"
  readOnlyRootFilesystem: false
  #Users and groups
  runAsUser:
    rule: RunAsAny
  supplementalGroups:
    rule: RunAsAny
  fsGroup:
    rule: RunAsAny
  #Privilege Escalation
  allowPrivilegeEscalation: false
  defaultAllowPrivilegeEscalation: false
  #Capabilities
  allowedCapabilities: ['NET_ADMIN', 'NET_RAW']
  defaultAddCapabilities: []
  requiredDropCapabilities: []
  #Host namespaces
  hostPID: false
  hostIPC: false
  hostNetwork: true
  hostPorts:
  - min: 0
    max: 65535
  # seLinux
  seLinux:
    # seLinux is unused in CaaSP
    rule: 'RunAsAny'
```

```yaml
---
kind: ClusterRole
apiVersion: rbac.authorization.Kubernetes.io/v1
metadata:
  name: flannel
rules:
- apiGroups: ['extensions']
  resources: ['podsecuritypolicies']
  verbs: ['use']
  resourceNames: ['psp.flannel.unprivileged']
- apiGroups:
  - ""
  resources:
  - pods
  verbs:
  - get
- apiGroups:
  - ""
  resources:
  - nodes
  verbs:
  - list
  - watch
- apiGroups:
  - ""
  resources:
  - nodes/status
  verbs:
  - patch
---
kind: ClusterRoleBinding
apiVersion: rbac.authorization.Kubernetes.io/v1
metadata:
  name: flannel
roleRef:
  apiGroup: rbac.authorization.Kubernetes.io
  kind: ClusterRole
  name: flannel
subjects:
- kind: ServiceAccount
  name: flannel
  namespace: kube-system
---
apiVersion: v1
```

```yaml
kind: ServiceAccount
metadata:
  name: flannel
  namespace: kube-system
---
kind: ConfigMap
apiVersion: v1
metadata:
  name: kube-flannel-cfg
  namespace: kube-system
  labels:
    tier: node
    app: flannel
data:
  cni-conf.json: |
    {
      "name": "cbr0",
      "cniVersion": "0.3.1",
      "plugins": [
        {
          "type": "flannel",
          "delegate": {
            "hairpinMode": true,
            "isDefaultGateway": true
          }
        },
        {
          "type": "portmap",
          "capabilities": {
            "portMappings": true
          }
        }
      ]
    }
  net-conf.json: |
    {
      "Network": "10.244.0.0/16",
      "Backend": {
        "Type": "vxlan"
      }
    }
---
apiVersion: apps/v1
kind: DaemonSet
```

```yaml
metadata:
  name: kube-flannel-ds
  namespace: kube-system
  labels:
    tier: node
    app: flannel
spec:
  selector:
    matchLabels:
      app: flannel
  template:
    metadata:
      labels:
        tier: node
        app: flannel
    spec:
      affinity:
        nodeAffinity:
          requiredDuringSchedulingIgnoredDuringExecution:
            nodeSelectorTerms:
            - matchExpressions:
              - key: kubernetes.io/os
                operator: In
                values:
                - linux
      hostNetwork: true
      priorityClassName: system-node-critical
      tolerations:
      - operator: Exists
        effect: NoSchedule
      serviceAccountName: flannel
      initContainers:
      - name: install-cni
        image: quay.io/coreos/flannel:v0.13.1-rc2
        command:
        - cp
        args:
        - -f
        - /etc/kube-flannel/cni-conf.json
        - /etc/cni/net.d/10-flannel.conflist
        volumeMounts:
        - name: cni
          mountPath: /etc/cni/net.d
        - name: flannel-cfg
```

```yaml
          mountPath: /etc/kube-flannel/
      containers:
      - name: kube-flannel
        image: quay.io/coreos/flannel:v0.13.1-rc2
        command:
        - /opt/bin/flanneld
        args:
        - --ip-masq
        - --kube-subnet-mgr
        resources:
          requests:
            cpu: "100m"
            memory: "50Mi"
          limits:
            cpu: "100m"
            memory: "50Mi"
        securityContext:
          privileged: false
          capabilities:
            add: ["NET_ADMIN", "NET_RAW"]
        env:
        - name: POD_NAME
          valueFrom:
            fieldRef:
              fieldPath: metadata.name
        - name: POD_NAMESPACE
          valueFrom:
            fieldRef:
              fieldPath: metadata.namespace
        volumeMounts:
        - name: run
          mountPath: /run/flannel
        - name: flannel-cfg
          mountPath: /etc/kube-flannel/
      volumes:
      - name: run
        hostPath:
          path: /run/flannel
      - name: cni
        hostPath:
          path: /etc/cni/net.d
      - name: flannel-cfg
        configMap:
          name: kube-flannel-cfg
```

(3) 查看 Kubernetes Flannel 网络插件是否部署成功,如图 11-3 所示,操作指令如下:

```
kubectl -n kube-system get pods|grep -aiE flannel
```

```
[root@master1 ~]# kubectl get nodes
NAME      STATUS   ROLES                  AGE   VERSION
master1   Ready    control-plane,master   15d   v1.20.4
node1     Ready    <none>                 15d   v1.20.4
[root@master1 ~]#
[root@master1 ~]#
[root@master1 ~]#
[root@master1 ~]# kubectl -n kube-system get pods|grep -aiE flannel
kube-flannel-ds-62zrs          1/1   Running   0   48s
kube-flannel-ds-p96c2          1/1   Running   0   48s
[root@master1 ~]#
[root@master1 ~]#
```

图 11-3　Kubernetes Flannel 网络插件部署

(4) 安装 Calico 网络插件。

Kubernetes 云计算平台网络通信除了使用 Flannel 之外,还可以使用 Calico(一种容器之间互通的网络方案)。两个网络插件都可以实现 Kubernetes 容器互通。实际生产环境中二选一即可。

在虚拟化平台中(如 OpenStack、Docker 等)都需要实现容器之间互连,但同时也需要对容器进行隔离控制,就像在互联网中的服务仅开放 80 端口、公有云的多租户一样,提供隔离和管控机制。

在多数虚拟化平台实现中,通常都使用二层隔离技术实现容器的网络,这些技术有一些弊端。例如,需要依赖 VLAN、Bridge 和隧道等技术,其中 Bridge 带来了复杂性,VLAN 隔离和 Tunnel 隧道则消耗更多的资源并对物理环境有要求,随着网络规模增大,整体会变得越加复杂。

我们尝试把 Host 当作 Internet 中的路由器,同样使用 BGP 同步路由,并使用 iptables 来做安全访问策略,最终设计出了 Calico 方案。

Calico 不使用隧道或 NAT 来实现转发,而是巧妙地把所有二、三层流量转换成三层流量,并通过 Host 上路由配置完成跨 Host 转发。

为了保证 Calico 正常工作,需要传递--pod-network-cidr=10.10.0.0/16 到 kubeadm init 或更新 calico.yaml 文件,以便与 Pod 网络相匹配。

(1) 部署 Calico 网络插件,同样需要从官网下载 calico.yaml 文件,操作指令如下:

```
kubectl apply -f https://docs.projectcalico.org/v3.10/manifests/calico.yaml
```

(2) 如果安装 Kubernetes Flannel 网络插件,必须通过 kubeadm init 配置-pod-network-cidr=10.10.0.0/16 参数。

(3) 验证 Calico 网络插件,安装 Calico Pod 网络后,确认 Coredns 及其他 Pod 全部运行正常,查看 master 节点状态为 Ready,如图 11-4 所示,操作指令如下:

```
kubectl get nodes
kubectl -n kube-system get pods
```

```
[root@node1 rpm]# kubectl get nodes
NAME    STATUS  ROLES                AGE    VERSION
node1   Ready   control-plane,master 8m22s  v1.20.4
[root@node1 rpm]#
[root@node1 rpm]# kubectl -n kube-system get pods
NAME                                          READY  STATUS             RESTARTS
calico-kube-controllers-7854b85cf7-v8qhr      0/1    ContainerCreating  0
calico-node-swfn4                             0/1    PodInitializing    0
coredns-7f89b7bc75-2gs55                      0/1    ContainerCreating  0
coredns-7f89b7bc75-gqdg9                      0/1    ContainerCreating  0
etcd-node1                                    1/1    Running            0
kube-apiserver-node1                          1/1    Running            0
kube-controller-manager-node1                 1/1    Running            0
```

图 11-4 Kubernetes Calico 网络插件部署

至此，Kubernetes 的 master1 节点和 node1 节点部署完成。接下来就可以管理和使用 Kubernetes 集群了。

11.9　Kubernetes 开启 IPVS 模式

IPVS（IP Virtual Server）实现了传输层负载均衡，也就是常说的 4 层 LAN 交换。Linux 内核默认集成 IPVS。IPVS 运行在主机上，在真实服务器集群前充当负载均衡器。

Kubernetes 默认使用 iptables 实现服务访问和代理，如果使用 IPVS 模式进行服务访问代理，kube-proxy 会监视 Kubernetes 集群中的对象和端点（endpoint），调用 netlink 接口以相应地创建 IPVS 规则并定期与 Kubernetes 中的 service 对象和 endpoints 同步 IPVS 规则，以确保 IPVS 状态与期望的一致。当访问 service 时，流量就会被重定向到后端的 Pod 上。

修改 kube-proxy 的 configmap，在 config.conf 中找到 mode 参数，改为 mode: "ipvs"并保存，操作指令如下：

```
kubectl -n kube-system get cm kube-proxy -o yaml | sed 's/mode: ""/mode: "ipvs"/g' | kubectl replace -f -
#或者手动修改
kubectl -n kube-system edit cm kube-proxy
kubectl -n kube-system get cm kube-proxy -o yaml | grep mode
   mode: "ipvs"
#重启 kube-proxy Pod
kubectl -n kube-system delete pods -l Kubernetes-app=kube-proxy
#确认 IPVS 模式开启成功
kubectl -n kube-system logs -f -l Kubernetes-app=kube-proxy | grep ipvs
#日志中打印出 Using ipvs Proxier,说明 IPVS 模式已经开启
#创建 service 之后,可以使用命令 ipvsadm -L -n 查看是否采用 IPVS 模式进行转发
ipvsadm -L -n
```

11.10 Kubernetes 集群故障排错

Kubernetes 集群可能出现 Pod 或 node 节点状态异常等情况，可以通过查看日志分析错误原因。常见排错操作方法和指令如下：

```
#查看 Pod 日志
kubectl -n kube-system logs -f <pod name>
#查看 Pod 运行状态及事件
kubectl -n kube-system describe pods <pod name>
node not ready                            #可以通过分析 node 相关日志排查
kubectl describe nodes <node name>
systemctl status docker
systemctl status kubelet
journalctl -xeu docker
journalctl -xeu kubelet
tail -f /var/log/messages
```

11.11 Kubernetes 集群节点移除

Kubernetes 云计算平台部署成功之后，在运行过程中，随着服务器时间寿命增加，会淘汰下架一些服务器，此时需要删除节点。以删除 node1 节点为例，在 master1 节点上执行如下指令：

```
kubectl drain node1 --delete-local-data --force --ignore-daemonsets
kubectl delete node node1
kubeadm reset -f
```

11.12 etcd 分布式案例操作

etcd 是一个分布式的、高可用的、一致的 key-value 存储数据库，基于 Go 语言实现，主要用于共享配置和服务发现。

在 Kubernetes 云计算平台中，Kubernetes 服务配置信息的管理共享和服务发现是一个很基本、很重要的问题。etcd 可以集中管理配置信息，服务端将配置信息存储于 etcd，客户端通过 etcd 得到服务配置信息，etcd 监听配置信息的改变，发现改变后通知客户端。

为了防止单点故障，还可以启动多个 etcd 组成集群。etcd 集群使用 raft 一致性算法处理日志复制，保证多节点数据的强一致性。

根据以上 Kubernetes 集群部署，etcd 默认也会被自动部署成功，可以使用以下指令测试 etcd 集群是否正常。

```
#查看etcd集群节点列表
etcdctl member list
#查看etcd集群节点状态
etcdctl cluster-health
#获取etcd中config key的值
etcdctl get /atomic.io/network/config
#查看etcd目录树
etcdctl ls /atomic.io/network/subnets
#删除etcd目录树
etcdctl rm  /atomic.io/network/  --recursive
#创建etcd config key和value
etcdctl mk /atomic.io/network/config '{"Network":"172.17.0.0/16"}'
```

Kubernetes 的 node 节点搭建和配置 Flannel 网络，etcd 中的/atomic.io/network/config 节点会被 node 节点上的 Flannel 用来创建 Docker IP 地址网段。查看 etcd 配置中心的网段信息，操作指令如下：

```
etcdctl ls /atomic.io/network/subnets
```

第 12 章 Kubernetes 企业网络 Flannel 实战

Flannel 是 CoreOS 团队针对 Kubernetes 设计的一个覆盖网络（Overlay Network）工具，其目的在于帮助每一个使用 Kuberentes 的 CoreOS 主机拥有一个完整的子网。

Flannel 通过给每台宿主机分配一个子网的方式为容器提供虚拟网络，它基于 Linux TUN/TAP，使用 UDP 封装 IP 包来创建 overlay 网络，并借助 etcd 维护网络的分配情况。

12.1 Flannel 工作原理

Flannel 是 CoreOS 团队针对 Kubernetes 设计的一个网络规划服务，简单来说，它的功能是让集群中的不同节点主机创建的 Docker 容器都具有全集群唯一的虚拟 IP 地址。

在默认的 Docker 配置中，每个 node 的 Docker 服务会分别负责所在节点容器的 IP 分配。node 内部的容器之间可以相互访问，但是跨主机网络相互间是不能通信的。

Flannel 的设计目的就是为集群中所有节点重新规划 IP 地址的使用规则，从而使得不同节点上的容器能够获得同属一个内网且不重复的 IP 地址，并让属于不同节点上的容器能够直接通过内网 IP 地址进行通信。

Flannel 使用 etcd 存储配置数据和子网分配信息。Flannel 启动之后，后台进程首先检索配置和正在使用的子网列表，然后选择一个可用的子网，并尝试注册它。etcd 也存储每个主机对应的 IP。

Flannel 使用 etcd 的 watch 机制监视/atomic.io/network/subnets 下面所有元素的变化信息，并且根据它来维护一个路由表。为了提高性能，Flannel 优化了 Universal TAP/TUN 设备，并对 TUN 和 UDP 之间的 IP 分片做了代理。

12.2　Flannel 架构介绍

Flannel 默认使用 8285 端口作为 UDP 封装报文的端口，VxLan 使用 8472 端口，如图 12-1 所示。

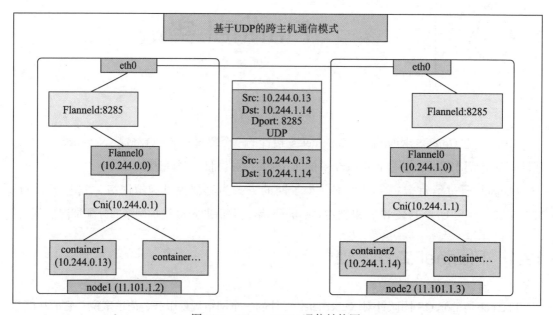

图 12-1　Flannel UDP 通信结构图

一个网络报文请求是怎么从一个容器发送到另外一个容器的呢？例如，从 master1 上的 container1 容器（IP 地址：10.244.0.13）访问 node1 上面的 container2 容器（IP 地址：10.244.1.14）。

（1）container1 容器 10.244.0.13 直接访问目标容器 container2 的 IP 地址 10.244.1.14，请求默认通过容器内部的 eth0 网卡发送出去。

（2）请求报文通过 Veth pair（虚拟设备对）被发送到 Docker 宿主机 VethXXX 设备上。

（3）VethXXX 设备是直接连接到虚拟交换机 Cni0 的，所以请求报文通过虚拟 Bridge Cni0 发送出去。

（4）查找 Docker 宿主机的路由表信息，同时外部容器 IP 的报文都会转发到 Flannel0 虚拟网卡（这是一个 P2P 的虚拟网卡），然后报文就被转发到监听在另一端的 flanneld 进程。

（5）因为 Flannel 在 etcd 中存储着子网和宿主机 IP 的对应关系，所以能够找到 10.244.1.14 对应的宿主机 IP 地址为 11.101.1.3，进而开始组装 UDP 数据包，并发送数据到目的主机。

（6）这个请求得以完成的原因是每个节点上都启动了一个 flanneld udp 进程，都监听着

8285 端口,所以 master1 通过 flanneld 进程把数据包通过宿主机的 Interface 网卡发送给 node1 的 flanneld 进程的相应端口即可。

(7)请求报文到达 node1 之后,继续往上传输到传输层,交给监听在 8285 端口的 flanneld 程序处理。

(8)请求数据被解包,然后发送给 Flannel0 虚拟网卡。

(9)查找 Kubernetes 主机节点的路由表,发现对应容器的报文要交给 Cni0。

(10)Cni0 找到连到自己的容器,把报文发送给 container2。

12.3 Kubernetes Dashboard UI 实战

Kubernetes 最重要的工作是对 Docker 容器集群统一的管理和调度。通常使用命令行操作 Kubernetes 集群及各个节点,非常不方便,如果使用 UI 界面可视化操作,更加方便管理和维护。以下为配置 Kubernetes dashboard 的完整过程。

(1)下载 dashboard 配置文件。

```
wget https://raw.githubusercontent.com/kubernetes/dashboard/v2.0.0-rc5/aio/deploy/recommended.yaml
\cp recommended.yaml recommended.yaml.bak
```

(2)修改文件 recommended.yaml 的 39 行内容。因为默认情况下 service 的类型是 cluster IP,所以需要更改为 NodePort 的方式,便于访问,也可映射到指定的端口。

```
spec:
  type: NodePort
  ports:
    - port: 443
      targetPort: 8443
      nodePort: 31001
  selector:
    Kubernetes -app: kubernetes-dashboard
```

(3)修改文件 recommended.yaml 的 195 行内容。因为默认情况下 Dashboard 为英文显示,故可以设置为中文。

```
env:
        - name: ACCEPT_LANGUAGE
          value: zh
```

(4)创建 Dashboard 服务,操作指令如下:

```
kubectl apply -f recommended.yaml
```

（5）查看 Dashboard 运行状态。

```
kubectl get pod -n kubernetes-dashboard
kubectl get svc -n kubernetes-dashboard
```

（6）基于 Token 的方式访问，设置和绑定 Dashboard 权限，获取 Token 值，如图 12-2 所示，操作指令如下：

```
#创建 Dashboard 的管理用户
kubectl create serviceaccount dashboard-admin -n kube-system
#将创建的 Dashboard 用户绑定为管理用户
kubectl create clusterrolebinding dashboard-cluster-admin --clusterrole=cluster-admin --serviceaccount=kube-system:dashboard-admin
#获取刚刚创建的用户对应的 Token 名称
kubectl get secrets -n kube-system | grep dashboard
#查看 Token 的详细信息
kubectl describe secrets -n kube-system $(kubectl get secrets -n kube-system | grep dashboard |awk '{print $1}')
```

(a)

(b)

图 12-2　Kubernetes 获取 Token 值

（a）获取 Kubernetes Token 密钥；（b）查看 Kubernetes admin 密钥信息

（7）通过浏览器访问 Dashboard Web（地址为 https://192.168.1.146:31001/），输入 Token 登录即可，如图 12-3 所示。

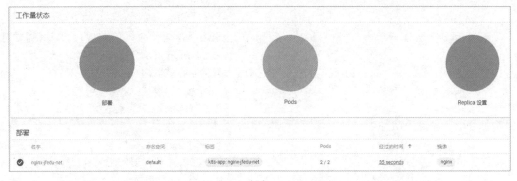

图 12-3　Kubernetes Web 界面展示

（a）打开 Kubernetes UI 登录界面；（b）查看 Kubernetes UI 容器 Pods 状态；
（c）查看 Kubernetes nginx-jfedu-net 部署状态

12.4 Kubernetes YAML 文件详解

Kubernetes YAML 配置文件主要分为基本标签、元数据标签、资源内容 3 部分，要想熟练地掌握 Kubernetes，必须要了解 YAML 配置文件中常见的参数和指令的含义。

（1）基本标签主要在文件起始位置。例如：

```
apiVersion: v1                               #版本号,例如 v1
kind: Namespace                              #类型或者控制器
```

（2）元数据标签主要在文件中部位置。例如：

```
metadata:                                    #数据标签
  name: nginx-deployment
  labels:                                    #子标签
    app: nginx                               #业务容器
```

（3）资源内容主要在文件末尾位置。例如：

```
spec:                                        #Pod 中容器的详细定义
  containers:                                #容器列表
  - name: nginx                              #容器名称
    image: nginx                             #容器的镜像名称
    imagePullPolicy: [Always | Never | IfNotPresent]
                    #获取镜像的策略,其中 Always 表示下载镜像;IfnotPresent 表示优
                    #先使用本地镜像,否则下载镜像;Never 表示仅使用本地镜像
    command: nginx  #容器的启动命令列表,如果不指定,则使用打包时使用的启动命令
    args: -g daemon off                      #容器的启动命令参数列表
    workingDir: /root/                       #容器的工作目录
    volumeMounts:                            #挂载到容器内部的存储卷配置
    - name: nginx                            #引用 Pod 定义的共享存储卷的名称,需要用
                                             # volumes[]部分定义的卷名
      mountPath: /usr/share/nginx/html       #存储卷在容器内 mount 的绝对路径
      readOnly: boolean                      #是否为只读模式
      ports:                                 #需要暴露的端口库号列表
      - name: nginx                          #端口号名称
        containerPort: 80                    #容器需要监听的端口号
        hostPort: 80                         #容器所在主机需要监听的端口号,默认为与
                                             #Container 相同
        protocol: TCP                        #端口协议,支持 TCP 和 UDP,默认为 TCP
      env:                                   #容器运行前需设置的环境变量列表
      - name: WEB                            #环境变量名称
        value: www.jfedu.net                 #环境变量的值
      resources:                             #资源限制和请求的设置
```

```
      limits:                    #资源限制的设置
        cpu: 1000m                #CPU 的限制,单位为 core 数,将用于 docker run --cpu-shares
                                  #参数
        memory: 1024m             #内存限制,单位可以为 Mib/Gib,将用于 docker run -memory
                                  #参数
      requests:                   #资源请求的设置
        cpu: 100m                 #CPU 请求,容器启动的初始可用数量
        memory: 1024m             #内存清楚,容器启动的初始可用数量
      livenessProbe:              #对 Pod 内各容器健康检查的设置,当探测无响应几次后将自动重启
                                  #该容器,检查方法有 exec、httpGet 和 tcpSocket,对一个容
                                  #器只需设置其中一种方法即可
        exec:                     #将 Pod 容器内检查方式设置为 exec
          command: [string]                    #exec 方式需要制定的命令或脚本
        httpGet:      #将 Pod 内各容器健康检查方法设置为 HttpGet,需要制定 path 和 port
          path: string
          port: number
          host: string
          scheme: string
          HttpHeaders:
          - name: string
            value: string
        tcpSocket:                #将 Pod 内各容器健康检查方式设置为 tcpSocket 方式
          port: number
        initialDelaySeconds: 0    #容器启动完成后首次探测的时间,单位为 s
        timeoutSeconds: 0 #对容器健康检查探测等待响应的超时时间,单位为 s,默认为 1s
        periodSeconds: 0   #对容器监控检查的定期探测时间设置,单位为 s,默认为 10s 一次
        successThreshold: 0
        failureThreshold: 0
        securityContext:
          privileged:false
  restartPolicy: [Always | Never | OnFailure]
                                  #Pod 的重启策略,其中,Always 表示一旦不管以何种方式终止
                                  #运行,kubelet 都将重启;OnFailure 表示只有 Pod 以非 0 退
                                  #出码退出才重启;Never 表示不再重启该 Pod
    nodeSelector: obeject         #设置 nodeSelector,表示将该 Pod 调度到包含这个 label
                                  #的 node 上,以 key: value 的格式指定
    imagePullSecrets:             #拉取镜像时使用的 secret 名称,以 key: secretkey 格式指定
    - name: string
    hostNetwork:false             #是否使用主机网络模式,默认为 false,如果设置为 true,表示
                                  #使用宿主机网络
    volumes:                      #在该 Pod 上定义共享存储卷列表
    - name: string                #共享存储卷名称 (volumes 类型有很多种)
```

```
        emptyDir: {}              #类型为 emtyDir 的存储卷,与 Pod 同生命周期的一个临时目录。
                                  #为空值
        hostPath: string          #类型为 hostPath 的存储卷,表示挂载 Pod 所在宿主机的目录
          path: string            #Pod 所在宿主机的目录,将被用于同期中 mount 的目录
        secret:                   #类型为 secret 的存储卷,挂载集群与定义的 secret 对象到容器
                                  #内部
          secretname: string
          items:
          - key: string
            path: string
        configMap:                #类型为 configMap 的存储卷,挂载预定义的 configMap 对象到
                                  #容器内部
          name: string
          items:
          - key: string
            path: string
```

12.5　kubectl 常见指令操作

Kubernetes 云计算平台部署和创建完成后,可以通过 kubectl 指令查看 Pod 和 service 的状态、信息,操作指令如下:

```
#查看 Kubernetes 集群所有的节点信息
kubectl get nodes
#删除 Kubernetes 集群中某个特定节点
kubectl delete nodes/10.0.0.123
#获取 Kubernetes 集群命名空间
kubectl get namespace
#获取 Kubernetes 所有命名空间有哪些部署
kubectl get deployment --all-namespaces
#查看 nginx 部署详细的信息
kubectl describe deployments/nginx -n default
#将 nginx 部署的镜像更新至 nginx1.19 版本
kubectl -n default set image deployments/nginx nginx=nginx:v1.19
#将 nginx 部署的 Pod 组容器副本数调整为 5 个
kubectl patch deployment nginx -p '{"spec":{"replicas":3}}' -n default
#获取 nginx 部署的 yaml 配置,输出到 nginx.yaml 文件
kubectl get deploy nginx -o yaml --export >nginx.yaml
#修改 nginx.yaml 文件,重新应用现有的 nginx 部署
kubectl apply -f nginx.yaml
#获取所有命名空间的详细信息、VIP、运行时间等
kubectl get svc --all-namespaces
```

```
#获取所有 Pod 所属的命名空间
kubectl get pods --all-namespaces
#获取所有命名空间的 Pod 详细 IP 信息
kubectl get pods -o wide --all-namespaces
#查看 dashboard 服务详细信息
kubectl describe service/kubernetes-dashboard --namespace="kube-system"
#获取 dashboard 容器详细信息
kubectl describe pod/kubernetes-dashboard-530803917-816df --namespace="kube-system"
#强制删除 dashboard 容器资源
kubectl delete pod/kubernetes-dashboard-530803917-816df --namespace="kube-system" --grace-period=0 --force
#强制删除一个 node 上的所有容器、服务、部署,不再接受新的 Pod 进程资源创建
kubectl drain 10.0.0.122 --force --ignore-daemonsets --delete-local-data
#恢复 node 上接受新的 Pod 进程资源创建
kubectl uncordon 10.0.0.122
```

12.6　Kubernetes 本地私有仓库实战

Docker 仓库主要用于存放 Docker 镜像。Docker 仓库分为公共仓库和私有仓库,基于 Registry 可以搭建本地私有仓库,使用私有仓库的优点如下:

（1）节省网络带宽,不用去 Docker 官网仓库下载每个镜像。

（2）从本地私有仓库中下载 Docker 镜像。

（3）组件公司内部私有仓库,方便各部门使用,服务器管理更加统一。

（4）可以基于 GIT 或者 SVN、Jenkins 更新本地 Docker 私有仓库镜像版本。

官方提供 Docker Registry 构建本地私有仓库,目前最新版本为 Registry v2。最新版的 Docker 已不再支持 Registry v1。Registry v2 使用 Go 语言编写,在性能和安全性上作了很多优化,重新设计了镜像的存储格式。以下为在 192.168.1.148 服务器上构建 Docker 本地私有仓库的方法及步骤。

（1）下载 Docker Registry 镜像。命令如下:

```
docker pull registry
```

（2）启动私有仓库容器。命令如下:

```
mkdir -p /data/registry/
docker run -itd -p 5000:5000 -v /data/registry:/var/lib/registry docker.io/registry
```

Docker 私有仓库创建和启动命令如图 12-4 所示。

```
[root@localhost ~]#
[root@localhost ~]# docker images
REPOSITORY              TAG                 IMAGE ID            CREAT
docker.io/registry      latest              9d0c4eabab4d        5 wee
[root@localhost ~]#
[root@localhost ~]# mkdir -p /data/registry/
[root@localhost ~]# docker run -itd -p 5000:5000 -v /data/regist
53c8771c78524659dcc94b30c452da54273d02772e410566308cc2d234d94d19

[root@localhost ~]#
[root@localhost ~]# docker ps -a
CONTAINER ID            IMAGE                       COMMAND
      PORTS                         NAMES
53c8771c7852            docker.io/registry          "/entrypoint.sh /etc/"
      0.0.0.0:5000->5000/tcp        compassionate_mcnulty
[root@localhost ~]#
```

图 12-4 Docker 私有仓库创建和启动命令

默认情况下，会将仓库存放于容器内的/tmp/registry 目录下，这样如果容器被删除，则存放于容器中的镜像也会丢失，所以一般情况下会指定本地一个目录挂载到容器内的/var/lib/registry 下。

（3）上传镜像至本地私有仓库。

客户端上传镜像至本地私有仓库。下面以 busybox 镜像为例，将 busybox 上传至私有仓库服务器。

```
docker   pull    busybox
docker   tag     busybox 192.168.1.148:5000/busybox
docker   push    192.168.1.148:5000/busybox
```

（4）检测本地私有仓库。

```
curl -XGET http://192.168.1.148:5000/v2/_catalog
curl -XGET http://192.168.1.148:5000/v2/busybox/tags/list
```

（5）在 Kubernetes 集群中其他节点（Docker 客户端）的主机上添加本地仓库地址，修改/etc/docker/daemon.json 文件，添加如下代码，同时重启 Docker 服务即可。

```
{
"insecure-registries":["192.168.1.148:5000"]
}
```

第 13 章 Kubernetes 核心组件 service 实战

13.1　Kubernetes service 概念

service 是 Kubernetes 最核心的概念，通过创建 service，可以为一组具有相同功能的 Pod 应用提供统一的访问入口，并且将请求进行负载分发到后端的各个容器应用上。

在 Kubernetes 中，在受到 RC 调控的时候，Pod 副本是变化的，对于 Pod 容器的 IP 也是变化的，如发生迁移或者伸缩的时候。这对于 Pod 的访问者来说是不可接受的。

Kubernetes 中的 service 是一种抽象概念，它定义了一个 Pod 逻辑集合以及访问它们的策略，service 与 Pod 的关联同样是基于 label 来完成的。service 的目标是提供一种桥梁，它会为访问者提供一个固定访问地址，用于在访问时重定向到相应的后端，这使得非 Kubernetes 原生应用程序在无须为 Kubemces 编写特定代码的前提下，可以轻松访问后端。

service 同 RC 一样，都是通过 label 来关联 Pod 的。当在 service 的 YAML 文件中定义了该 service 的 selector 中的 label 为 app:jfedu-app 时，这个 service 会将 Pod->metadata->labels 中 label 为 app:jfedu-app 的 Pod 作为分发请求的后端。

当 Pod 发生变化（增加、减少、重建等）时，service 会及时更新。这样一来，service 就可以作为 Pod 的访问入口，起到代理服务器的作用，而对于访问者来说，通过 service 进行访问，无须关注后端 Pod 容器是否有变化或更新。

13.2　Kubernetes service 实现方式

Kubernetes 分配给 service 的固定 IP 是一个虚拟 IP，在外部是无法寻址的。在真实的系统实现上，Kubernetes 通过 kube-proxy 组件实现虚拟 IP 路由及转发。所以在之前集群部署的环节，在每个 node 上均部署了 Proxy 这个组件，从而实现了 Kubernetes 层级的虚拟转发网络。

Kubernetes 为每个 service 分配一个唯一的 ClusterIP，所以当使用 ClusterIP：port 的组合访问一个 service 时，不管 port 是什么，这个组合是不可能发生重复的。另一方面，kube-proxy 为每个 service 真正打开的是一个绝对不会重复的随机端口，用户在 service 描述文件中指定的访问端口会被映射到这个随机端口上。这就是为什么用户可以在创建 service 时随意指定访问端口。

在 Kubernetes 集群中，Pod 的 IP 是在 docker0 网段动态分配的，当进行重启、扩容等操作时，IP 地址会随之变化。当某个 Pod（frontend）需要访问其依赖的另外一组 Pod（backend）时，如果 backend 的 IP 发生变化，如何保证 frontend 到 backend 的正常通信变得非常重要，此时需要借助 service 实现统一访问。

service 的 Virtual IP 是由 Kubernetes 虚拟出来的内部网络，外部是无法寻址的。但是有些服务又需要被外部访问，如 Web 前段，这时就需要加一层网络转发，即外网到内网的转发。Kubernetes 提供了 ClusterIP、nodePort、LoadBalancer、Ingress 四种方式。

（1）ClusterIP：在 Kubernetes 平台创建容器时，选择内部服务，默认会创建一个 service，其 IP 类型为 ClusterIP，该 IP（virtual IP Address，虚拟 IP 地址）只能在 Kubernetes 集群内部使用。

（2）NodePort：Kubernetes 会在每个 node 上暴露出一个端口：NodePort，外部网络可以通过（任一 node）[NodeIP]:[NodePort]访问到后端的 service。

（3）LoadBalancer：在 NodePort 基础上，Kubernetes 可以请求底层云平台创建一个负载均衡器，将每个 node 作为后端进行服务分发。该模式需要底层云平台（如 GCE）支持。

（4）Ingress：是一种 HTTP 方式的路由转发机制，由 Ingress Controller 和 HTTP 代理服务器组合而成。Ingress Controller 实时监控 Kubernetes API，实时更新 HTTP 代理服务器的转发规则。HTTP 代理服务器有 GCE Load-Balancer、HAProxy、Nginx 等开源方案。

13.3 service 实战：ClusterIP 案例演练

ClusterIP 模式通常又称为内部服务，当然外部服务方式也有 ClusterIP。默认创建一个 service 内部服务，Kubernetes 将会在集群中生成一个 VIP 地址，该 VIP 是不能寻址的，只能集群内部访问，其原理是通过 kube-proxy 调用 iptables 防火墙规则添加 NAT 映射，此时用户在 Kubernetes 内部通过 VIP 地址就可以访问到 service 均衡后端的 Pod 容器服务。

ClusterIP 模式在每个 node 节点上不会配置 IP 地址，同时端口也不会显示，只能通过 Web 界面或者其他命令行指令查看。

（1）ClusterIP 内部服务案例演练配置实战如图 13-1 所示。选择内部网络，操作界面如下：

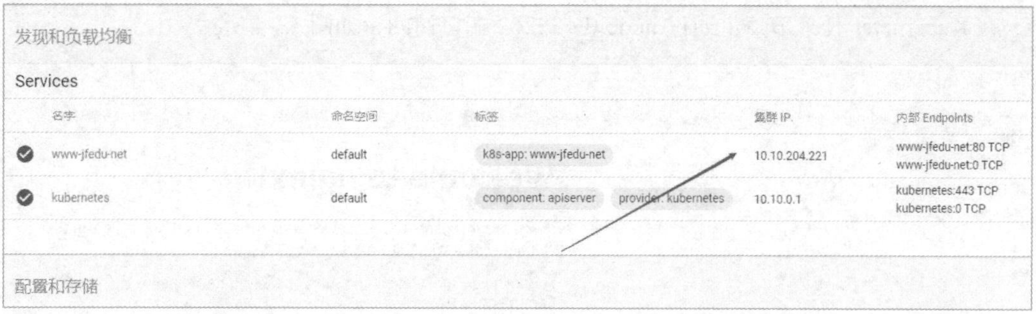

图 13-1　Kubernetes 创建容器设置为内部服务

（2）在 Kubernetes 任意节点上通过 Cluster IP+80 端口访问，如图 13-2 所示。

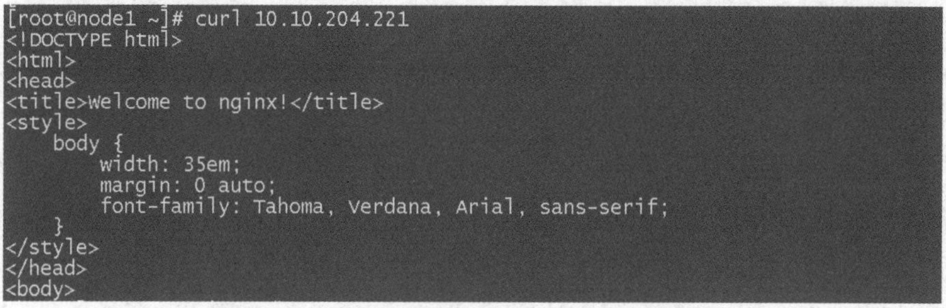

图 13-2　Kubernetes 内部服务 VIP 访问效果

13.4　service 实战：NodePort 案例演练

NodePort 模式下，Kubernetes 将会在每个 node 上打开一个端口，且每个 node 的端口都是一样的，通过 <NodeIP>:NodePort 的方式，Kubernetes 集群外部的程序可以访问 service。

NodePort 中每个 node 节点的端口有很多（1~65 535 个），Kubernetes 外部服务创建如图 13-3 所示。

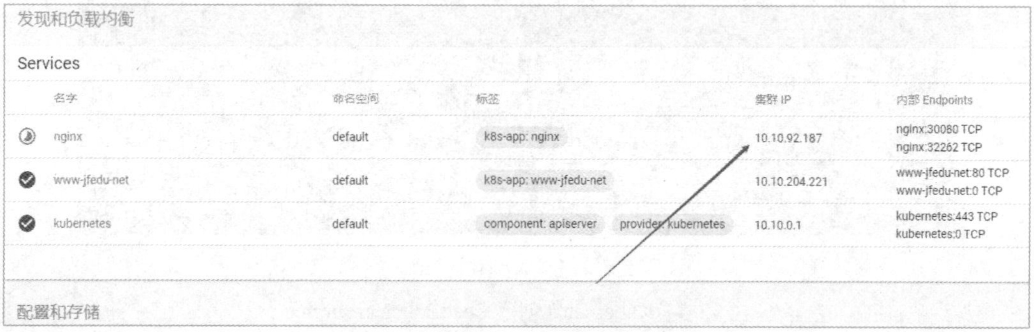

图 13-3　Kubernetes 外部服务创建

在 Kubernetes 任意节点上通过 node IP+32262 端口访问，如图 13-4 所示。

图 13-4　Kubernetes 外部服务访问和验证

13.5　service 实战：LoadBalancer 案例演练

LoadBalancer service 是 Kubernetes 深度结合云平台的一个组件；当使用 LoadBalancer service 暴露服务时，实际上是通过向底层云平台申请创建一个负载均衡器来向外暴露服务。

目前 LoadBalancer service 支持的云平台已经相对完善，如国外的 GCE、DigitalOcean，国内的阿里云，以及私有云 Openstack 等。由于 LoadBalancer service 深度结合了云平台，所以能在一些云平台上使用。

LoadBalancer 会分配 ClusterIP 和 NodePort，通过 Cloud Provider 实现 LoadBalancer 设备的配制，并且在 LoadBalancer 设备配置中将<NodeIP>:NodePort 作为 Pool Member，LoadBalancer 设备依据转发规则将流量转到节点的 NodePort，如图 13-5 所示。

创建 LoadBalancer service，操作指令如下，如图 13-5 所示。

```
kubectl expose deployment nginxv1 --port=8081 --target-port=80
--type=LoadBalancer
```

```
lhost ~]#
lhost ~]# kubectl expose deployment nginxv1 --port=8081 --target-port=80 --typ
ginxv1" exposed
lhost ~]#
lhost ~]#
lhost ~]#
lhost ~]#
lhost ~]# kubectl get svc
    CLUSTER-IP        EXTERNAL-IP      PORT(S)            AGE
    172.17.188.1      <none>           443/TCP            2d
    172.17.188.139    <pending>        8081:30608/TCP     21s
lhost ~]#
lhost ~]#
lhost ~]#
```

（a）

图 13-5　Kubernetes 创建 LoadBalancer service

（a）Kubernetes 暴露端口

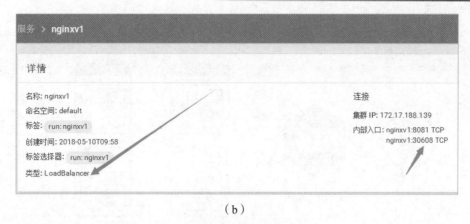

(b)

图 13-5　Kubernetes 创建 LoadBalancer service（续）

（b）查看 nginxv1 服务状态

13.6　service 实战：Ingress 案例演练

Ingress 是 Kubernetes 1.2 后才出现的，通过 Ingress 用户可以使用 Nginx 等开源的反向代理服务器实现对外暴露服务，后面 Traefik 用的也是 Ingress。使用 Ingress 通常需要以下 3 个组件。

（1）Ingress Controller：可以将 Ingress Controller 视作监视器，Ingress Controller 通过不断地与 Kubernetes API 交互，实时感知后端 service 和 Pod 的变化。当得到这些变化信息后，Ingress Controller 会跟 Ingress 生成相应的配置，然后将生成的配置刷新到反向代理服务器（Nginx）配置文件中，达到服务自动发现的作用。

（2）Ingress：Ingress 主要用来实现规则定义。例如，某个域名对应某个 service，即当某个域名的请求进来时转发给某个 service，这个规则将与 Ingress Controller 结合，然后 Ingress Controller 将其动态写入反向代理负载均衡器配置中，从而实现整体的服务发现和负载均衡。

（3）反向代理服务器（Nginx）：反向代理服务器种类很多，可以采用 Nginx、HAProxy、Apache 等，通常 Nginx 使用非常多。Nginx 的特点有轻量级、高性能、配置简单、管理便捷等，此处不需要另外部署一套 Nginx，因为在部署 Ingress Controller 服务时，会自动部署 Nginx。

用户访问某个网站域名时，请求首先会到达反向代理服务器，Ingress Controller 通过与 Ingress 交互得知某个域名对应哪个 service，再通过与 Kubernetes API 交互得知 service 地址等信息。综合以后，生成配置文件并实时写入反向代理服务器，然后反向代理服务器重新载入该规则便可实现服务发现，即动态映射，如图 13-6 所示。

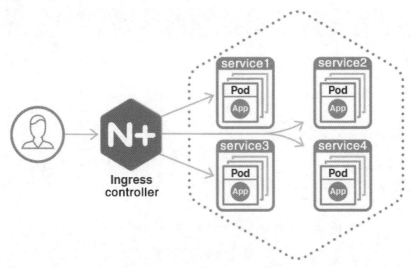

图 13-6 Nginx Ingress 内部结构图

Ingress Controller 通过和 Kubernetes API 交互，动态感知集群中 Ingress 规则变化，然后读取它，按照自定义的规则，规则就是写明了哪个域名对应哪个 service，生成一段 Nginx 配置。

再写到 Nginx-ingress-controller 的 Pod 中，这个 Ingress Controller 的 Pod 中运行着一个 Nginx 服务，控制器会把生成的 Nginx 配置写入/etc/nginx.conf 文件，然后重新载入使配置生效。以此达到域名分别配置和动态更新的目的。

无论如何请求，反向代理服务器的对外端口最终会暴露在固定的 node 上，同时以 Hostport 方式监听 80 端口，就解决了其他方式部署不确定反向代理服务器在哪儿的问题，同时访问每个 node 的 80 端口都能正确解析请求，如果前端再放置 Nginx，就又实现了一层负载均衡。

（1）部署 Ingress-nginx-controller 控制服务，从官网下载 YAML 配置文件，操作指令如下：

```
wget https://raw.githubusercontent.com/kubernetes/ingress-nginx/nginx-
0.30.0/deploy/static/mandatory.yaml
```

（2）修改文件 mandatory.yaml，在 213 行中加入如下代码：

```
hostNetwork: true    #在 Pod 中使用 hostNetwork:true 配置网络,Pod 中运行的应用程序
                     #可以直接看到宿主主机的网络接口,宿主主机所在的局域网上所有网络
                     #接口都可以访问该应用程序
```

（3）创建并应用 YAML 文件，操作指令如下：

```
kubectl apply -f mandatory.yaml
kubectl get pod -n ingress-nginx
```

（4）将 Ingress-nginx-controller 暴露为一个 service 资源对象，在命令行窗口打开文件 service-nodeport.yaml，YAML 代码如下：

```
apiVersion: v1
kind: Service
metadata:
  name: ingress-nginx
  namespace: ingress-nginx
  labels:
    app.kubernetes.io/name: ingress-nginx
    app.kubernetes.io/part-of: ingress-nginx
spec:
  type: NodePort
  ports:
    - name: http
      port: 80
      targetPort: 80
      protocol: TCP
    - name: https
      port: 443
      targetPort: 443
      protocol: TCP
  selector:
    app.kubernetes.io/name: ingress-nginx
    app.kubernetes.io/part-of: ingress-nginx
```

（5）执行上述 Ingress-nginx Service yaml 文件，操作指令如下：

```
kubectl apply -f service-nodeport.yaml
kubectl get svc -n ingress-nginx
```

（6）创建 ingress 规则 YAML 文件，关联 v1-jfedu-net 和 v2-jfedu-net 服务即可。ingress.yaml 配置文件代码如下：

```
apiVersion: extensions/v1beta1
kind: Ingress
metadata:
  name: v1-jfedu-net
spec:
  rules:
    - host: v1.jfedu.net
      http:
        paths:
        - path: /
          backend:
            serviceName: v1-jfedu-net
            servicePort: 80
---
apiVersion: extensions/v1beta1
kind: Ingress
metadata:
```

```
    name: v2-jfedu-net
spec:
  rules:
    - host: v2.jfedu.net
      http:
        paths:
          - path: /
            backend:
              serviceName: v2-jfedu-net
              servicePort: 80
```

（7）创建 Ingress 关联 v1-jfedu-net 和 v2-jfedu-net 服务，如图 13-7 所示，操作指令如下：

```
kubectl apply -f ingress.yaml
kubectl get ingresses
```

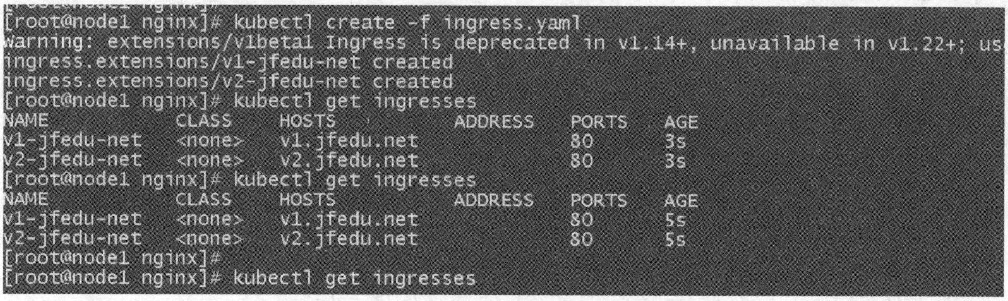

图 13-7　Kubernetes Nginx Ingress 配置实战

（8）分别创建两个应用部署和容器，并设置服务为内部服务，部署名称分别为 v1-jfedu-net 和 v2-jfedu-net，如图 13-8 所示。

（9）查看 Ingress server 服务对外监听的 node IP 端口为 80，在客户端 hosts 文件中添加 v1.jfedu.net、v2.jfedu.net 域名映射，通过浏览器访问，如图 13-9 所示。

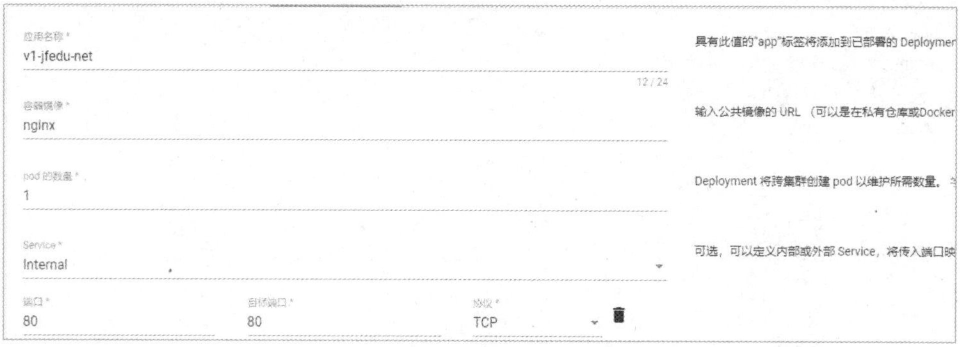

（a）

图 13-8　Kubernetes 创建应用和服务

（a）Kubernetes 创建 v1-jfedu-net 应用和服务

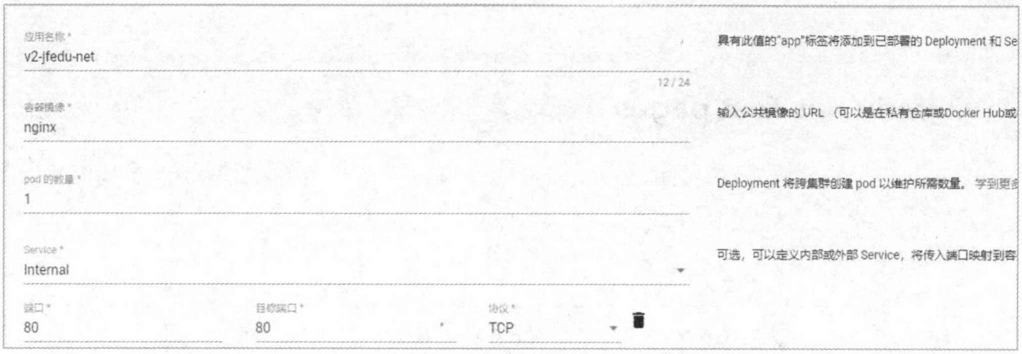

```
应用名称*
v2-jfedu-net                                         12 / 24

容器镜像*
nginx

pod 的数量*
1

Service*
Internal

端口*              目标端口*           协议*
80                 80                 TCP
```

（b）

图 13-8　Kubernetes 创建应用和服务（续）

（b）Kubernetes 创建 v2-jfedu-net 应用和服务

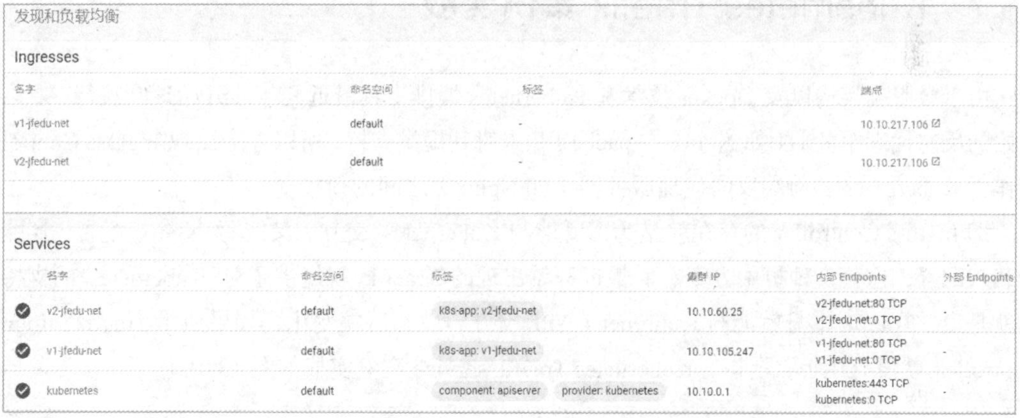

（a）

（b）

图 13-9　Kubernetes Nginx Ingress 案例实战

（a）查看 v1 和 v2 ingress 配置；（b）访问 v1.jfedu.net 测试页面

（c）

图 13-9　Kubernetes Nginx Ingress 案例实战（续）

（c）访问 v2.jfedu.net 测试页面

13.7　Kubernetes Traefik 案例实战

由于微服务架构以及 Docker 技术和 Kubernetes 编排工具最近几年才开始逐渐流行，所以一开始的反向代理服务器（如 Nginx、Apache）并未对其提供支持，所以才会出现 Ingress Controller 来作为 Kubernetes 和前端反向代理服务器（如 Nginx）之间的衔接。

即 Ingress Controller 的存在就是为了既能与 Kubernetes 交互，又能写 Nginx 配置，还能重新载入配置，这是一种折中方案。而最近开始出现的 Traefik 就提供了对 Kubernetes 的支持，也就是说，Traefik 本身就能与 Kubernetes API 交互，感知后端变化，如果使用 Traefik，Ingress Controller 就没有用了，所以，Kubernetes Traefik service 案例实战如图 13-10 所示。

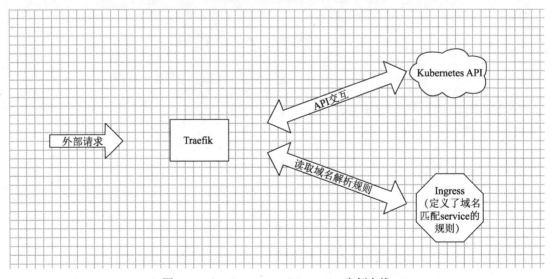

图 13-10　Kubernetes Traefik service 案例实战

Kubernetes Traefik 在企业生产环境中也被广泛采用，部署 Traefik service 也非常简单，操作的方法和步骤如下。

（1）创建 Traefik rbac 验证配置，rbac.yaml 文件内容如下：

```yaml
---
kind: ClusterRole
apiVersion: rbac.authorization.Kubernetes.io/v1beta1
metadata:
  name: traefik-ingress-controller
rules:
  - apiGroups:
      - ""
    resources:
      - services
      - endpoints
      - secrets
    verbs:
      - get
      - list
      - watch
  - apiGroups:
      - extensions
    resources:
      - ingresses
    verbs:
      - get
      - list
      - watch
  - apiGroups:
      - extensions
    resources:
      - ingresses/status
    verbs:
      - update
---
kind: ClusterRoleBinding
apiVersion: rbac.authorization.Kubernetes.io/v1beta1
metadata:
  name: traefik-ingress-controller
roleRef:
  apiGroup: rbac.authorization.Kubernetes.io
  kind: ClusterRole
  name: traefik-ingress-controller
subjects:
- kind: ServiceAccount
  name: traefik-ingress-controller
```

namespace: kube-system
```

（2）以 DaemonSet 的方式在每个 node 上启动一个 Traefik，并使用 hostPort 的方式让其监听每个 node 的 80 端口。创建部署 YAML 文件 traefik-depolyment.yaml，代码如下：

```yaml
---
apiVersion: v1
kind: ServiceAccount
metadata:
  name: traefik-ingress-controller
  namespace: kube-system
---
kind: DaemonSet
apiVersion: apps/v1
metadata:
  name: traefik-ingress-controller
  namespace: kube-system
  labels:
    Kubernetes-app: traefik-ingress-lb
spec:
  selector:
    matchLabels:
      Kubernetes-app: traefik-ingress-lb
      name: traefik-ingress-lb
  template:
    metadata:
      labels:
        Kubernetes-app: traefik-ingress-lb
        name: traefik-ingress-lb
    spec:
      serviceAccountName: traefik-ingress-controller
      terminationGracePeriodSeconds: 60
      containers:
      - image: traefik:v1.7
        name: traefik-ingress-lb
        ports:
        - name: http
          containerPort: 80
          hostPort: 80
        - name: admin
          containerPort: 8080
          hostPort: 8080
        securityContext:
          capabilities:
            drop:
            - ALL
            add:
```
```

```yaml
        - NET_BIND_SERVICE
      args:
      - --api
      - --kubernetes
      - --logLevel=INFO
---
kind: Service
apiVersion: v1
metadata:
  name: traefik-ingress-service
  namespace: kube-system
spec:
  selector:
    Kubernetes-app: traefik-ingress-lb
  ports:
    - protocol: TCP
      port: 80
      name: web
    - protocol: TCP
      port: 8080
      name: admin
```

（3）以 Deployment 的方式启动一个 Traefik，并使用 hostPort 的方式让其监听每个 node 的 80 端口。创建部署 YAML 文件 traefik-depolyment.yaml，代码如下：

```yaml
---
apiVersion: v1
kind: ServiceAccount
metadata:
  name: traefik-ingress-controller
  namespace: kube-system
---
kind: Deployment
apiVersion: apps/v1
metadata:
  name: traefik-ingress-controller
  namespace: kube-system
  labels:
    Kubernetes-app: traefik-ingress-lb
spec:
  replicas: 1
  selector:
    matchLabels:
      Kubernetes-app: traefik-ingress-lb
  template:
    metadata:
      labels:
```

```
      Kubernetes-app: traefik-ingress-lb
      name: traefik-ingress-lb
   spec:
     serviceAccountName: traefik-ingress-controller
     terminationGracePeriodSeconds: 60
     containers:
     - image: traefik:v1.7
       name: traefik-ingress-lb
       ports:
       - name: http
         containerPort: 80
       - name: admin
         containerPort: 8080
       args:
       - --api
       - --kubernetes
       - --logLevel=INFO
---
kind: Service
apiVersion: v1
metadata:
  name: traefik-ingress-service
  namespace: kube-system
spec:
  type: NodePort
  selector:
    Kubernetes-app: traefik-ingress-lb
  ports:
    - protocol: TCP
      port: 80
      name: web
      targetPort: 80
    - protocol: TCP
      port: 8080
      name: admin
```

其中 Traefik 监听 node 的 80 和 8080 端口，80 提供正常服务，8080 是其自带的用户界面。

（4）部署 Ingress 规则，Ingress Controller 是无须部署的，所以直接部署 Ingress，编写 traefik-ingress.yaml 文件，文件代码如下：

```
apiVersion: extensions/v1beta1
kind: Ingress
metadata:
  name: v1-jfedu-net
spec:
  rules:
```

```
    - host: v1.jfedu.net
      http:
        paths:
        - path: /
          backend:
            serviceName: v1-jfedu-net
            servicePort: 80
---
apiVersion: extensions/v1beta1
kind: Ingress
metadata:
  name: v2-jfedu-net
spec:
  rules:
    - host: v2.jfedu.net
      http:
        paths:
        - path: /
          backend:
            serviceName: v2-jfedu-net
            servicePort: 80
```

实际上，因为集群中已经存在了相应的名为 v1-jfedu-net 和 v2-jfedu-net 的 service，对应的 service 后端也有很多 Pod，所以这里就不再具体介绍部署实际业务容器（v1-jfedu-net、v2-jfedu-net）的过程了，测试的时候，只需要把这两个 service 替换成自己业务的 service 即可。

（5）部署 Traefik UI。Traefik 本身还提供了一套用户界面，同样以 Ingress 方式暴露，只需创建 traefik-ui.yaml 文件即可。traefik-ui.yaml 文件内容如下：

```
---
apiVersion: v1
kind: Service
metadata:
  name: traefik-web-ui
  namespace: kube-system
spec:
  selector:
    Kubernetes-app: traefik-ingress-lb
  ports:
  - name: web
    port: 80
    targetPort: 8080
---
apiVersion: extensions/v1beta1
kind: Ingress
```

```yaml
metadata:
  name: traefik-web-ui
  namespace: kube-system
spec:
  rules:
  - host: traefik-ui.minikube
    http:
      paths:
      - path: /
        backend:
          serviceName: traefik-web-ui
          servicePort: web
```

（6）创建两个应用部署和容器，并且设置服务为内部服务，部署名称分别为 v1-jfedu-net 和 v2-jfedu-net，操作如图 13-11 所示。

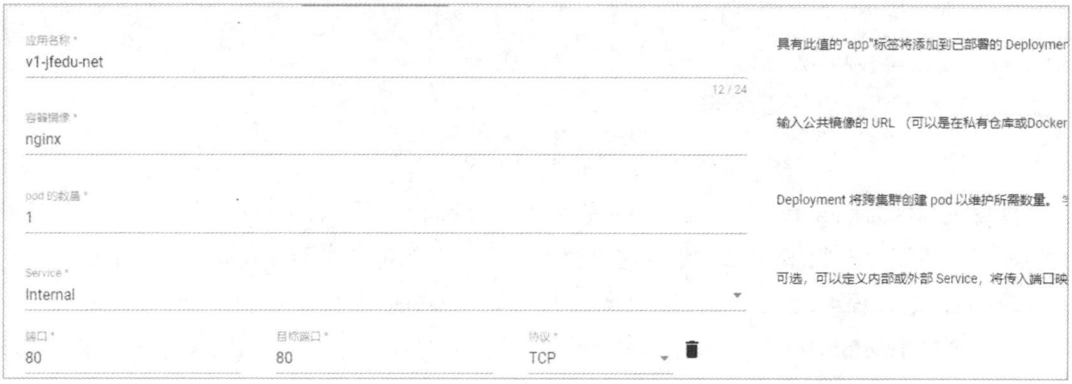

（a）

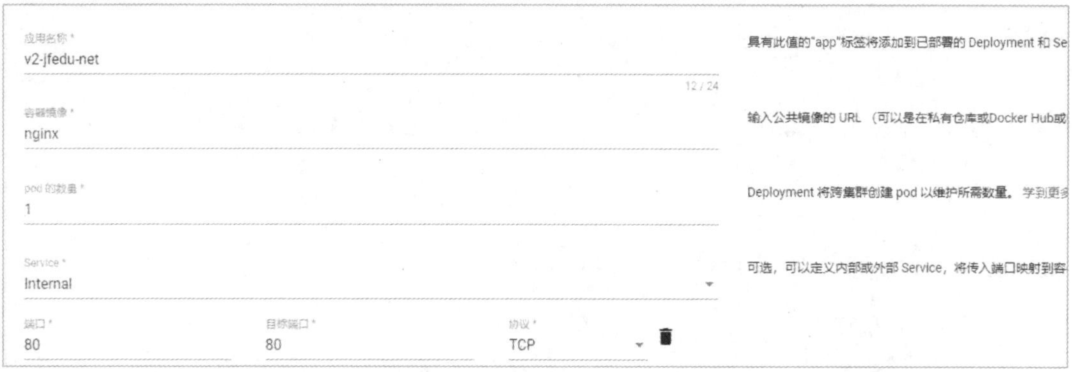

（b）

图 13-11　创建两个内部服务

（a）Kubernetes 创建内部服务 v1-jfedu-net；（b）Kubernetes 创建内部服务 v2-jfedu-net

（7）经过以上操作，Kubernetes Traefik service 所需配置文件均创建完成，执行以下指令使其生效即可，操作指令如下：

```
kubectl apply -f rbac.yaml
kubectl apply -f traefik-depolyment.yaml
kubectl apply -f traefik-ingress.yaml
kubectl apply -f traefik-ui.yaml
```

（8）查看 Traefik Web 服务对外监听的 node IP 端口为 32405（需要修改），在客户端 hosts 文件中添加 v1.jfedu.net、v2.jfedu.net 域名映射，通过浏览器访问，如图 13-12 所示。

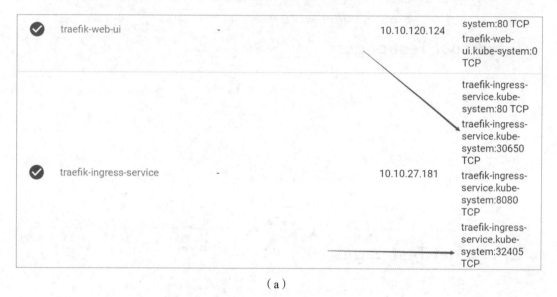

（a）

（b）

图 13-12　Kubernetes Traefik 案例实战 1

（a）查看 Traefik nodeport 端口；（b）访问 Traefik Web 界面

（9）查看以上 service，NodePort 端口为 30650，即访问虚拟主机需要带上该端口。如何将其端口修改为 80 端口呢？修改 master 节点上的 API server 配置文件/etc/kubernetes/manifests/kube-apiserver.yaml，在 command 字段 kube-apiserver 下添加如下代码：

```
- --service-node-port-range=1-65535
```

（10）分别设置两个部署应用的网站内容，在客户端添加 hosts 绑定到任意一个 node 节点 IP，浏览器访问两个域名，如图 13-13 所示。

图 13-13　Kubernetes Traefik 案例实战 2

（a）v1.jfedu.net 测试页面；（b）v2.jfedu.net 测试页面

第 14 章 Kubernetes 容器升级实战

14.1 Kubernetes 容器升级概念

传统的网站升级更新,通常是将服务全部下线,业务停止后再更新版本和配置,然后重新启动并提供服务。这样的模式已经完全不能满足发展需求了。

高并发、高可用系统普及的今天,服务的升级更新至少要做到"业务不中断"。而滚动更新(Rolling-update)恰好是满足这一需求的系统更新升级方案。

滚动更新就是针对多实例服务的一种不中断服务的更新升级方式。一般情况下,对于多实例服务,滚动更新采用对各个实例逐个进行单独更新,而非同一时刻对所有实例进行全部更新的方式。

对于 Kubernetes 集群部署的 service 来说,Rolling-update 就是指一次仅更新一个 Pod,然后逐个进行更新,而不是在同一时刻将该 service 下面的所有 Pod 关闭,然后更新。逐个更新可以避免业务中断。

Kubernetes 在 kubectl cli 工具中仅提供了对 Replication controller 的 Rolling-update 支持,通过命令 kubectl -help 查看指令信息,如图 14-1 所示。

14.2 Kubernetes 容器升级实现方式

(1)查看部署列表,获取部署应用名称。

```
kubectl get deployments -n default
```

(2)查看正在运行的 Pod。

```
kubectl get pods -n default
```

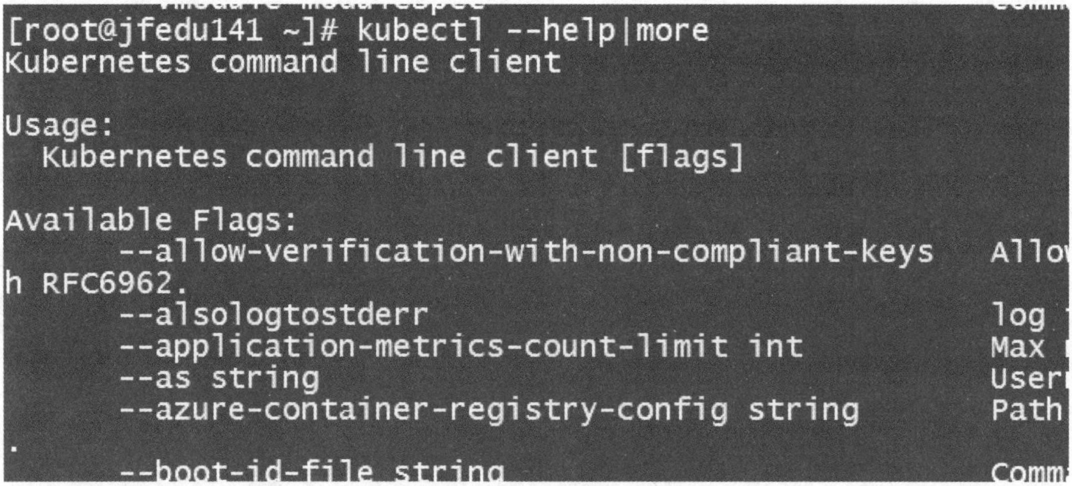

图 14-1 kubectl 指令帮助信息

（3）通过 Pod 描述，查看部署程序的当前映像版本。

```
kubectl describe pods -n default
```

（4）仓库源中提前制作最新更新的镜像，执行如下指令，升级镜像版本即可。升级之前一定要保证仓库源中有最新提交的镜像，代码如下，如图 14-2 所示。

```
kubectl -n default set image deployments/nginx-v1 nginx-v1=docker.io/nginx:v1
```

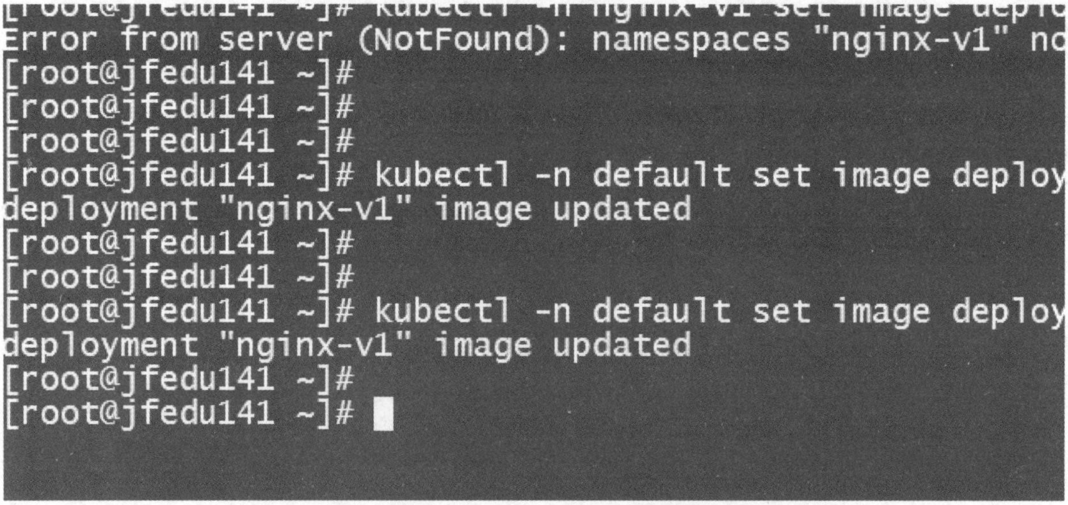

图 14-2 kubectl 更新部署应用镜像

14.3 Kubernetes 容器升级测试

（1）创建 nginx-v1 部署，查看 nginx-v1 部署的容器列表，如图 14-3 所示。

图 14-3 nginx-v1 部署容器列表

（a）创建 nginx-v1 部署；（b）查看 nginx-v1 部署的容器列表

（2）镜像更新完成之后，查看 nginx-v1 部署的容器列表，如图 14-4 所示。

图 14-4　nginx-v1 部署的容器列表

（a）查看 nginx-v1 部署和容器组；（b）查看 nginx-v1 使用的镜像

14.4　Kubernetes 容器升级验证

（1）检查 Kubernetes 更新 rollout 状态，操作指令如下：

```
kubectl -n default rollout status deployments/nginx-v1
```

（2）检查 Kubernetes Pod 详情，操作指令如下：

```
kubectl describe pods -n default
```

14.5 Kubernetes 容器升级回滚

（1）Kubernetes 部署镜像回滚，操作指令如下：

```
kubectl -n default rollout undo deployments/nginx-v1
```

（2）查看已经部署的版本，操作指令如下：

```
kubectl rollout history deploy/nginx-v1
```

（3）查看某个版本详细信息，如图 14-5 所示，操作指令如下：

```
kubectl rollout history deployment/nginx-v1 --revision=8
```

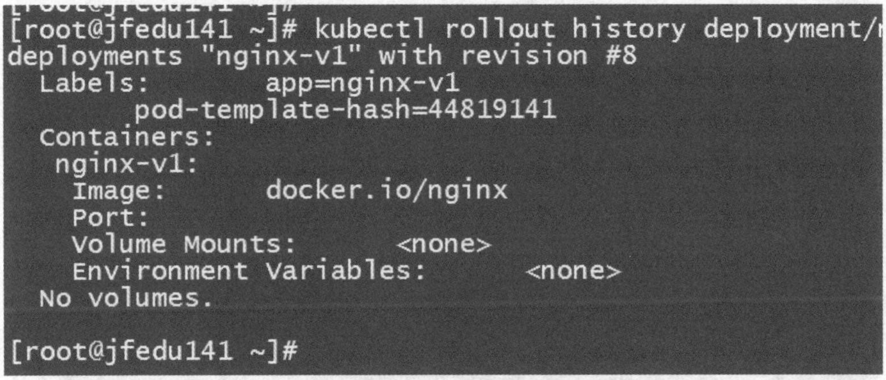

```
[root@jfedu141 ~]# kubectl rollout history deployment/n
deployments "nginx-v1" with revision #8
  Labels:         app=nginx-v1
                  pod-template-hash=44819141
  Containers:
   nginx-v1:
    Image:       docker.io/nginx
    Port:
    Volume Mounts:          <none>
    Environment Variables:  <none>
  No volumes.

[root@jfedu141 ~]#
```

图 14-5　Kubernetes 滚动升级查看版本详细信息

（4）Kubernetes 完美支持回滚至某个版本，还可以通过资源文件进行配置保留的历史版次量。Kubernetes 回滚某个版本，如图 14-6 所示，操作指令如下：

```
kubectl -n default rollout undo deployment/nginx-v1  --to-revision=8
```

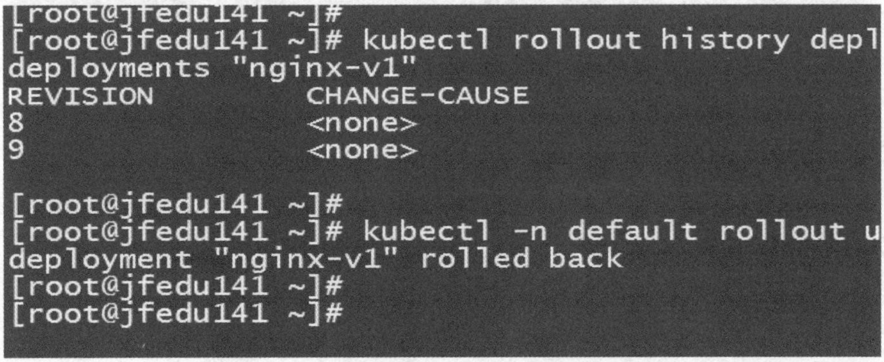

```
[root@jfedu141 ~]#
[root@jfedu141 ~]# kubectl rollout history depl
deployments "nginx-v1"
REVISION            CHANGE-CAUSE
8                   <none>
9                   <none>

[root@jfedu141 ~]#
[root@jfedu141 ~]# kubectl -n default rollout u
deployment "nginx-v1" rolled back
[root@jfedu141 ~]#
[root@jfedu141 ~]#
```

图 14-6　Kubernetes 镜像滚动至指定版本

14.6　Kubernetes 滚动升级和回滚原理

Kubernetes 精确地控制着整个发布过程，分批次有序地进行滚动更新，直到把所有旧的副本全部更新到新版本。实际上 Kubernetes 是通过两个参数精确控制每次滚动的 Pod 数量。

（1）maxSurge：滚动更新过程中运行操作期望副本数的最大 Pod 数量，可以为绝对数值（如5），也可以为百分数（如 10%），但不能为 0，默认为 25%。

（2）maxUnavailable：滚动更新过程中不可用的最大 Pod 数量，可以为绝对数值（如 5），也可以为百分数（如 10%），但不能为 0，默认为 25%。

如果未指定这两个可选参数，则 Kubernetes 会使用默认配置，查找默认配置指令，如图 14-7 所示，操作指令如下：

```
kubectl -n default get deployment nginx-v1 -o yaml
```

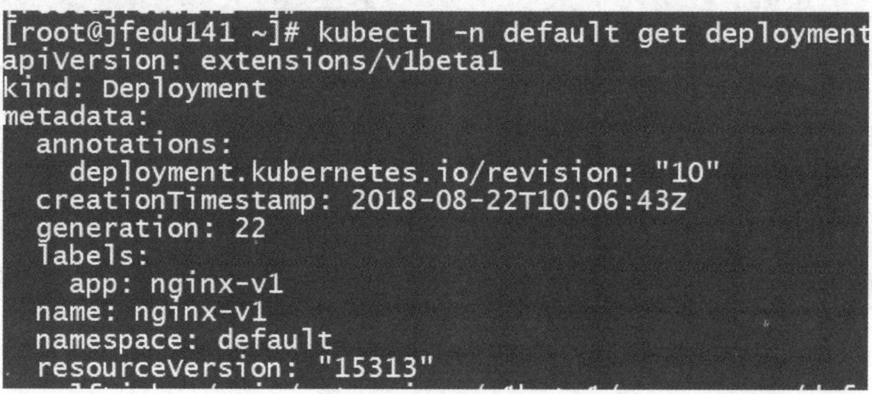

```
[root@jfedu141 ~]# kubectl -n default get deployment
apiVersion: extensions/v1beta1
kind: Deployment
metadata:
  annotations:
    deployment.kubernetes.io/revision: "10"
  creationTimestamp: 2018-08-22T10:06:43Z
  generation: 22
  labels:
    app: nginx-v1
  name: nginx-v1
  namespace: default
  resourceVersion: "15313"
```

（a）

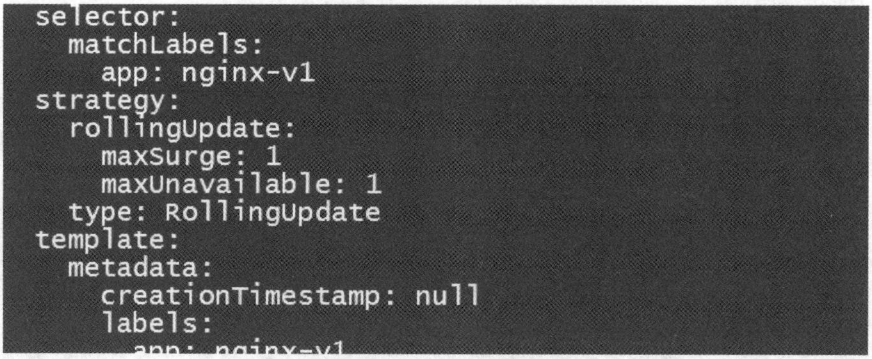

```
  selector:
    matchLabels:
      app: nginx-v1
  strategy:
    rollingUpdate:
      maxSurge: 1
      maxUnavailable: 1
    type: RollingUpdate
  template:
    metadata:
      creationTimestamp: null
      labels:
        app: nginx-v1
```

（b）

图 14-7　Kubernetes nginx-v1 部署 YAML 文件

（a）获取 nginx-v1 部署日志；（b）查看 nginx-v1 部署详情

（1）查看 Kubernetes nginx-v1 应用部署的概况，如图 14-8 所示。

```
[root@jfedu141 ~]#
[root@jfedu141 ~]# kubectl get deployments -n default
NAME        DESIRED   CURRENT   UP-TO-DATE   AVAILABLE   AGE
nginx-v1    10        10        10           8           5h
[root@jfedu141 ~]#
[root@jfedu141 ~]#
[root@jfedu141 ~]# kubectl get deployments -n default
NAME        DESIRED   CURRENT   UP-TO-DATE   AVAILABLE   AGE
nginx-v1    10        10        10           9           5h
[root@jfedu141 ~]#
[root@jfedu141 ~]#
[root@jfedu141 ~]# kubectl get deployments -n default
NAME        DESIRED   CURRENT   UP-TO-DATE   AVAILABLE   AGE
nginx-v1    10        10        10           10          5h
[root@jfedu141 ~]#
```

图 14-8　Kubernetes nginx-v1 应用部署概况

其中，部分参数含义如下。

① DESIRED：最终期望处于 READY 状态的副本数。

② CURRENT：当前的副本总数。

③ UP-TO-DATE：当前完成更新的副本数。

④ AVAILABLE：当前可用的副本数。

（2）查看 Kubernetes nginx-v1 应用镜像更新的效果，如图 14-9 所示，操作指令如下：

```
kubectl -n default describe deployment nginx-v1
```

```
# kubectl -n default describe deployment nginx-v1
        nginx-v1
        default
        Wed, 22 Aug 2018 18:06:43 +0800
        app=nginx-v1
        app=nginx-v1
        10 updated | 10 total | 10 available | 0 unavail
        RollingUpdate
        0
tegy:   1 max unavailable, 1 max surge

tatus   Reason
------  ------
rue     MinimumReplicasAvailable
```

图 14-9　Kubernetes nginx-v1 应用镜像更新效果

（3）整个滚动过程是通过控制两个副本集完成的，分别是新副本和旧副本，如图 14-10 所示。名称如下。

① 新副本集：nginx-v1-224846633。

② 旧副本集：nginx-v1-44819141。

副本集		
名称	标签	容器组
✓ nginx-v1-44819141	app: nginx-v1 pod-template-hash: 44819141	0 / 0
✓ nginx-v1-224846633	app: nginx-v1 pod-template-hash: 224846633	10 / 10

服务

图 14-10　Kubernetes nginx-v1 应用副本查看

（4）理想状态下的 Kubernetes 镜像更新滚动的过程如下：

① 创建 1 个新的副本集，并为其分配 3 个新版本的 Pod，使副本总数达到 11，一切正常。

② 通知旧副本集，销毁 2 个旧版本的 Pod，使可用副本总数保持到 8，一切正常。

③ 当 2 个副本销毁成功后，通知新副本集，再新增 2 个新版本的 Pod，使副本总数达到 11，一切正常。

④ 只要销毁成功，新副本集就会创造新的 Pod，一直循环，直到旧的副本集 Pod 数量为 0。

（5）滚动升级一个服务，实际是创建一个新的 RS，然后逐渐将新 RS 中的副本数增加到理想状态，将旧 RS 中的副本数减小到 0 的复合操作。

（6）无论是否理想，Kubernetes 最终都会使应用程序全部更新到期望状态，并保持最大的副本总数和可用副本总数的不变性。

第 15 章 Kubernetes+NFS 持久化存储实战

15.1 Kubernetes 服务运行状态

Kubernetes 运行的服务，从简单到复杂可以分成三类：无状态服务、普通有状态服务和有状态集群服务。下面分别介绍 Kubernetes 是如何运行这三类服务的。

（1）无状态服务。

Kubernetes 使用 RC（或更新的 ReplicaSet）保证一个服务的实例数量，如果某个 Pod 实例由于某种原因崩溃了，RC 会立刻用这个 Pod 的模板启动一个 Pod 替代它，由于是无状态服务，新启动的 Pod 与原来健康状态下的 Pod 一模一样。在 Pod 被重建后，它的 IP 地址可能发生变化，为了对外提供一个稳定的访问接口，Kubernetes 引入了 service。一个 service 后面可以挂多个 Pod，从而实现服务的高可用。

（2）普通有状态服务。

与无状态服务相比，它多了状态保存的需求。Kubernetes 提供了以 Volume 和 Persistent Volume 为基础的存储系统，可以实现服务的状态保存。

（3）有状态集群服务。

与普通有状态服务相比，它多了集群管理的需求。Kubernetes 为此开发了一套以 PetSet 为核心的全新特性，方便了有状态集群服务在 Kubernetes 上的部署和管理。具体来说是通过 Init Container 做集群的初始化工作，用无头服务（Headless Service）维持集群成员的稳定关系，用动态存储供给方便集群扩容，最后用 PetSet 综合管理整个集群。

15.2 Kubernetes 存储系统

Kubernetes 的存储系统从基础到高级又大致分为三个层次：普通存储卷、持久存储卷

（Persistent Volume）和动态存储供应（Dynamic Provisioning）。

（1）普通存储卷之单节点存储卷。

单节点存储卷是最简单的普通存储卷，它和 Docker 的存储卷类似，使用的是 Pod 所在 Kubernetes 节点的本地目录。具体有两种，一种是 emptyDir，是一个匿名的空目录，由 Kubernetes 在创建 Pod 时创建，删除 Pod 时删除；另外一种是 hostPath，与 emptyDir 的区别在于，后者在 Pod 之外独立存在，由用户指定路径名。这类和节点绑定的存储卷在 Pod 迁移到其他节点后数据就会丢失，所以只能用于存储临时数据或用于在同一个 Pod 里的容器之间共享数据。普通 Volume 目前支持的各种存储插件及情况如图 15-1 所示。

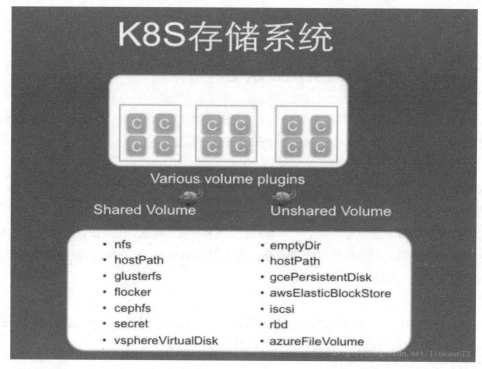

图 15-1　普通 Volume 目前支持的各种存储插件及情况

（2）持久存储卷。

普通存储卷和使用它的 Pod 之间是一种静态绑定关系，在定义 Pod 的文件里，同时定义了 Pod 使用的存储卷。存储卷是 Pod 的附属品，无法单独创建一个存储卷，因为它不是一个独立的 Kubernetes 资源对象。

而持久存储卷（Persistent Volume，PV）是一个 Kubernetes 资源对象，所以可以单独创建一个 PV。它不和 Pod 直接发生关系，而是通过 Persistent Volume Claim（PV 索取，PVC）实现动态绑定。Pod 定义里指定的是 PVC，然后 PVC 会根据 Pod 的要求自动绑定合适的 PV 给 Pod 使用。

(3)动态存储供应。

与普通存储卷类似,这里不再赘述。

15.3 Kubernetes 存储绑定的概念

用户根据所需存储空间大小和访问模式创建(或在动态部署中已创建)一个 PVC。

Kubernetes 的 master 节点循环监控新产生的 PVC,找到与之匹配的 PV(如果有),并把它们绑定在一起。

动态配置时,循环会一直将 PV 与这个 PVC 绑定,直到 PV 完全匹配 PVC,以避免 PVC 请求和得到的 PV 不一致。绑定一旦形成,PVC 绑定就是专有的,不管是使用何种模式绑定的。

如果找不到匹配的存储卷,用户请求会一直保持未绑定状态。在匹配的存储卷可用之后,用户请求将会被绑定。例如,一个配置很多 50Gi PV 的集群不会匹配到一个要求 100Gi 的 PVC。只有在 100Gi PV 被加到集群之后,这个 PVC 才可以被绑定。

15.4 PV 的访问模式

Kubernetes PV 访问模式通常分为 3 种:ReadWriteOnce 是最基本的方式,可读可写,但只支持被单个 Pod 挂载;ReadOnlyMany 以只读的方式被多个 Pod 挂载;ReadWriteMany 可以以读写的方式被多个 Pod 共享。不是每一种存储都支持这 3 种方式。

在 PVC 绑定 PV 时通常根据两个条件绑定:一个是存储的大小,另一个就是访问模式。Kubernetes PV 种类和访问类型如图 15-2 所示。访问模式全称及简写如下。

(1)ReadWriteOnce:RWO。

(2)ReadOnlyMany:ROX。

(3)ReadWriteMany:RWX。

PV 的生命周期首先是 Provision,即创建 PV。这里创建 PV 有两种方式,分别为静态创建和动态创建。

(1)静态创建:管理员手动创建一批 PV,组成一个 PV 池,供 PVC 绑定。

(2)动态创建:在现有 PV 不满足 PVC 的请求时,可以使用存储分类(StorageClass)。PV 先创建分类,PVC 请求已创建的某个类(StorageClass)的资源,这样就达到动态配置的效果,即通过一个叫 StorageClass 的对象由存储系统根据 PVC 的要求自动创建。

PV 创建完后状态会变成 Available,等待被 PVC 绑定。一旦被 PVC 绑定,PV 的状态会变成 Bound,就可以被定义了相应 PVC 的 Pod 使用。Pod 使用完后会释放 PV,PV 的状态变成 Released。变成 Released 的 PV 会根据定义的回收策略做相应的回收工作。

Volume Plugin	ReadWriteOnce	ReadOnlyMany	ReadWriteMany
AWSElasticBlockStore	✓	-	-
AzureFile	✓	✓	✓
AzureDisk	✓	-	-
CephFS	✓	✓	✓
Cinder	✓	-	-
FC	✓	✓	-
FlexVolume	✓	✓	-
Flocker	✓	-	-
GCEPersistentDisk	✓	✓	-
Glusterfs	✓	✓	✓
HostPath	✓	-	-
iSCSI	✓	✓	-
PhotonPersistentDisk	✓	-	-
Quobyte	✓	✓	✓
NFS	✓	✓	✓
RBD	✓	✓	-

图 15-2　Kubernetes PV 种类和访问类型

有 3 种回收策略：Retain、Delete 和 Recycle。Retain 策略就是保留现场，Kubernetes 什么也不做，等待用户手动处理 PV 中的数据，处理完后，再手动删除 PV。Delete 策略，Kubernetes 会自动删除该 PV 及其中的数据。Recycle 策略，Kubernetes 会将 PV 中的数据删除，然后把 PV 的状态变成 Available，又可以被新的 PVC 绑定使用。

在实际的使用场景中，PV 的创建和使用通常不是同一个用户完成。这里有一个典型的应用场景：管理员创建一个 PV 池，开发人员创建 Pod 和 PVC，PVC 里定义了 Pod 所需存储的大小和访问模式，然后 PVC 会到 PV 池里自动匹配最合适的 PV 给 Pod 使用。

15.5　Kubernetes+NFS 静态存储模式

下面为静态 PV 存储操作方式。

基于 Linux 平台构建 NFS 网络文件系统，操作指令如下：

```
#安装 NFS 文件服务
yum install nfs-utils -y
#配置共享目录和权限
vim /etc/exports
/data/ *(rw,async,no_root_squash)
#启动 NFS 服务
service nfs restart
```

创建 PV 文件，pv.yaml 文件内容如下：

```
cat>pv.yaml<<EOF
apiVersion: v1
kind: PersistentVolume
metadata:
  name: nfs-pv
  namespace: default
spec:
  capacity:
    storage: 10G
  accessModes:
    - ReadWriteMany
  nfs:
    # FIXME: use the right IP
    server: 192.168.1.147
    path: /data/
EOF
```

PV 配置参数如下：

Capacity 指定 PV 的容量为 100MB。

accessModes 指定访问模式为 ReadWritMany，支持的访问模式有：

（1）ReadWriteOnce 表示 PV 能以 read-write 模式挂载到单个节点。

（2）ReadOnlyMany 表示 PV 能以 read-only 模式挂载到多个节点。

（3）ReadWriteMany 表示 PV 能以 read-write 模式挂载到多个节点。

persistentVolumeReclaimPolicy 指定当前 PV 的回收策略为 Recycle，支持的策略有：

（1）Retain 表示需要管理员手动回收。

（2）Recycle 表示清除 PV 中的数据，效果相当于执行 rm -rf /thevolume/*。

（3）Delete 表示删除 Storage Provider 上的对应存储资源，如 AWS EBS、GCE PD、Azure、Disk、OpenStack Cinder Volume 等。

storageClassName 指定 PV 的 class 为 NFS。相当于为 PV 设置了一个分类，PVC 可以指定 class 申请相应 class 的 PV。

15.6　PVC 存储卷创建

（1）创建 PVC，pvc.yaml 文件内容如下：

```
cat>pvc.yaml<<EOF
apiVersion: v1
kind: PersistentVolumeClaim
metadata:
```

```
    name: nfs-pvc
    namespace: default
spec:
  accessModes:
    - ReadWriteMany
  storageClassName: ""
  resources:
    requests:
      storage: 10G
EOF
```

（2）创建 PV、PVC 之后，需要通过 kubectl apply -f 命令读取 YAML 配置，使其部署生效。最终效果如图 15-3 所示。

图 15-3　Kubernetes PV、PVC 案例实战

15.7　Nginx 整合 PV 存储卷

（1）创建 Nginx Pod 容器，使用 PVC，nginx.yaml 文件内容如下：

```
cat>nginx.yaml<<EOF
apiVersion: v1
kind: ReplicationController
metadata:
  name: nginx-v1
  labels:
    name: nginx-v1
  namespace: default
spec:
  replicas: 1
  selector:
    name: nginx-v1
  template:
    metadata:
```

```
      labels:
        name: nginx-v1
    spec:
      containers:
      - name: nginx-v1
        image: nginx
        volumeMounts:
        - mountPath: /usr/share/nginx/html
          name: nginx-data
        ports:
        - containerPort: 80
      volumes:
      - name: nginx-data
        persistentVolumeClaim:
          claimName: nfs-pvc
EOF
```

（2）登录 node 节点，查看 NFS PV 资源是否挂载，如图 15-4 所示。

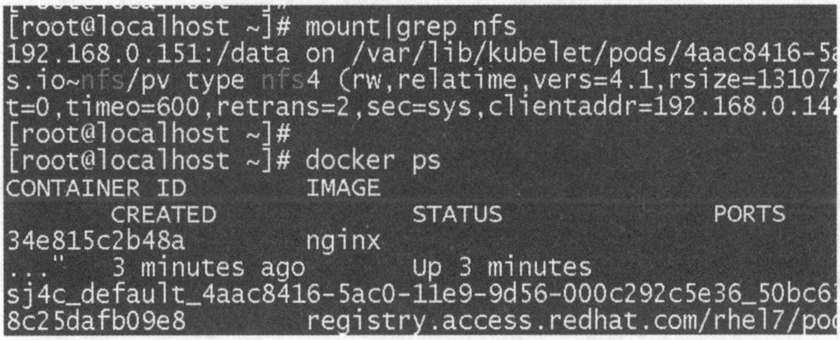

图 15-4　Kubernetes PV、PVC 案例测试 1

（3）测试 Kubernetes Pod NFS 与 NFS 服务器数据是否一致，如图 15-5 所示。

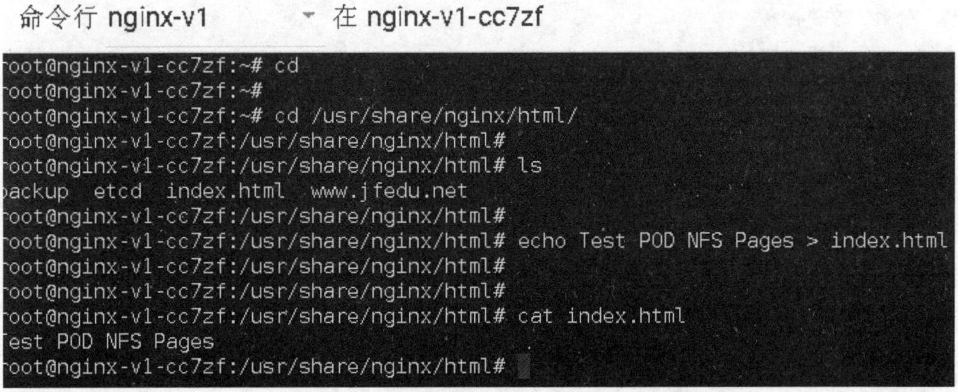

图 15-5　Kubernetes PV、PVC 案例测试 2

15.8　Kubernetes+NFS 动态存储模式

动态创建 PV，是指在现有 PV 不满足 PVC 的请求时，可以使用存储分类（StorageClass），具体过程是：PV 先创建分类，PVC 请求已创建的某个类（StorageClass）的资源，这样就达到动态配置的效果，即通过一个叫 StorageClass 的对象由存储系统根据 PVC 的要求自动创建。

其中动态方式是通过 StorageClass 完成的，这是一种新的存储供应方式。动态卷供给能力让管理员不必预先创建存储卷，而是随用户需求而创建，如图 15-6 所示。

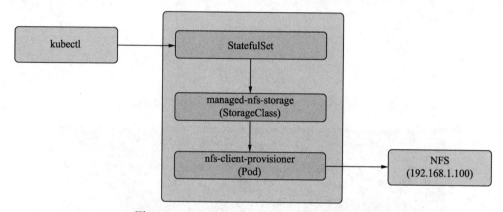

图 15-6　Kubernetes PVC 动态存储流程示意图

使用 StorageClass 有什么好处呢？StorageClass 除了由存储系统动态创建，节省了管理员的时间外，还可以封装不同类型的存储供 PVC 选用。

在 StorageClass 出现以前，PVC 绑定一个 PV 只能根据两个条件，一个是存储的大小，另一个是访问模式。在 StorageClass 出现后，等于增加了一个绑定维度。以下为动态 PV 操作方式。

基于 Linux 平台构建 NFS 网络文件系统，配置指令如下：

```
#安装 NFS 文件服务
yum install nfs-utils -y
#配置共享目录和权限
vim /etc/exports
/data/ *(rw,async,no_root_squash)
#启动 NFS 服务
service nfs restart
```

15.9　NFS 插件配置实战

（1）NFS 默认不支持动态存储，而是使用了第三方的 NFS 插件安装 NFS 插件，GitHub 地址如下：

```
https://github.com/kubernetes-incubator/external-storage/tree/master/nfs
-client/deploy
```

（2）下载 NFS 和动态 PV 配置文件，操作指令如下：

```
for file in class.yaml deployment.yaml rbac.yaml  ; do wget -c https://raw.
githubusercontent.com/kubernetes-incubator/external-storage/master/nfs-
client/deploy/$file ; done
```

（3）修改 deployment.yaml 文件内容，将 NFS 地址和路径修改正确，如图 15-7 所示。

```
containers:
  - name: nfs-client-provisioner
    image: quay.io/external_storage/nfs-client-provisioner:latest
    volumeMounts:
      - name: nfs-client-root
        mountPath: /persistentvolumes
    env:
      - name: PROVISIONER_NAME
        value: fuseim.pri/ifs
      - name: NFS_SERVER
        value: 192.168.1.147
      - name: NFS_PATH
        value: /data
volumes:
  - name: nfs-client-root
    nfs:
      server: 192.168.1.147
      path: /data
```

图 15-7 Kubernetes PVC 动态存储 YAML 文件修改

（4）依次应用 rbac.yaml、class.yaml 和 deployment.yaml 配置文件，操作指令如下：

```
kubectl create -f rbac.yaml
kubectl create -f class.yaml
kubectl create -f deployment.yaml
```

（5）创建 Kubernetes Pod 案例，自动获取 PV 资源即可。Pod 案例 YAML 文件代码如下：

```
apiVersion: v1
kind: Service
metadata:
  name: nginx
  labels:
    app: nginx
spec:
  ports:
  - port: 80
    name: nginx
  selector:
    app: nginx
  clusterIP: None
apiVersion: apps/v1
kind: StatefulSet
metadata:
  name: web
```

```yaml
spec:
  selector:
    matchLabels:
      app: nginx
  serviceName: "nginx"
  replicas: 1
  template:
    metadata:
      labels:
        app: nginx
    spec:
      imagePullSecrets:
      - name: huoban-harbor
      terminationGracePeriodSeconds: 10
      containers:
      - name: nginx
        image: nginx
        ports:
        - containerPort: 80
          name: web
        volumeMounts:
        - name: www
          mountPath: /usr/share/nginx/html
  volumeClaimTemplates:
  - metadata:
      name: www
    spec:
      accessModes: [ "ReadWriteOnce" ]
      storageClassName: "managed-nfs-storage"
      resources:
        requests:
          storage: 1Gi
```

（6）根据以上方法，如果 Pod 容器一直处于 Pending（挂起）状态，可以查看其 LOG 日志信息，报错信息为 Claim reference: selfLink was empty, can't make refere，解决方法如下：

```
#在/etc/kubernetes/manifests/kube-apiserver.yaml 配置文件中加入如下代码
- --feature-gates=RemoveSelfLink=false
#重新应用 apiserver.yaml 文件,同时重新创建所有 NFS yaml 文件即可
kubectl apply -f /etc/kubernetes/manifests/kube-apiserver.yaml
```

（7）在 NFS /data/目录下，能够看到 default 开头的动态 PV 创建的目录和相关的文件即可。

第 15 章　Kubernetes+NFS 持久化存储实战

（8）查看 Nginx Pod 对应的 service，显示没有 Cluster IP，如图 15-8 所示。

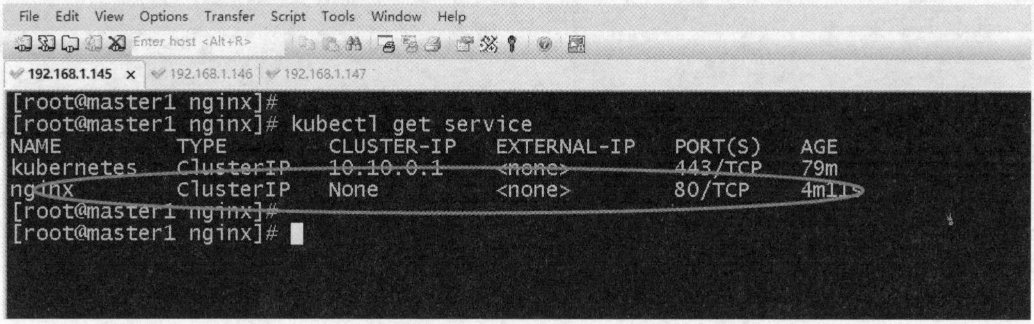

```
[root@master1 nginx]#
[root@master1 nginx]# kubectl get service
NAME         TYPE        CLUSTER-IP   EXTERNAL-IP   PORT(S)    AGE
kubernetes   ClusterIP   10.10.0.1    <none>        443/TCP    79m
nginx        ClusterIP   None         <none>        80/TCP     4m11s
[root@master1 nginx]#
[root@master1 nginx]#
```

图 15-8　Kubernetes Pod service 案例实战 1

（9）验证解析，每个 Pod 都拥有一个基于其顺序索引的、稳定的主机名，如图 15-9 所示，操作指令如下：

```
kubectl get pods --namespace=default
kubectl exec -it web-0 --namespace default /bin/bash
hostname
```

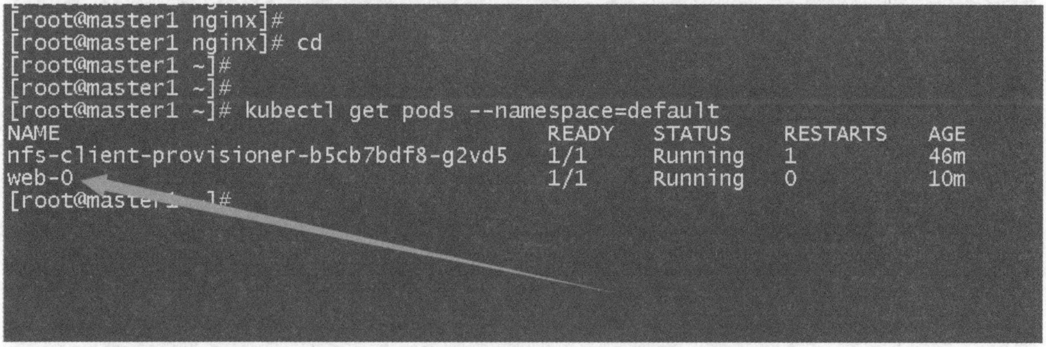

```
[root@master1 nginx]#
[root@master1 nginx]# cd
[root@master1 ~]#
[root@master1 ~]#
[root@master1 ~]# kubectl get pods --namespace=default
NAME                                        READY   STATUS    RESTARTS   AGE
nfs-client-provisioner-b5cb7bdf8-g2vd5      1/1     Running   1          46m
web-0                                       1/1     Running   0          10m
[root@master1 ~]#
```

图 15-9　Kubernetes Pod service 案例实战 2

（10）基于 kubectl run 运行一个提供 nslookup 命令的容器，该命令来自 dnsutils 包。通过对 Pod 的主机名执行 nslookup 命令，可以检查它们在集群内部的 DNS 地址，操作指令如下，结果如图 15-10 所示。

```
kubectl run -i --tty --image busybox:1.28.4 -n default dns-test --restart=Never --rm
nslookup web-0.nginx
```

（11）进入容器内部，访问 web-0.nginx.default.svc.cluster.local 域名，操作指令如下，结果如图 15-11 所示。

```
curl web-0.nginx.default.svc.cluster.local
```

图 15-10　Kubernetes Pod 案例测试 1

图 15-11　Kubernetes Pod 案例测试 2

（12）删除 web-0 Pod 容器，然后系统会自动创建一台，此时再次查看其域名对应的 IP 地址，如果通过 IP 访问还是访问到之前内容，即证明持久化操作成功，如图 15-12 所示。

图 15-12　Kubernetes Pod 案例测试 3

不管 web-0 重新调度去哪个 node 上，它都会继续监听各自的主机名，因为与其 PVC 相关联的 PV 被重新挂载到了对应的 VolumeMount 上。其 PV 将会被挂载到合适的挂载点上。

第 16 章 Kubernetes+CephFS 持久化存储实战

CephFS 模式下的 Kubernetes 服务运行状态、Kubernetes 存储系统、Kubernetes 存储绑定的概念、PV 的访问模式与 NFS 模式下的相同，见 15.1 节~15.4 节。

16.1 Kubernetes+CephFS 静态存储模式

Kubernetes 使用 CephFS 共享静态存储模式，需要先创建静态 PV，再手动创建 PVC，同时 PVC 绑定 PV 之后，方可创建部署业务使用 PV 资源。

16.2 PV 存储卷创建

（1）创建 Kubernetes CephFS 密钥，操作指令如下：

```
ceph auth get-key client.admin > /tmp/secret
kubectl create namespace cephfs
kubectl create secret generic ceph-admin-secret --from-file=/tmp/secret
```

（2）创建 PV，创建 pv.yaml 文件，操作指令如下：

```
cat>pv.yaml<<EOF
apiVersion: v1
kind: PersistentVolume
metadata:
  name: cephfs-pv1
spec:

  capacity:
    storage: 1Gi
  accessModes:
```

```
    - ReadWriteMany
  cephfs:
    monitors:
      - 192.168.1.145:6789
    user: admin
    secretRef:
      name: ceph-admin-secret
    readOnly: false
  persistentVolumeReclaimPolicy: Recycle
EOF
```

(3) Kubernetes PV 配置参数如下：

Capacity 指定 PV 的容量为 100MB。

accessModes 指定访问模式为 ReadWriteMany，支持的访问模式有：

(1) ReadWriteOnce 表示 PV 能以 read-write 模式挂载到单个节点。

(2) ReadOnlyMany 表示 PV 能以 read-only 模式挂载到多个节点。

(3) ReadWriteMany 表示 PV 能以 read-write 模式挂载到多个节点。

persistentVolumeReclaimPolicy 指定当前 PV 的回收策略为 Recycle，支持的策略有：

(1) Retain 表示需要管理员手动回收。

(2) Recycle 表示清除 PV 中的数据，效果相当于执行 rm -rf /thevolume/*。

(3) Delete 表示删除 Storage Provider 上的对应存储资源，如 AWS EBS、GCE PD、Azure、Disk、OpenStack Cinder Volume 等。

storageClassName 指定 PV 的 class 为 NFS。相当于为 PV 设置了一个分类，PVC 可以指定 class 申请相应 class 的 PV。

16.3 PVC 存储卷创建

Kubernetes CephFS PV 资源创建如图 16-1 所示。pvc.yaml 文件内容如下：

```
cat>pvc.yaml<<EOF
kind: PersistentVolumeClaim
apiVersion: v1
metadata:
  name: cephfs-pv-claim1
spec:
  accessModes:
    - ReadWriteMany
  resources:
    requests:
```

```
    storage: 1Gi
EOF
```

图 16-1 Kubernetes CephFS PV 资源创建

16.4 Nginx 整合 CephFS PV 存储卷

（1）创建 Nginx Pod 容器使用 PVC，nginx.yaml 文件内容如下：

```
cat>nginx.yaml<<EOF
apiVersion: v1
kind: ReplicationController
metadata:
  name: nginx-v1
  labels:
    name: nginx-v1
  namespace: default
spec:
  replicas: 1
  selector:
    name: nginx-v1
  template:
    metadata:
      labels:
        name: nginx-v1
    spec:
      containers:
      - name: nginx-v1
        image: nginx
        volumeMounts:
        - mountPath: /usr/share/nginx/html
          name: nginx-data
        ports:
        - containerPort: 80
      volumes:
      - name: nginx-data
        persistentVolumeClaim:
```

```
        claimName: cephfs-pv-claim1
EOF
```

（2）登录 node 节点，查看 CephFS PV 资源是否挂载，如图 16-2 所示。

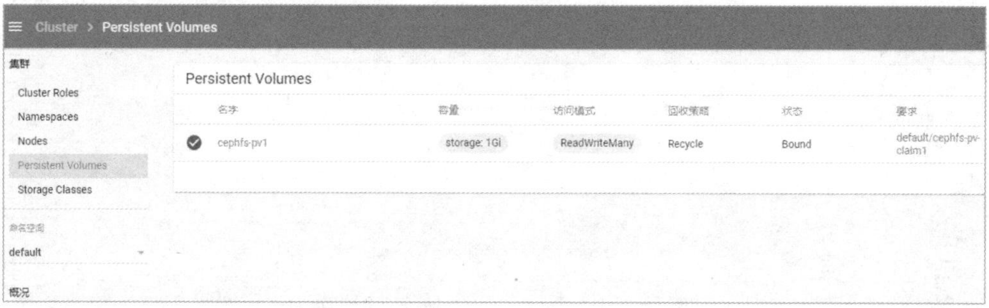

（a）

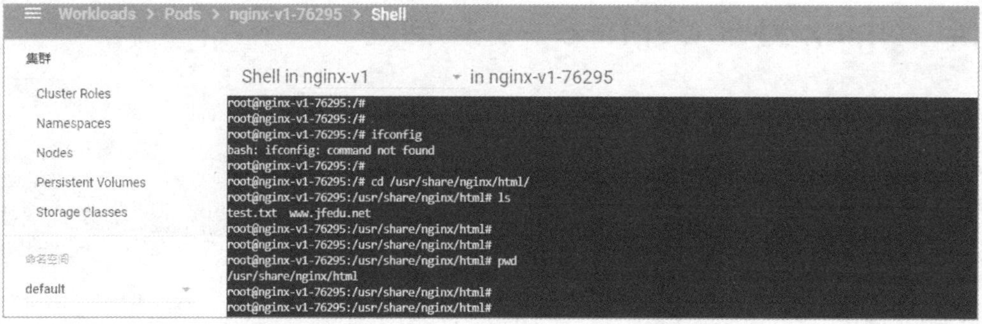

（b）

图 16-2　Kubernetes CephFS PV 存储实战 1

（a）查看 CephFS PV 运行状态；（b）查看 nginx-v1-76295 容器 html 目录内容

（3）测试 Kubernetes Pod CephFS 和 CephFS 服务器数据是否一致，如图 16-3 所示。

（a）

图 16-3　Kubernetes CephFS PV 存储实战 2

（a）查看 Kubernetes Pod 容器运行状态

第 16 章　Kubernetes+CephFS 持久化存储实战

```
[root@node3 ~]# ceph-fuse -m 192.168.1.145:6789 /mnt/
2021-07-27 14:46:11.408402 7f5b1968ef00 -1 asok(0x56357cda4000) AdminSocketConfigObs
ain socket to '/var/run/ceph/ceph-client.admin.asok': (17) File exists
ceph-fuse[44918]: starting ceph client
2021-07-27 14:46:11.408930 7f5b1968ef00 -1 init, newargv = 0x56357cd9e780 newargc=11
ceph-fuse[44918]: starting fuse
[root@node3 ~]#
[root@node3 ~]# cd /mnt/
[root@node3 mnt]# ls
2021  test.txt
[root@node3 mnt]# ls -l
total 1
-rw-r--r-- 1 root root 0 Jul 27 14:44 2021
-rw-r--r-- 1 root root 0 Jul 26 17:44 test.txt
drwxr-xr-x 1 root root 0 Jul 26 17:44
[root@node3 mnt]#
```

（b）

图 16-3　Kubernetes CephFS PV 存储实战 2（续）

（b）查看 ceph 挂载目录内容

16.5　Kubernetes+CephFS 动态存储模式

Kubernetes 使用 CephFS 共享动态存储模式，需要动态创建 PV，是指在现有 PV 不满足 PVC 的请求时，可以使用存储分类（StorageClass），描述具体过程为：PV 先创建分类，PVC 请求已创建的某个类（StorageClass）的资源，这样就达到动态配置的效果，即通过一个叫 StorageClass 的对象由存储系统根据 PVC 的要求自动创建。

其中动态方式是通过 StorageClass 完成的，这是一种新的存储供应方式。动态卷供给能力让管理员不必预先创建存储卷，而是随用户需求进行创建，如图 16-4 所示。

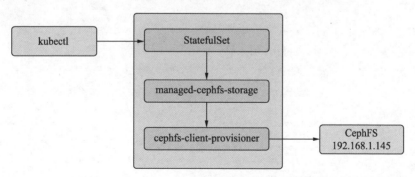

图 16-4　Kubernetes CephFS PVC 动态存储流程示意图

使用 StorageClass 有什么好处呢？除了由存储系统动态创建，节省了管理员的时间外，还可以封装不同类型的存储供 PVC 选用。

在 StorageClass 出现以前，PVC 绑定一个 PV 只能根据两个条件，一个是存储的大小，另一个是访问模式。在 StorageClass 出现后，等于增加了一个绑定维度。

16.6　CephFS 动态插件配置实战

（1）CephFS 默认不支持动态存储，而是需要安装第三方插件 Cephfs-provisioner，GitHub 地址如下：

```
https://kubernetes.io/docs/concepts/storage/storage-classes/
```

（2）下载 Cephfs-provisioner 与动态 PV 配置文件。

```
#CephFS 需要使用两个 Pool 来分别存储数据和元数据
ceph osd pool create fs_data 128
ceph osd pool create fs_metadata 128
ceph osd lspools
#创建一个 CephFS
ceph fs new cephfs fs_metadata fs_data
#查看 CephFS
ceph fs ls
```

（3）执行以上 YAML 文件，使其生效，并查看其 Pod 信息。

```
kubectl apply -f external-storage-cephfs-provisioner.yaml
```

（4）查看 Pod 状态，如图 16-5 所示，操作指令如下：

```
kubectl get pod -n kube-system |grep -aiE provisioner
```

图 16-5　查看 Pod 状态

（5）创建 CephFS secret 密钥，操作指令如下：

```
#查看 key 在 ceph 的 mon 或者 admin 节点
ceph auth get-key client.admin
ceph auth get-key client.kube
#创建 admin secret
#将 CEPH_ADMIN_SECRET 替换为 client.admin 获取到的 key
```

```
export CEPH_ADMIN_SECRET='AQCXgP5giXmCARAAzz1mWKqJ+dTFdLArr1Ee+Q=='
kubectl create secret generic ceph-secret --type="kubernetes.io/rbd" \
--from-literal=key=$CEPH_ADMIN_SECRET \
--namespace=kube-system
#查看secret
kubectl get secret ceph-user-secret -o yaml
kubectl get secret ceph-secret -n kube-system -o yaml
```

（6）配置 StorageClass，操作方法如下：

```
#如果使用kubeadm创建的集群provisioner,则使用如下方式
# provisioner: ceph.com/rbd
cat >storageclass-cephfs.yaml<<EOF
kind: StorageClass
apiVersion: storage.Kubernetes.io/v1
metadata:
  name: dynamic-cephfs
provisioner: ceph.com/cephfs
parameters:
    monitors: 192.168.1.145:6789
    adminId: admin
    adminSecretName: ceph-secret
    adminSecretNamespace: "kube-system"
    claimRoot: /volumes/kubernetes
EOF
kubectl apply -f storageclass-ceph-rdb.yaml
kubectl get sc
```

（7）创建 CephFS PVC 资源索取，操作指令如下：

```
cat>cephfs-pvc.yaml<<EOF
kind: PersistentVolumeClaim
apiVersion: v1
metadata:
  name: cephfs-claim
spec:
  accessModes:
    - ReadWriteOnce
  storageClassName: dynamic-cephfs
  resources:
    requests:
      storage: 2Gi
EOF
kubectl apply -f cephfs-pvc.yaml
```

（8）创建 Nginx Pod 容器，使用 CephFS 资源池，如图 16-6 所示，操作指令如下：

```
cat>nginx-pod.yaml<<EOF
apiVersion: v1
```

```yaml
kind: Pod
metadata:
  name: nginx-pod1
  labels:
    name: nginx-pod1
spec:
  containers:
  - name: nginx-pod1
    image: docker.io/library/nginx:latest
    ports:
    - name: web
      containerPort: 80
    volumeMounts:
    - name: cephfs
      mountPath: /usr/share/nginx/html
  volumes:
  - name: cephfs
    persistentVolumeClaim:
      claimName: cephfs-claim
EOF
```

图 16-6　Kubernetes CephFS PVC 动态存储实战

第 17 章 Kubernetes+Ceph RBD 持久化存储实战

Ceph RBD 模式下的 Kubernetes 服务运行状态、Kubernetes 存储系统、Kubernetes 存储绑定概念、PV 访问模式与 NFS 模式和 CephFS 模式下的相同，详见 15.1 节 ~ 15.4 节。

17.1 Kubernetes+Ceph RBD 静态存储模式

Kubernetes 使用 Ceph RBD 共享静态存储模式，需要先创建静态 PV，再手动创建 PVC，同时 PVC 绑定 PV 之后，方可创建部署业务使用 PV 资源。

17.2 PV 存储卷创建

（1）创建 Kubernetes Ceph RBD 密钥，操作指令如下：

```
ceph auth get-key client.admin > /tmp/secret
kubectl create namespace ceph rbd
kubectl create secret generic ceph-admin-secret --from-file=/tmp/secret
```

（2）创建 Ceph pool 和 Image，操作指令如下：

```
ceph osd pool create kube-nginx 128 128
rbd create kube-nginx/rbd0 -s 10G --image-feature layering
```

（3）创建 PV，pv.yaml 文件内容如下：

```
cat>pv.yaml<<EOF

apiVersion: v1
kind: PersistentVolume
metadata:
  name: rbd-pv1
spec:
```

```
      capacity:
        storage: 1Gi
      accessModes:
        - ReadWriteOnce
      rbd:
        monitors:
          - 192.168.1.145:6789
        pool: kube-nginx
        image: rbd0
        user: admin
        secretRef:
          name: ceph-admin-secret
      persistentVolumeReclaimPolicy: Recycle
EOF
```

（4）PV 配置参数如下：

Capacity 指定 PV 的容量为 100MB。

accessModes 指定访问模式为 ReadWriteOnce，支持的访问模式有：

（1）ReadWriteOnce 表示 PV 能以 read-write 模式挂载到单个节点。

（2）ReadOnlyMany 表示 PV 能以 read-only 模式挂载到多个节点。

（3）ReadWriteMany 表示 PV 能以 read-write 模式挂载到多个节点。

persistentVolumeReclaimPolicy 指定当前 PV 的回收策略为 Recycle，支持的策略有：

（1）Retain 表示需要管理员手动回收。

（2）Recycle 表示清除 PV 中的数据，效果相当于执行 rm -rf /thevolume/*。

（3）Delete 表示删除 Storage Provider 上的对应存储资源，如 AWS EBS、GCE PD、Azure Disk、OpenStack Cinder Volume 等。

storageClassName 指定 PV 的 class 为 NFS。相当于为 PV 设置了一个分类，PVC 可以指定 class 申请相应 class 的 PV。

17.3　PVC 存储卷创建

Kubernetes Ceph RBD PV 创建如图 17-1 所示，pvc.yaml 文件内容如下：

```
cat>pvc.yaml<<EOF
kind: PersistentVolumeClaim
apiVersion: v1
metadata:
  name: rbd-pv-claim1
spec:
  accessModes:
    - ReadWriteOnce
```

```
    resources:
      requests:
        storage: 1Gi
EOF
```

图 17-1　Kubernetes Ceph RBD PV 创建

17.4　Nginx 整合 Ceph PV 存储卷

（1）创建 Nginx Pod 容器使用 PVC，nginx.yaml 文件内容如下：

```
cat>nginx.yaml<<EOF
apiVersion: v1
kind: ReplicationController
metadata:
  name: nginx-v2
  labels:
    name: nginx-v2
  namespace: default
spec:
  replicas: 1
  selector:
    name: nginx-v2
  template:
    metadata:
      labels:
        name: nginx-v2
    spec:
      containers:
      - name: nginx-v2
        image: nginx
        volumeMounts:
        - mountPath: /usr/share/nginx/html
          name: nginx-data
        ports:
        - containerPort: 80
```

```
        volumes:
        - name: nginx-data
          persistentVolumeClaim:
            claimName: rbd-pv-claim1
EOF
```

(2) 登录 node 节点，查看 Ceph PV 资源是否挂载，如图 17-2 所示。

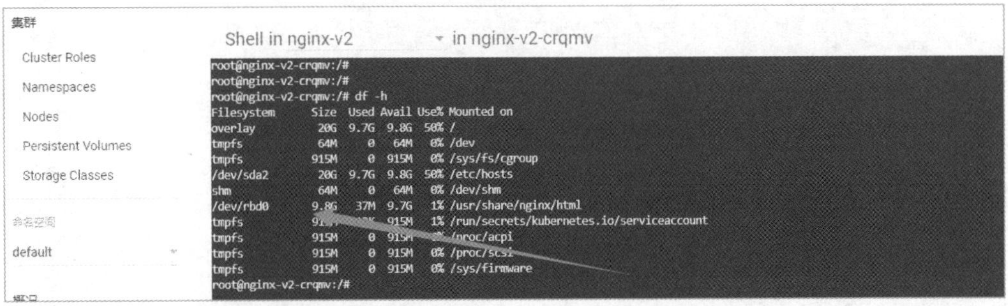

图 17-2　Kubernetes Ceph RBD 模式实战 1

(3) 测试 Kubernetes Pod Ceph 和 Ceph 服务器数据是否一致，如图 17-3 所示。

(a)

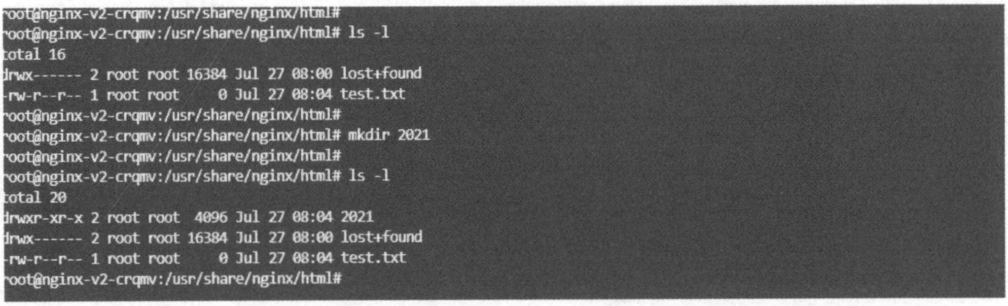

(b)

图 17-3　Kubernetes Ceph RBD 模式实战 2

(a) 查看 nginx-v2 pods 容器状态；(b) 查看 nginx-v2 pods 容器 html 数据

17.5　Kubernetes+Ceph RBD 动态存储模式

Kubernetes 使用 Ceph RBD 共享动态存储模式，需要动态创建 PV，是指在现有 PV 不满足 PVC 的请求时，可以使用存储分类（StorageClass），具体过程是：PV 先创建分类，PVC 请求已创建的某个类（StorageClass）的资源，这样就达到动态配置的效果，即通过一个叫 StorageClass 的对象由存储系统根据 PVC 的要求自动创建。

其中动态方式是通过 StorageClass 来完成的，这是一种新的存储供应方式。动态卷供给能力让管理员不必预先创建存储卷，而是随用户需求进行创建，如图 17-4 所示。

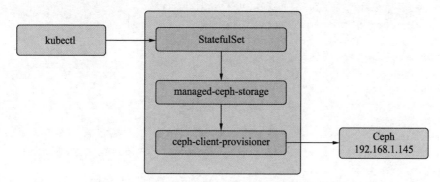

图 17-4　Kubernetes Ceph PVC 动态存储流程示意图

使用 StorageClass 有什么好处呢？除了由存储系统动态创建，节省了管理员的时间外，还可以封装不同类型的存储供 PVC 选用。

在 StorageClass 出现以前，PVC 绑定一个 PV 只能根据两个条件，一个是存储的大小，另一个是访问模式。在 StorageClass 出现后，等于增加了一个绑定维度。

17.6　Ceph RBD 插件配置实战

（1）Ceph 默认不支持动态存储，使用了第三方的 rbd-provisioner 插件安装 rbd-provisioner 插件，GitHub 地址如下：

https://kubernetes.io/docs/concepts/storage/storage-classes/

（2）下载 rbd-provisioner 和动态 PV 配置文件。

```
cat>external-storage-rbd-provisioner.yaml<<EOF
apiVersion: v1
kind: ServiceAccount
metadata:
  name: rbd-provisioner
```

```yaml
  namespace: kube-system
---
kind: ClusterRole
apiVersion: rbac.authorization.Kubernetes.io/v1
metadata:
  name: rbd-provisioner
rules:
  - apiGroups: [""]
    resources: ["persistentvolumes"]
    verbs: ["get", "list", "watch", "create", "delete"]
  - apiGroups: [""]
    resources: ["persistentvolumeclaims"]
    verbs: ["get", "list","watch", "update"]
  - apiGroups: ["storage. Kubernetes.io"]
    resources: ["storageclasses"]
    verbs: ["get", "list", "watch"]
  - apiGroups: [""]
    resources: ["events"]
    verbs: ["create", "update", "patch"]
  - apiGroups: [""]
    resources: ["endpoints"]
    verbs: ["get", "list", "watch", "create", "update", "patch"]
  - apiGroups: [""]
    resources: ["services"]
    resourceNames: ["kube-dns"]
    verbs: ["list", "get"]
---
kind: ClusterRoleBinding
apiVersion: rbac.authorization.Kubernetes.io/v1
metadata:
  name: rbd-provisioner
subjects:
  - kind: ServiceAccount
    name: rbd-provisioner
    namespace: kube-system
roleRef:
  kind: ClusterRole
  name: rbd-provisioner
  apiGroup: rbac.authorization.Kubernetes.io

---
apiVersion: rbac.authorization.Kubernetes.io/v1
kind: Role
metadata:
  name: rbd-provisioner
  namespace: kube-system
```

```yaml
rules:
- apiGroups: [""]
  resources: ["secrets"]
  verbs: ["get"]
---
apiVersion: rbac.authorization.Kubernetes.io/v1
kind: RoleBinding
metadata:
  name: rbd-provisioner
  namespace: kube-system
roleRef:
  apiGroup: rbac.authorization.Kubernetes.io
  kind: Role
  name: rbd-provisioner
subjects:
- kind: ServiceAccount
  name: rbd-provisioner
  namespace: kube-system

---
apiVersion: apps/v1
kind: Deployment
metadata:
  name: rbd-provisioner
  namespace: kube-system
spec:
  replicas: 1
  selector:
    matchLabels:
      app: rbd-provisioner
  strategy:
    type: Recreate
  template:
    metadata:
      labels:
        app: rbd-provisioner
    spec:
      containers:
      - name: rbd-provisioner
        image: "quay.io/external_storage/rbd-provisioner:latest"
        env:
        - name: PROVISIONER_NAME
          value: ceph.com/rbd
      serviceAccount: rbd-provisioner
EOF
```

（3）执行以上 YAML 文件，使其生效，并查看其 Pod 信息。

```
kubectl apply -f external-storage-rbd-provisioner.yaml
```

（4）查看 Pod 状态，如图 17-5 所示，操作命令如下：

```
kubectl get pod -n kube-system |grep -aiE provisioner
```

命名空间				
kube-system	rbd-provisioner	-		1/1
概况	coredns	k8s-app: kube-dns		2/2
工作量				
Cron Jobs	Pods			
Daemon Sets	名字	标签	节点	状态
Deployments	rbd-provisioner-76f6bc6669-49dlw	app: rbd-provisioner	node2	Running
Jobs		pod-template-hash: 76f6bc6669		

图 17-5　查看 Pod 状态

（5）创建 RBD OSD 资源池，操作指令如下：

```
#Kubernetes 集群中所有节点安装 ceph-common
#后期需要使用 rdb 命令 map 附加 rbd 创建的 image
yum install -y ceph-common
#在 ceph 的 mon 或者 admin 节点创建 osd pool
ceph osd pool create kube 4096
ceph osd pool ls
#在 ceph 的 mon 或者 admin 节点创建 Kubernetes 访问 ceph 的用户
ceph auth get-or-create client.kube mon 'allow r' osd 'allow class-read
object_prefix rbd_children,allow rwx pool=kube'-o ceph.client.kube.keyring
#查看 key 在 ceph 的 mon 或者 admin 节点
ceph auth get-key client.admin
ceph auth get-key client.kube
#创建 admin secret
#将 CEPH_ADMIN_SECRET 替换为 client.admin 获取到的 key
export CEPH_ADMIN_SECRET='AQCXgP5giXmCARAAzz1mWKqJ+dTFdLArr1Ee+Q=='
kubectl create secret generic ceph-secret --type="kubernetes.io/rbd" \
--from-literal=key=$CEPH_ADMIN_SECRET \
--namespace=kube-system
#在 default 命名空间创建 pvc,用于访问 ceph 的 secret
#将 CEPH_KUBE_SECRET 替换为 client.kube 获取到的 key
export CEPH_KUBE_SECRET='AQAQ/P9gcQSvARAAek8WxixlCnb4vMChm3eLhA=='
kubectl create secret generic ceph-user-secret --type="kubernetes.io/rbd" \
--from-literal=key=$CEPH_KUBE_SECRET \
--namespace=default
```

```
#查看 secret
kubectl get secret ceph-user-secret -o yaml
kubectl get secret ceph-secret -n kube-system -o yaml
```

(6）配置 StorageClass，操作方法如下：

```
#如果使用 kubeadm 创建的集群 provisioner,则使用如下方式
# provisioner: ceph.com/rbd
cat >storageclass-ceph-rdb.yaml<<EOF
kind: StorageClass
apiVersion: storage.Kubernetes.io/v1
metadata:
  name: dynamic-ceph-rdb
provisioner: ceph.com/rbd
# provisioner: kubernetes.io/rbd
parameters:
  monitors: 192.168.1.145:6789
  adminId: admin
  adminSecretName: ceph-secret
  adminSecretNamespace: kube-system
  pool: kube
  userId: kube
  userSecretName: ceph-user-secret
  fsType: ext4
  imageFormat: "2"
  imageFeatures: "layering"
EOF
kubectl apply -f storageclass-ceph-rdb.yaml
kubectl get sc
```

（7）创建 Ceph RBD PVC 资源索取，操作指令如下：

```
cat>ceph-rdb-pvc.yaml<<EOF
kind: PersistentVolumeClaim
apiVersion: v1
metadata:
  name: ceph-rdb-claim
spec:
  accessModes:
    - ReadWriteOnce
  storageClassName: dynamic-ceph-rdb
  resources:
    requests:
      storage: 2Gi
EOF
kubectl apply -f ceph-rdb-pvc.yaml
```

（8）创建 Nginx Pod 容器，使用 Ceph RBD 资源池，如图 17-6 所示，操作指令如下：

```yaml
apiVersion: v1
kind: Pod
metadata:
  name: nginx-pod1
  labels:
    name: nginx-pod1
spec:
  containers:
  - name: nginx-pod1
    image: docker.io/library/nginx:latest
    ports:
    - name: web
      containerPort: 80
    volumeMounts:
    - name: ceph-rdb
      mountPath: /usr/share/nginx/html
  volumes:
  - name: ceph-rdb
    persistentVolumeClaim:
      claimName: ceph-rdb-claim
```

图 17-6　Kubernetes Ceph RBD 实战

（a）查看 nginx-pod1 容器状态；（b）查看 nginx-pod1 容器内部数据

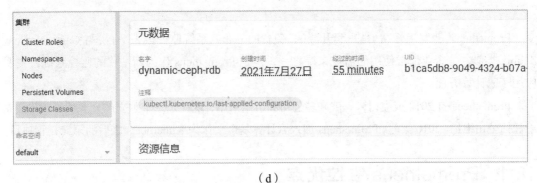

图 17-6　Kubernetes Ceph RBD 实战（续）

（c）查看 PVC 元数据状态；（d）查看 Kubernetes RBD 元数据信息

第 18 章 Prometheus 监控 Kubernetes 实战

Prometheus（普罗米修斯）是一套开源的、免费的分布式系统监控报警平台，与 Cacti、Nagios、Zabbix 类似，是企业最常使用的监控系统之一，但是 Prometheus 作为新一代的监控系统，主要应用于云计算方面。

Prometheus 自 2012 成立以来，被许多公司和组织采用，现在是一个独立的开源项目，并独立于任何公司维护。2016 年起，Prometheus 加入云计算基金会作为 Kubernetes 之后的第二托管项目。

18.1 Prometheus 监控优点

Prometheus 相比传统监控系统（Cacti、Nagios、Zabbix），有如下优点。

（1）易管理性：Prometheus 核心部分只有一个单独的二进制文件，可直接在本地工作，不依赖于分布式存储。

（2）业务数据相关性：监控服务的运行状态，基于 Prometheus 丰富的 Client 库，用户可以轻松地在应用程序中添加对 Prometheus 的支持，从而获取服务和应用内部真正的运行状态。

（3）性能高效性：单一 Prometheus 可以处理数以百万的监控指标，每秒处理数十万的数据点。

（4）易于伸缩性：使用功能分区（sharing）+集群（federation）对 Prometheus 进行扩展，形成一个逻辑集群。

（5）良好的可视化：Prometheus 除了自带 Prometheus UI，还提供了一个独立的基于 Ruby On Rails 的 Dashboard 解决方案 Promdash。另外，最新的 Grafana 可视化工具也提供了完整的 Prometheus 支持，基于 Prometheus 提供的 API 还可以实现自己的监控可视化 UI。

18.2 Prometheus 监控特点

Prometheus 监控特点如下：

（1）由度量名和键值对标识的时间序列数据的多维数据模型。
（2）灵活的查询语言。
（3）不依赖于分布式存储，单服务器节点是自治的。
（4）通过 HTTP 上的拉模型实现时间序列收集。
（5）通过中间网关支持推送时间序列。
（6）通过服务发现或静态配置发现目标。
（7）图形和仪表板支持的多种模式。

18.3　Prometheus 组件实战

Prometheus 生态由多个组件组成，并且这些组件大部分是可选的。

1. Prometheus server

Prometheus server（Prometheus 服务端）是 Prometheus 组件中的核心部分，负责实现对监控数据的获取、存储及查询。Prometheus server 可以通过静态配置管理监控目标，也可以配合使用 service discovery 的方式动态管理监控目标，并从这些监控目标中获取数据。

其次，Prometheus server 需要对采集到的数据进行存储，Prometheus server 本身就是一个实时数据库，将采集到的监控数据按照时间序列的方式存储在本地磁盘中。Prometheus server 对外提供了自定义的 PromQL，实现对数据的查询以及分析。

另外，Prometheus server 的联邦集群能力可以使其从其他 Prometheus server 实例中获取数据。

2. Exporter 监控客户端

Exporter 将监控数据采集的端点通过 HTTP 服务的形式暴露给 Prometheus server，Prometheus server 通过访问该 Exporter 提供的 Endpoint（端点），即可以获取需要采集的监控数据。可以将 Exporter 分为两类。

（1）直接采集：这一类 Exporter 直接内置了对 Prometheus 监控的支持，如 cAdvisor、Kubernetes、etcd、Gokit 等，都直接内置了用于向 Prometheus 暴露监控数据的端点。

（2）间接采集：原有监控目标并不直接支持 Prometheus，需要通过 Prometheus 提供的 Client Library 编写该监控目标的监控采集程序，如 MySQL Exporter、JMX Exporter、Consul Exporter 等。

3. Alertmanager 报警模块

在 Prometheus server 中支持基于 PromQL 创建告警规则，如果满足 PromQL 定义的规则，则会产生一条告警。在 Alertmanager 从 Prometheus server 端接收到告警后，会进行去除重复数据、分组，并检查告警的接收方式和接收人，然后发出报警。常见的接收方式有电子邮件、pagerduty、webhook 等。

4. PushGateway 网关

Prometheus 数据采集基于 Prometheus server 从 Exporter 拉取数据，因此当网络环境不允许 Prometheus server 和 Exporter 进行通信时，可以使用 PushGateway 进行中转。通过 PushGateway 将内部网络的监控数据主动推送到 Gateway 中，Prometheus server 采用针对 Exporter 同样的方式，将监控数据从 PushGateway 拉取到 Prometheus server。

5. Web UI 平台

Prometheus 的 Web 接口可用于简单可视化，以及语句执行或者服务状态监控。

18.4　Prometheus 体系结构

Prometheus 从 jobs 获取度量数据，也可以直接或通过推送网关获取临时 jobs 的度量数据。它在本地存储所有被获取的样本，并在这些数据运行规则中，对现有数据进行聚合和记录新的时间序列或生成警报。

通过 Grafana 或其他 API 消费者，可以可视化地查看收集到的数据。Prometheus 的整体架构和生态组件如图 18-1 所示。

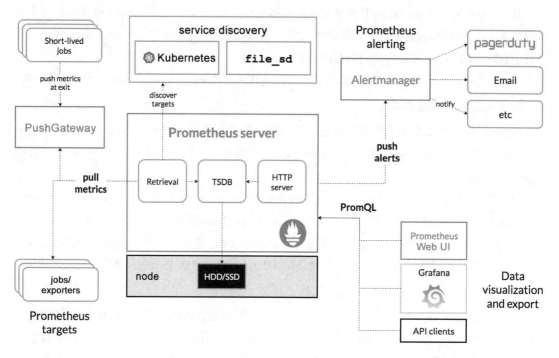

图 18-1　Prometheus 的整体架构和生态组件

18.5 Prometheus 工作流程

（1）Prometheus 服务器定期从配置好的 jobs 或 Exporter 中获取度量数据，或者接收来自推送网关发送过来的度量数据。
（2）Prometheus 服务器在本地存储收集到的度量数据，并对这些数据进行聚合。
（3）运行已定义好的 alert.rules，记录新的时间序列或者向告警管理器推送警报。
（4）告警管理器根据配置文件，对接收到的警报进行处理，并通过 Email、微信、钉钉等途径发出告警。
（5）Grafana 等图形工具获取到监控数据，并以图形化的方式进行展示。

18.6 Prometheus 和 Kubernetes 背景

在 Kubernetes 集群中部署 node-exporter、Prometheus、Grafana，同时使用 Prometheus 对 Kubernetes 整个集群进行监控。实现方法和原理如下：
（1）node-exporter 负责收集节点上的 metrics 监控数据，将数据推送给 Prometheus server 端。
（2）Prometheus server 负责存储这些监控数据。
（3）Grafana 将这些数据通过网页以图形的形式展现给用户。

18.7 Kubernetes 集群部署 node-exporter

（1）在 master 和 node 节点下载 Prometheus 相关镜像，操作指令如下：

```
docker pull prom/node-exporter
docker pull prom/prometheus:v2.26.0
docker pull grafana/grafana
```

（2）基于 DaemonSet 方式部署 node-exporter 组件，每个节点只部署一个 node-exporter 实例，操作指令如下：

```
cat>node-exporter.yaml<<EOF
apiVersion: apps/v1
kind: DaemonSet
metadata:
  name: node-exporter
  namespace: kube-system
  labels:
    Kubernetes-app: node-exporter
spec:
```

```
    selector:
      matchLabels:
        Kubernetes-app: node-exporter
    template:
      metadata:
        labels:
          Kubernetes-app: node-exporter
      spec:
        containers:
        - image: prom/node-exporter
          name: node-exporter
          ports:
          - containerPort: 9100
            protocol: TCP
            name: http
---
apiVersion: v1
kind: Service
metadata:
  labels:
    Kubernetes-app: node-exporter
  name: node-exporter
  namespace: kube-system
spec:
  ports:
  - name: http
    port: 9100
    nodePort: 31672
    protocol: TCP
  type: NodePort
  selector:
    Kubernetes-app: node-exporter
EOF
kubectl apply -f node-exporter.yaml
```

18.8 Kubernetes 集群部署 Prometheus

（1）部署 Prometheus 相关服务组件，可以从京峰官网下载 YAML 配置文件，操作指令如下：

```
mkdir prometheus/
cd prometheus/
for i in alertmanager-configmap.yaml alertmanager-deployment.yaml
alertmanager-pvc.yaml configmap.yaml grafana-deploy.yaml grafana-service.
yaml node-exporter.yaml prometheus.deploy.yml prometheus-rules.yaml
prometheus.svc.yaml rbac-setup.yaml ;do wget -c http://bbs.jingfengjiaoyu.
```

```
com/download/docker/prometheus/$i ;done
#批量应用以上 YAML 脚本
for i in alertmanager-configmap.yaml alertmanager-deployment.yaml
alertmanager-pvc.yaml configmap.yaml grafana-deploy.yaml grafana-service.
yaml node-exporter.yaml prometheus.deploy.yml prometheus-rules.yaml
prometheus.svc.yaml rbac-setup.yaml ;do kubectl apply -f $i ;sleep 3 ;done
```

（2）部署 Prometheus 相关服务组件，部署 Rbac 认证文件 rbac-setup.yaml，操作指令如下：

```
apiVersion: rbac.authorization.Kubernetes.io/v1
kind: ClusterRole
metadata:
  name: prometheus
rules:
- apiGroups: [""]
  resources:
  - nodes
  - nodes/proxy
  - services
  - endpoints
  - pods
  verbs: ["get", "list", "watch"]
- apiGroups:
  - extensions
  resources:
  - ingresses
  verbs: ["get", "list", "watch"]
- nonResourceURLs: ["/metrics"]
  verbs: ["get"]
---
apiVersion: v1
kind: ServiceAccount
metadata:
  name: prometheus
  namespace: kube-system
---
apiVersion: rbac.authorization.Kubernetes.io/v1
kind: ClusterRoleBinding
metadata:
  name: prometheus
roleRef:
  apiGroup: rbac.authorization.Kubernetes.io
  kind: ClusterRole
  name: prometheus
subjects:
- kind: ServiceAccount
  name: prometheus
  namespace: kube-system
```

（3）部署 Prometheus 相关服务组件，部署 Prometheus 主程序，保存为文件 prometheus.deploy.yml，操作指令如下：

```yaml
---
apiVersion: apps/v1
kind: Deployment
metadata:
  labels:
    name: prometheus-deployment
  name: prometheus
  namespace: kube-system
spec:
  replicas: 1
  selector:
    matchLabels:
      app: prometheus
  template:
    metadata:
      labels:
        app: prometheus
    spec:
      containers:
      - image: prom/prometheus:v2.26.0
        name: prometheus
        command:
        - "/bin/prometheus"
        args:
        - "--config.file=/etc/prometheus/prometheus.yml"
        - "--storage.tsdb.path=/prometheus"
        - "--storage.tsdb.retention=24h"
        ports:
        - containerPort: 9090
          protocol: TCP
        volumeMounts:
        - mountPath: "/prometheus"
          name: data
        - mountPath: "/etc/prometheus"
          name: config-volume
        resources:
          requests:
            cpu: 100m
            memory: 100Mi
          limits:
            cpu: 500m
            memory: 2500Mi
      serviceAccountName: prometheus
```

```
        volumes:
        - name: data
          emptyDir: {}
        - name: config-volume
          configMap:
            name: prometheus-config
```

（4）部署 Prometheus 相关服务组件，部署 Prometheus service，保存为文件 prometheus.svc.yml，操作指令如下：

```
kind: Service
apiVersion: v1
metadata:
  labels:
    app: prometheus
  name: prometheus
  namespace: kube-system
spec:
  type: NodePort
  ports:
  - port: 9090
    targetPort: 9090
    nodePort: 30003
  selector:
    app: prometheus
```

（5）以 ConfigMap 的形式管理 Prometheus 组件的配置文件 configmap.yaml，操作指令如下：

```
apiVersion: v1
kind: ConfigMap
metadata:
  name: prometheus-config
  namespace: kube-system
data:
  prometheus.yml: |
    global:
      scrape_interval:     15s
      evaluation_interval: 15s
    scrape_configs:

    - job_name: 'kubernetes-apiservers'
      kubernetes_sd_configs:
      - role: endpoints
      scheme: https
      tls_config:
        ca_file: /var/run/secrets/kubernetes.io/serviceaccount/ca.crt
      bearer_token_file: /var/run/secrets/kubernetes.io/serviceaccount/token
```

```yaml
      relabel_configs:
      - source_labels: [__meta_kubernetes_namespace, __meta_kubernetes_service_name, __meta_kubernetes_endpoint_port_name]
        action: keep
        regex: default;kubernetes;https

    - job_name: 'kubernetes-nodes'
      kubernetes_sd_configs:
      - role: node
      scheme: https
      tls_config:
        ca_file: /var/run/secrets/kubernetes.io/serviceaccount/ca.crt
      bearer_token_file: /var/run/secrets/kubernetes.io/serviceaccount/token
      relabel_configs:
      - action: labelmap
        regex: __meta_kubernetes_node_label_(.+)
      - target_label: __address__
        replacement: kubernetes.default.svc:443
      - source_labels: [__meta_kubernetes_node_name]
        regex: (.+)
        target_label: __metrics_path__
        replacement: /api/v1/nodes/${1}/proxy/metrics

    - job_name: 'kubernetes-cadvisor'
      kubernetes_sd_configs:
      - role: node
      scheme: https
      tls_config:
        ca_file: /var/run/secrets/kubernetes.io/serviceaccount/ca.crt
      bearer_token_file: /var/run/secrets/kubernetes.io/serviceaccount/token
      relabel_configs:
      - action: labelmap
        regex: __meta_kubernetes_node_label_(.+)
      - target_label: __address__
        replacement: kubernetes.default.svc:443
      - source_labels: [__meta_kubernetes_node_name]
        regex: (.+)
        target_label: __metrics_path__
        replacement: /api/v1/nodes/${1}/proxy/metrics/cadvisor

    - job_name: 'kubernetes-service-endpoints'
      kubernetes_sd_configs:
      - role: endpoints
      relabel_configs:
```

```yaml
      - source_labels: [__meta_kubernetes_service_annotation_prometheus_io_scrape]
        action: keep
        regex: true
      - source_labels: [__meta_kubernetes_service_annotation_prometheus_io_scheme]
        action: replace
        target_label: __scheme__
        regex: (https?)
      - source_labels: [__meta_kubernetes_service_annotation_prometheus_io_path]
        action: replace
        target_label: __metrics_path__
        regex: (.+)
      - source_labels: [__address__, __meta_kubernetes_service_annotation_prometheus_io_port]
        action: replace
        target_label: __address__
        regex: ([^:]+)(?::\d+)?;(\d+)
        replacement: $1:$2
      - action: labelmap
        regex: __meta_kubernetes_service_label_(.+)
      - source_labels: [__meta_kubernetes_namespace]
        action: replace
        target_label: kubernetes_namespace
      - source_labels: [__meta_kubernetes_service_name]
        action: replace
        target_label: kubernetes_name

    - job_name: 'kubernetes-services'
      kubernetes_sd_configs:
      - role: service
      metrics_path: /probe
      params:
        module: [http_2xx]
      relabel_configs:
      - source_labels: [__meta_kubernetes_service_annotation_prometheus_io_probe]
        action: keep
        regex: true
      - source_labels: [__address__]
        target_label: __param_target
      - target_label: __address__
        replacement: blackbox-exporter.example.com:9115
      - source_labels: [__param_target]
        target_label: instance
```

```yaml
      - action: labelmap
        regex: __meta_kubernetes_service_label_(.+)
      - source_labels: [__meta_kubernetes_namespace]
        target_label: kubernetes_namespace
      - source_labels: [__meta_kubernetes_service_name]
        target_label: kubernetes_name

    - job_name: 'kubernetes-ingresses'
      kubernetes_sd_configs:
      - role: ingress
      relabel_configs:
      - source_labels: [__meta_kubernetes_ingress_annotation_prometheus_io_probe]
        action: keep
        regex: true
      - source_labels: [__meta_kubernetes_ingress_scheme,__address__,__meta_kubernetes_ingress_path]
        regex: (.+);(.+);(.+)
        replacement: ${1}://${2}${3}
        target_label: __param_target
      - target_label: __address__
        replacement: blackbox-exporter.example.com:9115
      - source_labels: [__param_target]
        target_label: instance
      - action: labelmap
        regex: __meta_kubernetes_ingress_label_(.+)
      - source_labels: [__meta_kubernetes_namespace]
        target_label: kubernetes_namespace
      - source_labels: [__meta_kubernetes_ingress_name]
        target_label: kubernetes_name

    - job_name: 'kubernetes-pods'
      kubernetes_sd_configs:
      - role: pod
      relabel_configs:
      - source_labels: [__meta_kubernetes_pod_annotation_prometheus_io_scrape]
        action: keep
        regex: true
      - source_labels: [__meta_kubernetes_pod_annotation_prometheus_io_path]
        action: replace
        target_label: __metrics_path__
        regex: (.+)
      - source_labels: [__address__, __meta_kubernetes_pod_annotation_prometheus_io_port]
```

```
        action: replace
        regex: ([^:]+)(?::\d+)?;(\d+)
        replacement: $1:$2
        target_label: __address__
      - action: labelmap
        regex: __meta_kubernetes_pod_label_(.+)
      - source_labels: [__meta_kubernetes_namespace]
        action: replace
        target_label: kubernetes_namespace
      - source_labels: [__meta_kubernetes_pod_name]
        action: replace
        target_label: kubernetes_pod_name
```

18.9　Kubernetes 集群部署 Grafana

（1）部署 Grafana Web 图形展示界面，操作指令如下：

```
cat>grafana-deploy.yaml<<EOF
apiVersion: apps/v1
kind: Deployment
metadata:
  name: grafana-core
  namespace: kube-system
  labels:
    app: grafana
    component: core
spec:
  replicas: 1
  selector:
    matchLabels:
      app: grafana
  template:
    metadata:
      labels:
        app: grafana
        component: core
    spec:
      containers:
      - image: grafana/grafana:4.2.0
        name: grafana-core
        imagePullPolicy: IfNotPresent
        # env:
        resources:
        # keep request = limit to keep this container in guaranteed class
          limits:
```

```
                    cpu: 100m
                    memory: 500Mi
                requests:
                    cpu: 100m
                    memory: 500Mi
            env:
            # The following env variables set up basic auth twith the default
admin user and admin password
                - name: GF_AUTH_BASIC_ENABLED
                  value: "true"
                - name: GF_AUTH_ANONYMOUS_ENABLED
                  value: "false"
                # - name: GF_AUTH_ANONYMOUS_ORG_ROLE
                # value: Admin
                # does not really work, because of template variables in exported
dashboards
                # - name: GF_DASHBOARDS_JSON_ENABLED
                # value: "true"
            readinessProbe:
              httpGet:
                path: /login
                port: 3000
              # initialDelaySeconds: 30
              # timeoutSeconds: 1
            volumeMounts:
            - name: grafana-persistent-storage
              mountPath: /var
        volumes:
        - name: grafana-persistent-storage
          emptyDir: {}
EOF
kubectl apply -f grafana-deploy.yaml
```

（2）部署 Grafana service 和对外暴露 node port 端口，操作指令如下：

```
cat>grafana-service.yaml<<EOF
apiVersion: v1
kind: Service
metadata:
  name: grafana
  namespace: kube-system
  labels:
    app: grafana
    component: core
spec:
```

```
    type: NodePort
    ports:
      - port: 3000
    selector:
      app: grafana
      component: core
EOF
kubectl apply -f grafana-service.yaml
```

18.10　Kubernetes 配置和整合 Prometheus

（1）查看 node-exporter Pod 是否运行，通过浏览器访问任意 node 节点 IP 地址（http://192.1.146:31672/metrics），如图 18-2 所示。

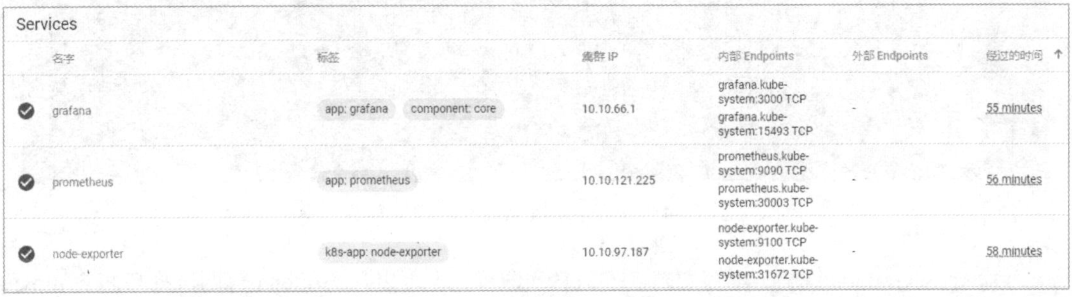

图 18-2　Prometheus 客户端 Exporter

（2）Prometheus 对应的 NodePort 端口为 30003，通过访问 node 节点 IP 地址（http://192.168.1.146:30003/targets），可以看到 Prometheus 已经成功连接上了 Kubernetes 的 API server（状态全部为 UP），如图 18-3 所示。

图 18-3　Prometheus 平台实战操作

（3）通过端口进行 Grafana 访问，默认用户名、密码均为 admin，通过浏览器访问 node 节点 IP 地址（http://10.10.121.225:9090），添加数据源，如图 18-4 所示。

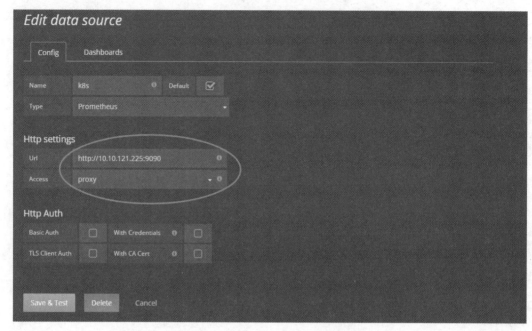

图 18-4　Grafana+Prometheus 整合操作

（4）导入 Grafana 模板，填写 ID（315|3119）即可，也可以从 Grafana 官网选择模板：https://grafana.com/grafana/dashboards，如图 18-5 所示。

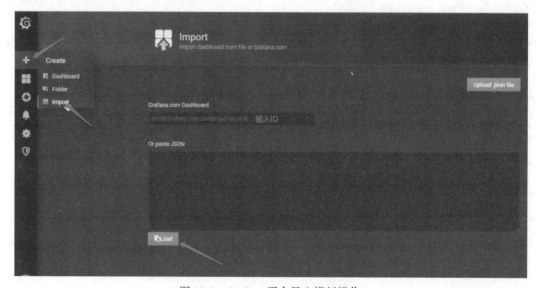

图 18-5　Grafana 平台导入模板操作

（5）最后查看 Prometheus 监控 Kubernetes node 数据展示信息，如图 18-6 所示。

图 18-6　Prometheus 监控 Kubernetes node 数据展示界面

18.11　Kubernetes+Prometheus 报警设置

Prometheus 触发一条告警的过程：Prometheus→触发阈值→超出持续时间→Alertmanager→分组|抑制|静默→媒体类型→邮件、钉钉、微信等。

（1）分组（group）。

将类似性质的警报合并为单个通知。例如，Web 服务是一组，CPU 是一组，不用发多个 CPU 超出范围的报警，只发单个 CPU 组的报警即可。

（2）静默（silences）。

是一种简单的特定时间静音的机制。例如，服务器要升级维护，可以先设置这个时间段告警静默，关闭其间的报警，否则在代码升级时，会触发一些报警。

（3）抑制（inhibition）。

当警报发出后，停止重复发送由此警报引发的其他警报，即合并一个故障引起的多个报警事件，可以消除冗余告警。例如，一个交换机上有 3 个节点，但由于交换机故障，导致 3 个节点网络不通，所以只发交换机的报警，而不发 3 个节点的网络报警，因为是交换机故障导致节点不能正常通信，不必发一堆报警，这也称为报警依赖。

18.12　Kubernetes Alertmanager 实战

（1）部署 Alertmanager，通过 Kubernetes+Prometheus 整合方式部署，需要部署文件 alertmanager-configmap.yaml，代码如下：

```yaml
apiVersion: v1
kind: ConfigMap
metadata:
  #配置文件名称
  name: alertmanager
  namespace: kube-system
  labels:
    kubernetes.io/cluster-service: "true"
    addonmanager.kubernetes.io/mode: EnsureExists
data:
  alertmanager.yml: |
    global:
      resolve_timeout: 5m
      #告警自定义邮件
      smtp_smarthost: 'smtp.163.com:25'
      smtp_from: 'wgkgood@163.com'
      smtp_auth_username: 'wgkgood@163.com'
      smtp_auth_password: 'FNGDCYDBFSXRLUUC'

    receivers:
    - name: default-receiver
      email_configs:
      - to: "wgkgood@163.com"

    route:
      group_interval: 1m
      group_wait: 10s
      receiver: default-receiver
      repeat_interval: 1m
```

（2）部署文件 alertmanager-deployment.yaml，代码如下：

```yaml
apiVersion: apps/v1
kind: Deployment
metadata:
  labels:
    name: alertmanager
  name: alertmanager
  namespace: kube-system
spec:
  replicas: 1
  selector:
    matchLabels:
      app: alertmanager
  template:
    metadata:
      labels:
```

```yaml
      app: alertmanager
    spec:
      containers:
      - image: prom/alertmanager:v0.16.1
        name: alertmanager
        ports:
        - containerPort: 9093
          protocol: TCP
        volumeMounts:
        - mountPath: "/alertmanager"
          name: data
        - mountPath: "/etc/alertmanager"
          name: config-volume
        resources:
          requests:
            cpu: 50m
            memory: 50Mi
          limits:
            cpu: 200m
            memory: 200Mi
      volumes:
      - name: data
        emptyDir: {}
      - name: config-volume
        configMap:
          name: alertmanager
---
apiVersion: v1
kind: Service
metadata:
  labels:
    app: alertmanager
  annotations:
    prometheus.io/scrape: 'true'
  name: alertmanager
  namespace: kube-system
spec:
  type: NodePort
  ports:
  - port: 9093
    targetPort: 9093
    nodePort: 31113
  selector:
    app: alertmanager
```

（3）部署文件 alertmanager-pvc.yaml，代码如下：

```yaml
apiVersion: v1
kind: PersistentVolumeClaim
metadata:
  name: alertmanager
  namespace: kube-system
  labels:
    kubernetes.io/cluster-service: "true"
    addonmanager.kubernetes.io/mode: EnsureExists
spec:
  #使用自己的动态 PV
  storageClassName: managed-nfs-storage
  accessModes:
    - ReadWriteOnce
  resources:
    requests:
      storage: "2Gi"
```

（4）部署文件 prometheus-rules.yaml，代码如下：

```yaml
apiVersion: v1
kind: ConfigMap
metadata:
  name: prometheus-rules
  namespace: kube-system
data:
  #通用角色
  general.rules: |
    groups:
    - name: general.rules
      rules:
      - alert: InstanceDown
        expr: up == 0
        for: 1m
        labels:
          severity: error
        annotations:
          summary: "Instance {{ $labels.instance }} 停止工作"
          description: "{{ $labels.instance }} job {{ $labels.job }} 已经停止 5 分钟以上."
  #node 对所有资源的监控
  node.rules: |
    groups:
    - name: node.rules
      rules:
      - alert: NodeFilesystemUsage
```

```yaml
    expr: 100 - (node_filesystem_free_bytes{fstype=~"ext4|xfs"} / node_filesystem_size_bytes{fstype=~"ext4|xfs"} * 100) > 80
    for: 1m
    labels:
      severity: warning
    annotations:
      summary: "Instance {{ $labels.instance }} : {{ $labels.mountpoint }} 分区使用率过高"
      description: "{{ $labels.instance }}: {{ $labels.mountpoint }} 分区使用大于 80% (当前值: {{ $value }})"

  - alert: NodeMemoryUsage
    expr: 100 - (node_memory_MemFree_bytes+node_memory_Cached_bytes+node_memory_Buffers_bytes) / node_memory_MemTotal_bytes * 100 > 80
    for: 1m
    labels:
      severity: warning
    annotations:
      summary: "Instance {{ $labels.instance }} 内存使用率过高"
      description: "{{ $labels.instance }}内存使用大于80% (当前值: {{ $value }})"

  - alert: NodeCPUUsage
    expr: 100 - (avg(irate(node_cpu_seconds_total{mode="idle"}[5m])) by (instance) * 100) > 60
    for: 1m
    labels:
      severity: warning
    annotations:
      summary: "Instance {{ $labels.instance }} CPU 使用率过高"
      description: "{{ $labels.instance }}CPU使用大于60% (当前值: {{ $value }})"
```

18.13 Alertmanager 实战部署

（1）部署 Alertmanager，单独部署在一台服务器上，默认监听端口是 9093，部署方法如下：

```
cd /usr/local/src
tar xvf alertmanager-0.20.0.linux-amd64.tar.gz
ln -s /usr/local/src/alertmanager-0.20.0.linux-amd64 /usr/local/alertmanager
cd /usr/local/alertmanager
```

（2）修改 Alertmanager 配置文件 vim alertmanager.yml，配置文件代码如下：

```
global:
  resolve_timeout: 5m                              #超时时间
```

```
    smtp_smarthost: 'smtp.163.com'           #smtp 服务器地址
    smtp_from: 'wgkgood@163.com'             #发件人
    smtp_auth_username: 'wgkgood'            #登录认证的用户名
    smtp_auth_password: 'jfedu666'           #登录的授权码
    smtp_hello: '@163.com'
    smtp_require_tls: false    #是否使用 tls
    route:                     #route 用来设置报警的分发策略,由谁去发
    group_by: ['alertname']    #采用哪个标签作为分组依据
    group_wait: 10s            #组告警等待时间,也就是告警产生后等待 10s,一个组内的
                               #告警 10s 后一起发送出去
    group_interval: 10s        #两组告警的间隔时间
    repeat_interval: 2m        #重复告警的间隔时间,减少相同邮件的发送频率
    receiver: 'web.hook'       #设置接收人;真正发送邮件不是由 route 完成,而是由
                               #receiver 发送邮件 receivers:- name: 'web.hook'
    #webhook_configs:          #调用指定的 API 把邮件发送出去
    #- url: 'http://127.0.0.1:5001/'
    email_configs:             #通过邮件的方式发送
  - to: 'wgkgood@163.com'      #接收人
    - source_match:            #源匹配级别;以下设置的级别的报警不会发送
        severity: 'critical'
      target_match:            #目标匹配级别
        severity: 'warning'
      equal: ['alertname', 'dev', 'instance']
```

(3)将 Alertmanager 设置为系统服务,操作方法和指令如下:

```
vim /etc/systemd/system/alertmanager.service
[Unit]Description=Prometheus lertManagerDocumentation=https://prometheus.
io/docs/introduction/overview/After=network.target
[Service]Restart=on-failureExecStart=/usr/local/alertmanager/alertmanager
--config.file=/usr/local/alertmanager/alertmanager.yml
[Install]WantedBy=multi-user.target
systemctl start alertmanager.service
systemctl enable alertmanager.service
```

(4) Prometheus 添加规则,绑定 Alertmanager 服务,操作指令如下:

```
vim /usr/local/prometheus/prometheus.yml
alerting:                                  #当触发告警时,把告警发送给下面所配置的服务
  alertmanagers:                           #当出告警时,把告警通知发送给 alertmanager
  - static_configs:
    - targets:
      - 192.168.3.146:9093                 #指定 alertmanager 地址及端口
    rule_files:                            #报警规则文件
    - "/usr/local/prometheus/rules.yml"    #指定报警的 rules.yml 文件所在路径
```

（5）创建 Prometheus 报警规则，新建 rules.yml 文件，内容如下：

```
vim /usr/local/prometheus/rules.yml
groups:
  - name: linux_pod.rules                    #指定名称
    rules:
    - alert: Pod_all_cpu_usage               #相当于 Zabbix 中的监控项；也是邮件的标题
      expr: (sum by(name)(rate(container_cpu_usage_seconds_total{image!=""}
[5m]))*100) > 75
                                             #PromQL 语句查询到所有 Pod 的 CPU 利用率
                                             #与后面的值作对比,查询到的是浮点数,需要
                                             #乘以 100,转换成整数
      for: 5m                                #每 5min 获取一次 Pod 的 CPU 利用率
      labels:
        severity: critical
        service: pods
      annotations:                           #此为当前所有容器的 CPU 利用率
        description: 容器 {{ $labels.name }} CPU 资源利用率大于 75%,(current
value is {{ $value }})                       #报警的描述信息内容
        summary: Dev CPU 负载告警
    - alert: Pod_all_memory_usage
      expr: sort_desc(avg by(name)(irate(container_memory_usage_bytes
{name!=""} [5m]))*100) > 1024^3*2
                                             #通过 PromQL 语句获取到所有 Pod 中的内存
                                             #利用率,将后面的单位 GB 转换成字节
      for: 10m
      labels:
        severity: critical
      annotations:
        description: 容器 {{ $labels.name }} Memory 资源利用率大于 2GB(当前
已用内存是: {{ $value }})
        summary: Dev Memory 负载告警
    - alert: Pod_all_network_receive_usage
      expr: sum by (name)(irate(container_network_receive_bytes_total
{container_name="POD"}[1m])) > 1024*1024*50
      for: 10m                               #因为获取的所有 Pod 网络利用率是字节,所以
                                             #把后面对比的 MB 转换成字节
      labels:
        severity: critical
      annotations:
        description: 容器 {{ $labels.name }} network_receive 资源利用率大于
50M , (current value is {{ $value }})
```

以上为 PromQL 语句通过 Grafana 找到相应的监控项，单击 edit 找到相应的 PromQL 语句即可。

（6）访问 Prometheus 的 Web 界面，以确认规则是否构建成功，如图 18-7 所示。

图 18-7 Prometheus 规则的定义

（7）检查 rule.yml 文件语法是否正确，如图 18-8 所示，操作指令如下：

```
promtool check rules rules.yml
```

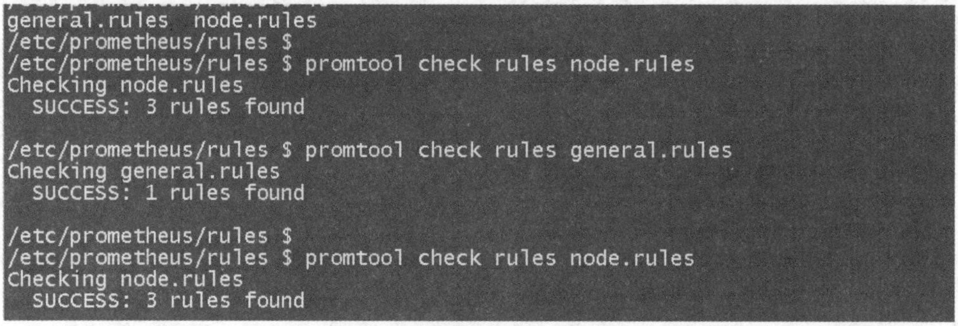

图 18-8 Prometheus 规则的检测

（8）列出当前 Alertmanager 服务器的所有告警，如图 18-9 所示，操作指令如下：

```
amtool alert --alertmanager.url=http://10.244.1.14:9093/
```

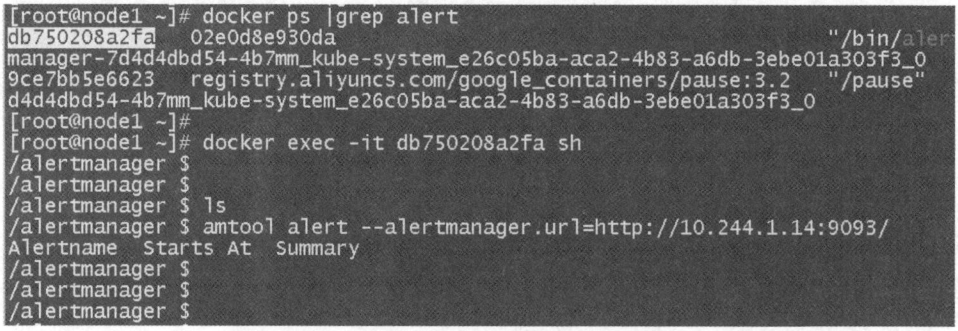

图 18-9 Prometheus Alertmanager 报警列表 1

（9）增加一台 node 节点，增加成功后将其关机，模拟宕机，测试邮件报警实战，如图 18-10 所示。

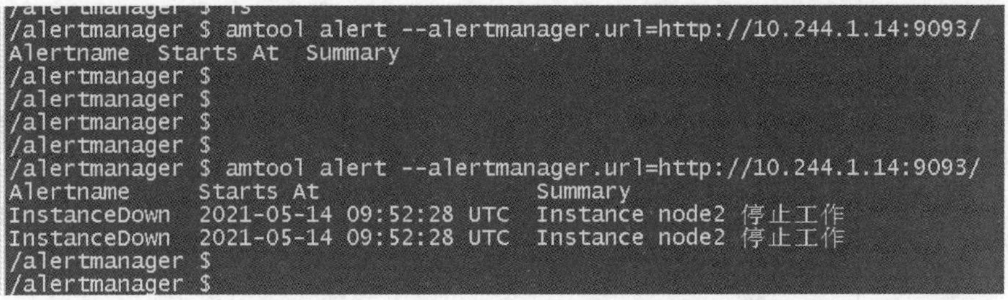
/alertmanager $ amtool alert --alertmanager.url=http://10.244.1.14:9093/
Alertname Starts At Summary
/alertmanager $
/alertmanager $
/alertmanager $
/alertmanager $
/alertmanager $ amtool alert --alertmanager.url=http://10.244.1.14:9093/
Alertname Starts At Summary
InstanceDown 2021-05-14 09:52:28 UTC Instance node2 停止工作
InstanceDown 2021-05-14 09:52:28 UTC Instance node2 停止工作
/alertmanager $
/alertmanager $

图 18-10 Prometheus Alertmanager 报警列表 2

（10）访问 Alertmanager Web 界面，查看报警界面，如图 18-11 所示。

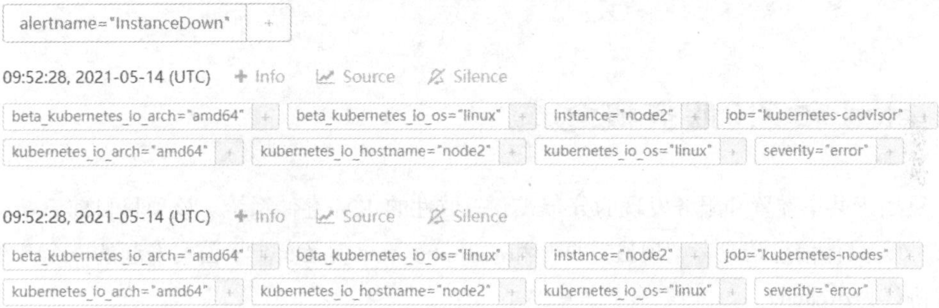
alertname="InstanceDown"
09:52:28, 2021-05-14 (UTC) + Info Source Silence
beta_kubernetes_io_arch="amd64" beta_kubernetes_io_os="linux" instance="node2" job="kubernetes-cadvisor"
kubernetes_io_arch="amd64" kubernetes_io_hostname="node2" kubernetes_io_os="linux" severity="error"
09:52:28, 2021-05-14 (UTC) + Info Source Silence
beta_kubernetes_io_arch="amd64" beta_kubernetes_io_os="linux" instance="node2" job="kubernetes-nodes"
kubernetes_io_arch="amd64" kubernetes_io_hostname="node2" kubernetes_io_os="linux" severity="error"

图 18-11 Prometheus Alertmanager 报警界面

（11）登录 163 邮件服务器，查看是否收到 node2 报警信息，如图 18-12 所示。

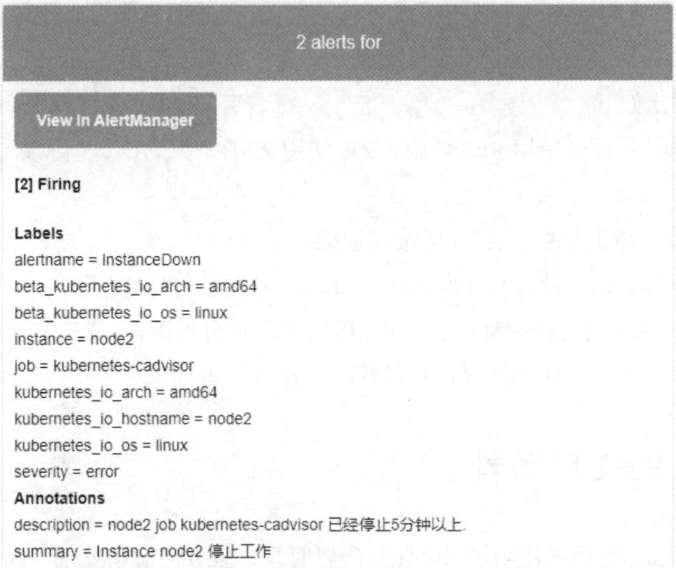
2 alerts for
View in AlertManager
[2] Firing
Labels
alertname = InstanceDown
beta_kubernetes_io_arch = amd64
beta_kubernetes_io_os = linux
instance = node2
job = kubernetes-cadvisor
kubernetes_io_arch = amd64
kubernetes_io_hostname = node2
kubernetes_io_os = linux
severity = error
Annotations
description = node2 job kubernetes-cadvisor 已经停止5分钟以上.
summary = Instance node2 停止工作

图 18-12 Prometheus Alertmanager 报警信息

第 19 章　Kubernetes etcd 服务实战

19.1　etcd 和 ZK 服务概念

etcd 是用于共享配置和服务发现的分布式、一致性的 KV 存储系统。该项目目前最新稳定版本为 2.3.0，具体信息请参考"项目首页"和 Github。etcd 是 CoreOS 公司发起的一个开源项目，授权协议为 Apache。

市场上能提供配置共享和服务发现的系统比较多，其中最为大家熟知的是 ZooKeeper（以下简称 ZK），而 etcd 可以算得上是后起之秀了。在项目实现的一致性协议易理解性、运维、安全等多个维度上，etcd 相比 ZK 都占据优势。

ZK 作为典型代表与 etcd 进行比较，而不考虑将 Consul 项目作为比较对象，原因是 Consul 的可靠性和稳定性还需要时间验证（项目发起方自身服务都未使用 Consul）。

etcd 和 ZK 对比如下。

（1）一致性协议：etcd 使用 Raft 协议，ZK 使用 ZAB（类 PAXOS 协议），前者容易理解，方便工程实现。

（2）运维方面：etcd 方便运维，ZK 难以运维。

（3）项目活跃度：etcd 社区与开发活跃，ZK 已经快被用户弃用了。

（4）API：etcd 提供 HTTP+JSON、gRPC 接口、跨平台跨语言，ZK 需要使用其客户端。

（5）访问安全方面：etcd 支持 HTTPS 访问，ZK 在这方面缺失。

19.2　etcd 的使用场景

和 ZK 类似，etcd 有很多使用场景，包括配置管理、服务注册于发现、选主、应用调度、分布式队列、分布式锁。

19.3 etcd 读写性能

按照官网给出的 Benchmark，在 2 个 CPU、1.8GB 内存、SSD 磁盘这样的配置下，单节点的写性能可以达到 16 000QPS，而先写后读也能达到 12 000QPS。这个性能是相当可观的。

19.4 etcd 工作原理

etcd 使用 Raft 协议维护集群内各个节点状态的一致性。简单地说，etcd 集群是一个分布式系统，由多个节点相互通信构成整体对外服务，每个节点都存储了完整的数据，并通过 Raft 协议保证每个节点维护的数据是一致的，如图 19-1 所示。

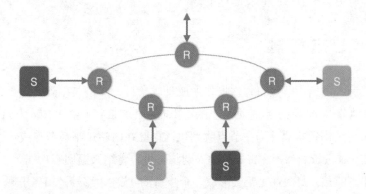

图 19-1　etcd 内部结构图

每个 etcd 节点都维护了一个状态机，且任意时刻至多存在一个有效的主节点。主节点用于处理所有来自客户端的写操作，通过 Raft 协议保证写操作对状态机的改动会可靠地同步到其他节点。

etcd 工作原理核心部分在于 Raft 协议。本节接下来将简要介绍 Raft 协议，具体细节请参考相关论文。

Raft 协议主要分为 3 部分：etcd 选主、etcd 日志复制、etcd 安全性。

19.5 etcd 选主

Raft 协议是用于维护一组服务节点数据一致性的协议。这一组服务节点构成一个集群，且有一个主节点对外提供服务。当集群初始化，或者主节点崩溃后，将面临选主问题。集群中的每个节点在任意时刻处于 Leader、Follower、Candidate 这三个角色之一。选举特点如下：

（1）在集群初始化时，每个节点都是 Follower 角色。

（2）集群中至多存在 1 个有效的主节点，通过心跳与其他节点同步数据。

（3）当 Follower 在一定时间内没有收到来自主节点的心跳时，会将自己角色改变为 Candidate，并发起一次选主投票；当收到包括自己在内超过半数节点赞成后，选举成功；当收到票数不足半数选举失败，或者选举超时。若本轮未选出主节点，将进行下一轮选举（出现这种情况，是由于多个节点同时选举，所有节点均为获得过半选票）。

（4）Candidate 节点收到来自主节点的信息后，会立即终止选举过程，进入 Follower 角色。

（5）为了避免陷入选主失败循环，每个节点未收到心跳发起选举的时间是一定范围内的随机值，这样能够避免 2 个节点同时发起选主。

19.6 etcd 日志复制

所谓日志复制，是指主节点将每次操作形成日志条目，并持久化到本地磁盘，然后通过网络 I/O 发送给其他节点。其他节点根据日志的逻辑时钟（TERM）和日志编号（INDEX）判断是否将该日志记录持久化到本地。当主节点收到包括自己在内超过半数节点成功返回时，认为该日志是可提交的（committed），并将日志输入到状态机，将结果返回客户端。

这里需要注意的是，每次选主都会形成一个唯一的 TERM 编号，相当于逻辑时钟。每一条日志都有全局唯一的编号，如图 19-2 所示。

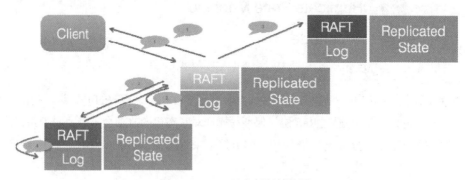

图 19-2　etcd 日志复制结构图

主节点通过网络 I/O 向其他节点追加日志。某节点收到日志追加的消息，将首先判断该日志的 TERM 是否过期，以及该日志条目的 INDEX 是否比当前和提交的日志 INDEX 更早。若已过期，或者比提交的日志更早，则拒绝追加，并返回该节点当前已提交的日志编号；否则，将进行日志追加，并返回成功信息。

当主节点收到其他节点关于日志追加的回复后，若发现被拒绝，则根据该节点返回的已提交日志编号。

主节点向其他节点同步日志，还做了拥塞控制。具体地说，主节点发现日志复制的目标节点拒绝了某次日志追加消息，将进入日志探测阶段，逐条发送日志，直到目标节点接受日志，然后进入快速复制阶段，可进行批量日志追加。

按照日志复制的逻辑可以看到，集群中慢节点不影响整个集群的性能。另外一个特点是，数据只从主节点复制到 Follower 节点，这样大大简化了逻辑流程。

19.7　etcd 安全性

到目前为止，选主以及日志复制并不能保证节点间数据一致。试想，当某个节点挂掉了一段时间后再次重启，并当选为主节点，而在其挂掉这段时间内，集群若有超过半数节点存活，集群会正常工作，那么会有日志提交。因为这些提交的日志无法传递给挂掉的节点，所以当挂掉的节点再次当选主节点时，它将缺失部分已提交的日志。

在这样的场景下，按 Raft 协议，它将自己的日志复制给其他节点，会覆盖集群已经提交的日志。这显然是不可接受的。

其他协议解决这个问题的办法是，新当选的主节点会询问其他节点，和自己的数据对比，确定集群已提交数据，然后将缺失的数据同步过来。这个方案有明显缺陷，增加了集群恢复服务的时间（集群在选举阶段不可服务），且增加了协议的复杂度。

Raft 解决的办法是，在选主逻辑中，对能够成为主节点的节点加以限制，确保选出的节点包含了集群已经提交的所有日志。如果新选出的主节点已经包含了集群所有提交的日志，就不需要和其他节点比对数据了，从而简化了流程，缩短了集群恢复服务的时间。

这里存在一个问题，这样限制之后，还能否选出主节点呢？答案是：只要仍然有超过半数节点存活，这样的主节点一定能够选出。因为已经提交的日志必然被集群中超过半数节点持久化，显然前一个主节点提交的最后一条日志也被集群中大部分节点持久化。

当主节点挂掉后，集群中仍有大部分节点存活，那么这些存活的节点中一定存在一个节点包含了已经提交的日志。

19.8　etcd 使用案例

据公开资料显示，至少有 CoreOS、Google Kubernetes、Cloud Foundry，以及在 Github 上超过 500 个项目在使用 etcd。

19.9　etcd 接口使用

etcd 主要的通信协议是 HTTP，在最新版本中支持 Google gRPC 方式访问。具体支持接口情况如下：

（1）etcd 是一个高可靠的 KV 存储系统，支持 PUT/GET/DELETE 接口。
（2）为了支持服务注册与发现，支持 watch 接口（通过 http long poll 实现）。
（3）支持 key 持有 TTL 属性。
（4）支持 CAS（compare and swap）操作。
（5）支持多 key 的事务操作。
（6）支持目录操作。

第 20 章 Kubernetes+HAProxy 高可用集群

20.1 Kubernetes 高可用集群概念

Kubernetes Apiserver 提供了 Kubernetes 各类资源对象（Pod、RC、service 等）的增删改查及 watch 等 HTTP Rest 接口，是整个系统的数据总线和数据中心。

Kubernetes API server 的功能如下：

（1）提供了集群管理的 Rest API 接口（包括认证授权、数据校验以及集群状态变更）。

（2）提供其他模块之间的数据交互和通信的枢纽（其他模块通过 API server 查询或修改数据，只有 API server 才直接操作 etcd）。

（3）是资源配额控制的入口。

（4）拥有完备的集群安全机制。

API server 是用户和 Kubernetes 集群交互的入口，封装了核心对象的增删改查操作，提供了 RESTFul 风格的 API 接口，通过 etcd 实现持久化并维护对象的一致性。在整个 Kubernetes 集群中，API server 服务至关重要，一旦宕机，整个 Kubernetes 平台将无法使用，所以保障企业高可用是运维必备的工作。

20.2 Kubernetes 高可用工作原理

运行 Keepalived 和 HAProxy 的节点称为 LB(Load balancer,负载均衡)节点。因为 Keepalived 是一主多备运行模式，所以至少有 2 个 LB 节点。此处采用 master 节点的 2 台机器，HAProxy 监听的端口（8443）需要与 kube-apiserver 的端口 6443 不同，以避免冲突。

下面是基于 Keepalived 和 HAProxy 实现 kube-apiserver 高可用的步骤。

（1）Keepalived 提供 kube-apiserver 对外服务的 VIP。

（2）HAProxy 监听 VIP，后端连接所有 kube-apiserver 实例，提供健康检查和负载均衡功能。

Keepalived 在运行过程中周期检查本机的 HAProxy 进程状态，如果检测到 HAProxy 进程异常，则触发重新选主的过程，VIP 将飘移到新选出来的主节点，从而实现 VIP 的高可用。

最终实现所有组件（如 kubectl、kube-controller-manager、kube-scheduler、kube-proxy 等）都通过 VIP+HAProxy 监听的 8443 端口访问 kube-apiserver 服务，如图 20-1 所示。

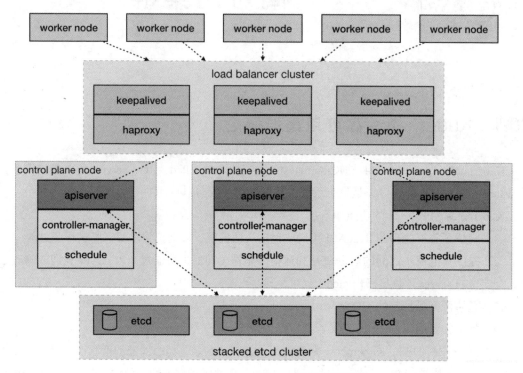

图 20-1　Kubernetes master 高可用结构图

20.3　HAProxy 安装配置

HAProxy 安装配置步骤相对比较简单，跟其他源码软件安装方法大致相同，以下为 HAProxy 配置方法及步骤。

（1）HAProxy 编译及安装，操作指令如下，如图 20-2 所示。

```
cd /usr/src
wget -c https://www.haproxy.org/download/2.1/src/haproxy-2.1.12.tar.gz
yum install wget gcc -y
tar xzf haproxy-2.1.12.tar.gz
cd haproxy-2.1.12
```

第 20 章　Kubernetes+HAProxy 高可用集群

```
make  TARGET=linux310  PREFIX=/usr/local/haproxy/
make  install  PREFIX=/usr/local/haproxy
```

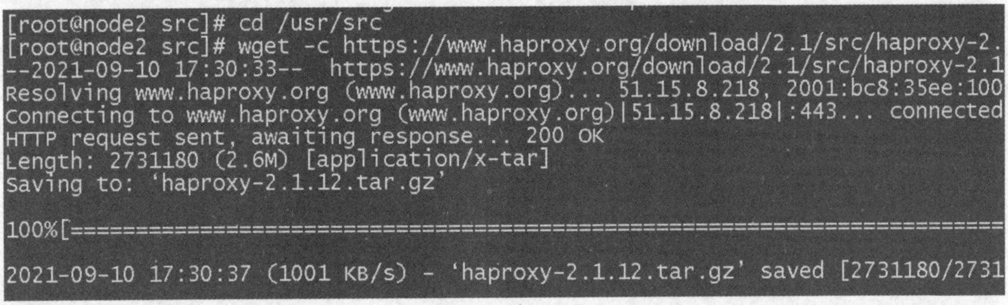

```
[root@node2 src]# cd /usr/src
[root@node2 src]# wget -c https://www.haproxy.org/download/2.1/src/haproxy-2.
--2021-09-10 17:30:33--  https://www.haproxy.org/download/2.1/src/haproxy-2.1
Resolving www.haproxy.org (www.haproxy.org)... 51.15.8.218, 2001:bc8:35ee:100
Connecting to www.haproxy.org (www.haproxy.org)|51.15.8.218|:443... connected
HTTP request sent, awaiting response... 200 OK
Length: 2731180 (2.6M) [application/x-tar]
Saving to: 'haproxy-2.1.12.tar.gz'

100%[===================================================================
2021-09-10 17:30:37 (1001 KB/s) - 'haproxy-2.1.12.tar.gz' saved [2731180/2731
```

(a)

```
[root@node2 haproxy-2.1.12]# make  install  PREFIX=/usr/local/haproxy
install: creating directory '/usr/local/haproxy'
install: creating directory '/usr/local/haproxy/sbin'
'haproxy' -> '/usr/local/haproxy/sbin/haproxy'
install: creating directory '/usr/local/haproxy/share'
install: creating directory '/usr/local/haproxy/share/man'
install: creating directory '/usr/local/haproxy/share/man/man1'
'doc/haproxy.1' -> '/usr/local/haproxy/share/man/man1/haproxy.1'
install: creating directory '/usr/local/haproxy/doc'
install: creating directory '/usr/local/haproxy/doc/haproxy'
'doc/configuration.txt' -> '/usr/local/haproxy/doc/haproxy/configuratio
'doc/management.txt' -> '/usr/local/haproxy/doc/haproxy/management.txt'
```

(b)

图 20-2　Kubernetes HAProxy 编译及安装

(a) 下载 HAProxy 软件包；(b) 编译安装 HAProxy 软件服务

(2) 配置 HAProxy 服务，操作指令如下，如图 20-3 所示。

```
useradd -s /sbin/nologin haproxy -M
cd /usr/local/haproxy ;mkdir -p etc/
touch /usr/local/haproxy/etc/haproxy.cfg
cd /usr/local/haproxy/etc/
```

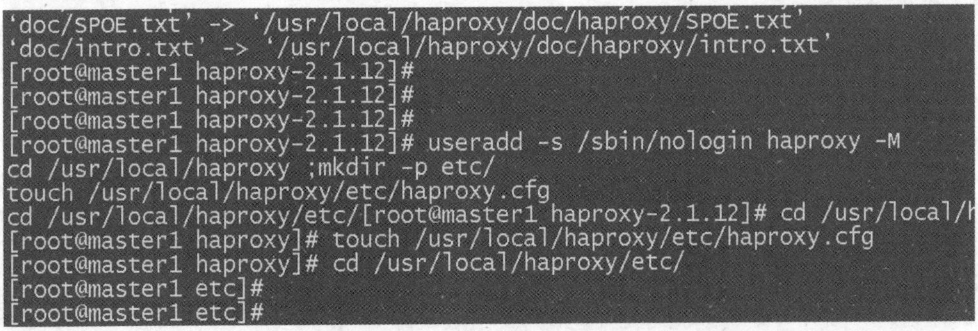

```
'doc/SPOE.txt' -> '/usr/local/haproxy/doc/haproxy/SPOE.txt'
'doc/intro.txt' -> '/usr/local/haproxy/doc/haproxy/intro.txt'
[root@master1 haproxy-2.1.12]#
[root@master1 haproxy-2.1.12]#
[root@master1 haproxy-2.1.12]#
[root@master1 haproxy-2.1.12]# useradd -s /sbin/nologin haproxy -M
cd /usr/local/haproxy ;mkdir -p etc/
touch /usr/local/haproxy/etc/haproxy.cfg
cd /usr/local/haproxy/etc/[root@master1 haproxy-2.1.12]# cd /usr/local/h
[root@master1 haproxy]# touch /usr/local/haproxy/etc/haproxy.cfg
[root@master1 haproxy]# cd /usr/local/haproxy/etc/
[root@master1 etc]#
[root@master1 etc]#
```

图 20-3　配置 HAProxy 服务

（3）haproxy.cfg 配置文件内容如下，如图 20-4 所示。

```
global
    log /dev/log    local0
    log /dev/log    local1 notice
    chroot /usr/local/haproxy
    stats socket /usr/local/haproxy/haproxy-admin.sock mode 660 level admin
    stats timeout 30s
    user haproxy
    group haproxy
    daemon
    nbproc 1
defaults
    log     global
    timeout connect 5000
    timeout client  10m
    timeout server  10m
listen admin_stats
    bind 0.0.0.0:10080
    mode http
    log 127.0.0.1 local0 err
    stats refresh 30s
    stats uri /admin
    stats realm welcome login\ Haproxy
    stats auth admin:123456
    stats hide-version
    stats admin if TRUE
listen kube-master
    bind 0.0.0.0:8443
    mode tcp
    option tcplog
    balance source
    server master1 192.168.1.145:6443 check inter 2000 fall 2 rise 2 weight 1
    server master2 192.168.1.146:6443 check inter 2000 fall 2 rise 2 weight 1
    server master3 192.168.1.147:6443 check inter 2000 fall 2 rise 2 weight 1
```

```
global
    log /dev/log    local0
    log /dev/log    local1 notice
    chroot /usr/local/haproxy
    stats socket /usr/local/haproxy/haproxy-admin.sock mode 660 level admin
    stats timeout 30s
    user haproxy
    group haproxy
    daemon
    nbproc 1
defaults
    log     global
    timeout connect 5000
```

图 20-4 Kubernetes HAProxy 部署实战

（4）启动 HAProxy 服务，操作指令如下，如图 20-5 所示。

```
/usr/local/haproxy/sbin/haproxy -f /usr/local/haproxy/etc/haproxy.cfg
ps -ef|grep -aiE haproxy
netstat -tnlp|grep -aiE haproxy
```

```
"haproxy.cfg" 33L, 925C written
[root@master1 etc]# /usr/local/haproxy/sbin/haproxy  -f   /usr/local/hapro
[root@master1 etc]# ps -ef|grep -aiE haproxy
haproxy   50632     1  0 17:34 ?        00:00:00 /usr/local/haproxy/sbin/hap
root      50714 29572  0 17:34 pts/0    00:00:00 grep --color=auto -aiE hapro
[root@master1 etc]#
[root@master1 etc]# netstat -tnlp|grep -aiE haproxy
tcp        0      0 0.0.0.0:8443            0.0.0.0:*               LISTEN
tcp        0      0 0.0.0.0:10080           0.0.0.0:*               LISTEN
[root@master1 etc]#
[root@master1 etc]#
[root@master1 etc]#
```

图 20-5　启动 HAProxy 服务

（5）启动 HAProxy，报错如下：

```
[WARNING] 217/202150 (2857) : Proxy 'chinaapp.sinaapp.com': in multi-process
mode, stats will be limited to process assigned to the current request.
```

解决方法：修改源码配置 src/cfgparse.c，找到以下行，调整 nbproc > 1 数值即可。

```
if (nbproc > 1) {
            if (curproxy->uri_auth) {
-                   Warning("Proxy '%s': in multi-process mode, stats will
be limited to process assigned to the current request.\n",
+                   Warning("Proxy '%s': in multi-process mode, stats will
be limited to the process assigned to the current request.\n",
```

20.4　配置 Keepalived 服务

Keepalived 是一个类似于工作在 layer 3、layer 4 和 layer 7 交换机制的软件，该软件有两种功能，分别是健康检查、VRRP 冗余协议。Keepalived 是模块化设计，不同模块负责不同的功能。部署 Keepalived 命令如下，如图 20-6 所示。

```
cd /usr/src;
yum install openssl-devel popt* kernel kernel-devel -y
wget -c http://www.keepalived.org/software/keepalived-1.2.1.tar.gz
tar xzf keepalived-1.2.1.tar.gz
cd keepalived-1.2.1 &&
./configure&&make &&make install
DIR=/usr/local/;\cp $DIR/etc/rc.d/init.d/keepalived /etc/rc.d/init.d/
\cp $DIR/etc/sysconfig/keepalived /etc/sysconfig/
```

```
mkdir -p /etc/keepalived  && \cp $DIR/sbin/keepalived /usr/sbin/
cd /etc/keepalived/;touch keepalived.conf
```

```
make[1]: Leaving directory `/usr/src/keepalived-1.2.1/keepalived'
make -C genhash install
make[1]: Entering directory `/usr/src/keepalived-1.2.1/genhash'
install -d /usr/local/bin
install -m 755 ../bin/genhash /usr/local/bin/
install -d /usr/local/share/man/man1
install -m 644 ../doc/man/man1/genhash.1 /usr/local/share/man/man1
make[1]: Leaving directory `/usr/src/keepalived-1.2.1/genhash'
[root@node2 keepalived-1.2.1]# DIR=/usr/local/ ;\cp $DIR/etc/rc.d/in
[root@node2 keepalived-1.2.1]# \cp $DIR/etc/sysconfig/keepalived /et
[root@node2 keepalived-1.2.1]# mkdir -p /etc/keepalived  && \cp $DIR
[root@node2 keepalived-1.2.1]#
```

图 20-6　配置 Keepalived 服务

20.5　Keepalived master 配置实战

HAProxy+Keepalived master 端 keepalived.conf 配置文件如下：

```
! Configuration File for keepalived
global_defs {
notification_email {
    wgkgood@139.com
}
  notification_email_from wgkgood@139.com
  smtp_server 127.0.0.1
  smtp_connect_timeout 30
  router_id LVS_DEVEL
}
vrrp_script chk_haproxy {
   script "/data/sh/check_haproxy.sh"
   interval 2
   weight 2
}
# VIP1
vrrp_instance VI_1 {
    state MASTER
    interface ens33
    virtual_router_id 151
    priority 100
    advert_int 5
    nopreempt
    authentication {
       auth_type PASS
       auth_pass 2222
```

```
    }
    virtual_ipaddress {
        192.168.1.188
    }
    track_script {
    chk_haproxy
    }
}
```

20.6 Keepalived Backup 配置实战

（1）HAProxy+Keepalived Backup 端 master2，配置 keepalived.conf，同时修改优先级为 90，操作指令如下：

```
! Configuration File for keepalived
global_defs {
notification_email {
    wgkgood@139.com
}
    notification_email_from wgkgood@139.com
    smtp_server 127.0.0.1
    smtp_connect_timeout 30
    router_id LVS_DEVEL
}
vrrp_script chk_haproxy {
    script "/data/sh/check_haproxy.sh"
    interval 2
    weight 2
}
# VIP1
vrrp_instance VI_1 {
    state BACKUP
    interface ens33
    virtual_router_id 151
    priority 90
    advert_int 5
    nopreempt
    authentication {
        auth_type PASS
        auth_pass 2222
    }
    virtual_ipaddress {
        192.168.1.188
    }
    track_script {
```

```
        chk_haproxy
    }
}
```

（2）HAProxy+Keepalived Backup 端 master3，配置 keepalived.conf，同时修改优先级为 80，操作指令如下：

```
! Configuration File for keepalived
global_defs {
notification_email {
    wgkgood@139.com
}
  notification_email_from wgkgood@139.com
  smtp_server 127.0.0.1
  smtp_connect_timeout 30
  router_id LVS_DEVEL
}
vrrp_script chk_haproxy {
   script "/data/sh/check_haproxy.sh"
   interval 2
   weight 2
}
# VIP1
vrrp_instance VI_1 {
    state  BACKUP
    interface ens33
    virtual_router_id 151
    priority 80
    advert_int 5
    nopreempt
    authentication {
       auth_type  PASS
       auth_pass  2222
    }
    virtual_ipaddress {
       192.168.1.188
    }
    track_script {
    chk_haproxy
    }
}
```

20.7 创建 HAProxy 检查脚本

在/data/sh/目录下创建 HAProxy 服务检查脚本，脚本名称为 check_haproxy.sh，同时设置执行权限。脚本内容如下：

```
mkdir -p /data/sh/
cd /data/sh/
touch check_haproxy.sh
chmod +x check_haproxy.sh
cat>/data/sh/check_haproxy.sh<<EOF
#!/bin/bash
#auto check haprox process
#2021-11-12  jfedu.net
CHECK_NUM=\$(ps -ef|grep -aiE haproxy|grep -aicvE "check|grep")
if
   [[ \$CHECK_NUM -eq 0 ]];then
   /etc/init.d/keepalived stop
fi
EOF
```

20.8　HAProxy+Keepalived 验证

启动所有节点 Keepalived 服务，然后查看 145 的 Keepalived 后台日志，且访问 188 VIP 正常，确认其可提供服务，即证明 HAProxy+Keepalived 高可用架构配置完毕，如图 20-7 所示。

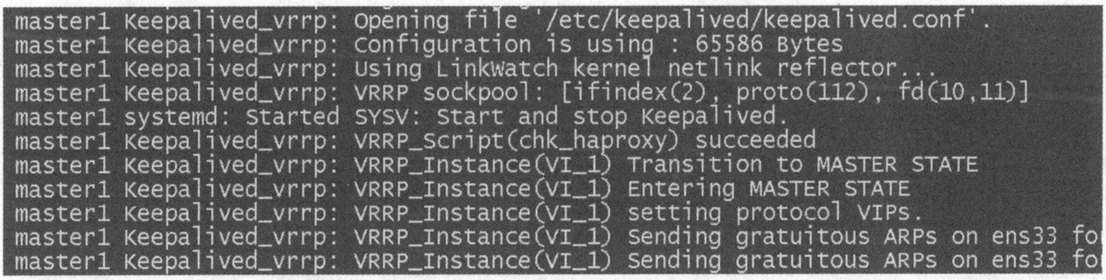

（a）

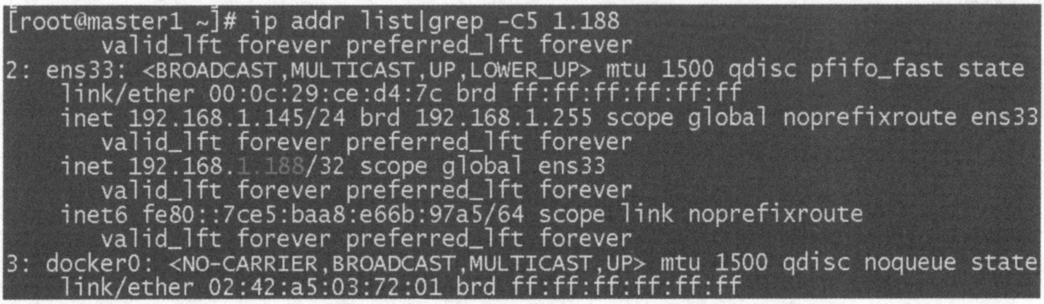

（b）

图 20-7　HAProxy+Keepalived 网站架构

（a）查看 keepalived 状态；（b）查看 1.188 VIP 是否存在

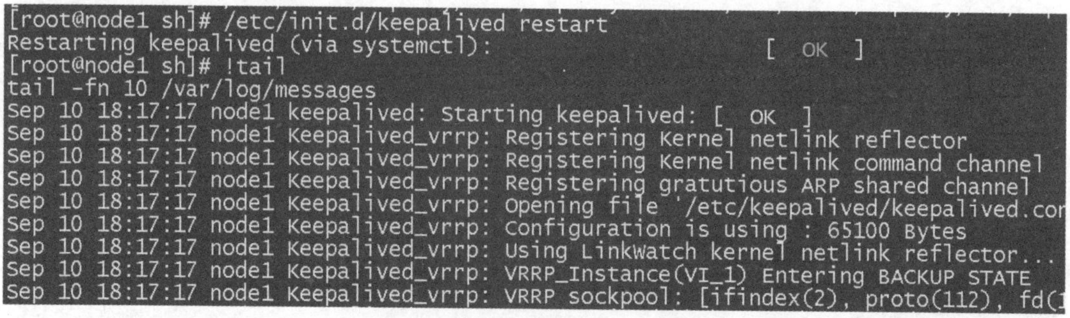

```
[root@node1 sh]# /etc/init.d/keepalived restart
Restarting keepalived (via systemctl):           [ OK ]
[root@node1 sh]# !tail
tail -fn 10 /var/log/messages
Sep 10 18:17:17 node1 keepalived: Starting keepalived: [ OK ]
Sep 10 18:17:17 node1 Keepalived_vrrp: Registering Kernel netlink reflector
Sep 10 18:17:17 node1 Keepalived_vrrp: Registering Kernel netlink command channel
Sep 10 18:17:17 node1 Keepalived_vrrp: Registering gratuitous ARP shared channel
Sep 10 18:17:17 node1 Keepalived_vrrp: Opening file '/etc/keepalived/keepalived.co
Sep 10 18:17:17 node1 Keepalived_vrrp: Configuration is using : 65100 Bytes
Sep 10 18:17:17 node1 Keepalived_vrrp: Using LinkWatch kernel netlink reflector...
Sep 10 18:17:17 node1 Keepalived_vrrp: VRRP_Instance(VI_1) Entering BACKUP STATE
Sep 10 18:17:17 node1 Keepalived_vrrp: VRRP sockpool: [ifindex(2), proto(112), fd(
```

(c)

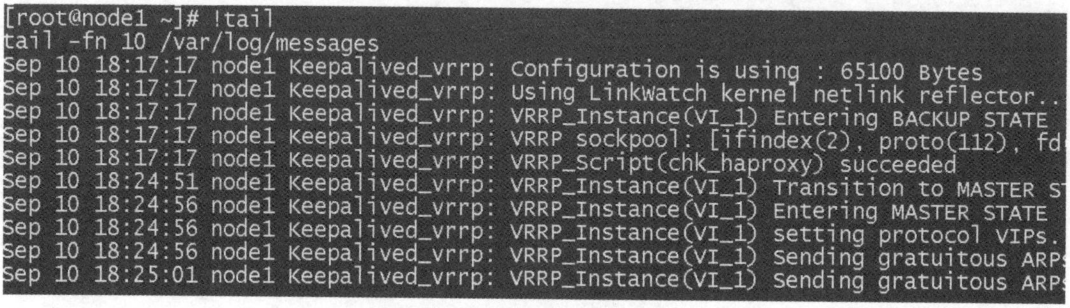

```
[root@node1 ~]# !tail
tail -fn 10 /var/log/messages
Sep 10 18:17:17 node1 Keepalived_vrrp: Configuration is using : 65100 Bytes
Sep 10 18:17:17 node1 Keepalived_vrrp: Using LinkWatch kernel netlink reflector..
Sep 10 18:17:17 node1 Keepalived_vrrp: VRRP_Instance(VI_1) Entering BACKUP STATE
Sep 10 18:17:17 node1 Keepalived_vrrp: VRRP sockpool: [ifindex(2), proto(112), fd
Sep 10 18:17:17 node1 Keepalived_vrrp: VRRP_Script(chk_haproxy) succeeded
Sep 10 18:24:51 node1 Keepalived_vrrp: VRRP_Instance(VI_1) Transition to MASTER S
Sep 10 18:24:56 node1 Keepalived_vrrp: VRRP_Instance(VI_1) Entering MASTER STATE
Sep 10 18:24:56 node1 Keepalived_vrrp: VRRP_Instance(VI_1) setting protocol VIPs.
Sep 10 18:24:56 node1 Keepalived_vrrp: VRRP_Instance(VI_1) Sending gratuitous ARP
Sep 10 18:25:01 node1 Keepalived_vrrp: VRRP_Instance(VI_1) Sending gratuitous ARP
```

(d)

图 20-7　HAProxy+Keepalived 网站架构（续）

（c）重启 Keepalived 服务并且查看日志；（d）查看内核日志发现 Keepalived 状态变成了 master

20.9　初始化 master 集群

（1）Kubernetes 集群引入 HAProxy 高可用集群，可以生成 Kubernetes 集群 init 配置文件，操作指令如下：

```
kubeadm config print init-defaults >kubeadmin-init.yaml
```

（2）根据以上指令生成配置文件之后，修改 kubeadmin-init.yaml 配置文件，添加集群 controlPlaneEndpoint: "192.168.1.188:8443"，最终代码如下：

```
apiVersion: kubeadm.Kubernetes.io/v1beta2
bootstrapTokens:
- groups:
  - system:bootstrappers:kubeadm:default-node-token
  token: abcdef.0123456789abcdef
  ttl: 24h0m0s
  usages:
  - signing
```

```
  - authentication
kind: InitConfiguration
localAPIEndpoint:
  advertiseAddress: 192.168.1.145
  bindPort: 6443
nodeRegistration:
  criSocket: /var/run/dockershim.sock
  name: master1
  taints:
  - effect: NoSchedule
    key: node-role.kubernetes.io/master
---
apiServer:
  timeoutForControlPlane: 4m0s
apiVersion: kubeadm.Kubernetes.io/v1beta2
certificatesDir: /etc/kubernetes/pki
clusterName: kubernetes
controlPlaneEndpoint: "192.168.1.188:8443"
controllerManager: {}
dns:
  type: CoreDNS
etcd:
  local:
    dataDir: /var/lib/etcd
imageRepository: registry.aliyuncs.com/google_containers
kind: ClusterConfiguration
kubernetesVersion: v1.20.0
networking:
  dnsDomain: cluster.local
  podSubnet: 10.244.0.0/16
  serviceSubnet: 10.10.0.0/16
scheduler: {}
```

（3）执行以下命令初始化集群即可，如图 20-8 所示，操作指令如下：

```
kubeadm init --config kubeadmin-init.yaml --upload-certs
```

（4）如果集群只有 3 台 master，没有 node 节点，可以去除 master 节点上的污点标记，使其可以分配 Pod 资源，操作指令如下：

```
kubectl taint nodes --all node-role.kubernetes.io/master-
```

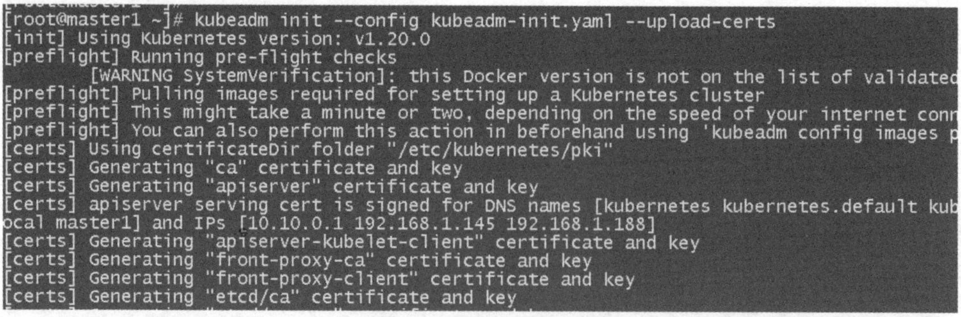
```
[root@master1 ~]# kubeadm init --config kubeadm-init.yaml --upload-certs
[init] Using Kubernetes version: v1.20.0
[preflight] Running pre-flight checks
        [WARNING SystemVerification]: this Docker version is not on the list of validated
[preflight] Pulling images required for setting up a Kubernetes cluster
[preflight] This might take a minute or two, depending on the speed of your internet conn
[preflight] You can also perform this action in beforehand using 'kubeadm config images p
[certs] Using certificateDir folder "/etc/kubernetes/pki"
[certs] Generating "ca" certificate and key
[certs] Generating "apiserver" certificate and key
[certs] apiserver serving cert is signed for DNS names [kubernetes kubernetes.default kub
ocal master1] and IPs [10.10.0.1 192.168.1.145 192.168.1.188]
[certs] Generating "apiserver-kubelet-client" certificate and key
[certs] Generating "front-proxy-ca" certificate and key
[certs] Generating "front-proxy-client" certificate and key
[certs] Generating "etcd/ca" certificate and key
```

（a）

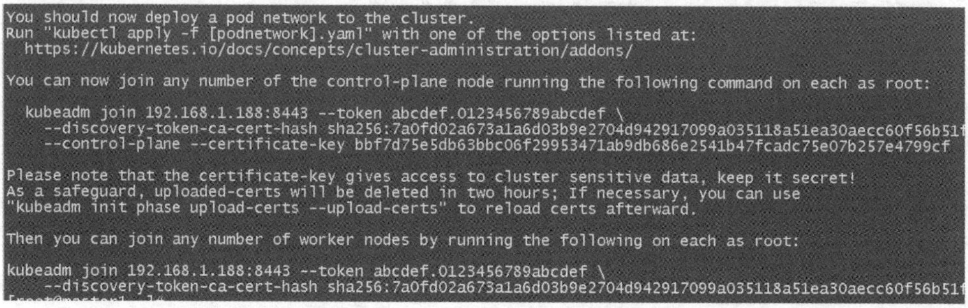
```
You should now deploy a pod network to the cluster.
Run "kubectl apply -f [podnetwork].yaml" with one of the options listed at:
  https://kubernetes.io/docs/concepts/cluster-administration/addons/

You can now join any number of the control-plane node running the following command on each as root:

  kubeadm join 192.168.1.188:8443 --token abcdef.0123456789abcdef \
    --discovery-token-ca-cert-hash sha256:7a0fd02a673a1a6d03b9e2704d942917099a035118a51ea30aecc60f56b51f
    --control-plane --certificate-key bbf7d75e5db63bbc06f29953471ab9db686e2541b47fcadc75e07b257e4799cf

Please note that the certificate-key gives access to cluster sensitive data, keep it secret!
As a safeguard, uploaded-certs will be deleted in two hours; If necessary, you can use
"kubeadm init phase upload-certs --upload-certs" to reload certs afterward.

Then you can join any number of worker nodes by running the following on each as root:

kubeadm join 192.168.1.188:8443 --token abcdef.0123456789abcdef \
    --discovery-token-ca-cert-hash sha256:7a0fd02a673a1a6d03b9e2704d942917099a035118a51ea30aecc60f56b51f
```

（b）

图 20-8　Kubernetes 多 master 集群初始化

（a）初始化 Kubernetes 集群；（b）查看 Kubernetes 初始化详细日志

20.10　Kubernetes Dashboard UI 实战

Kubernetes 实现的最重要的工作是对 Docker 容器集群进行统一管理和调度，通常使用命令行操作 Kubernetes 集群及各个节点。命令行操作非常不方便，如果使用 UI 界面进行可视化操作，会更加方便管理和维护。以下为配置 Kubernetes Dashboard 的完整过程：

（1）下载 Dashboard 配置文件。

```
wget https://raw.githubusercontent.com/kubernetes/dashboard/v2.0.0-rc5/
aio/deploy/recommended.yaml
\cp recommended.yaml recommended.yaml.bak
```

（2）修改文件 recommended.yaml 的 39 行内容。因为默认情况下 service 的类型是 cluster IP，需更改为 NodePort 的方式，便于访问，也可映射到指定的端口。

```
spec:
  type: NodePort
  ports:
    - port: 443
      targetPort: 8443
```

```
    nodePort: 31001
  selector:
    Kubernetes-app: kubernetes-dashboard
```

（3）修改文件 recommended.yaml 的 195 行内容，因为默认情况下 Dashboard 为英文显示，可以设置为中文。

```
env:
        - name: ACCEPT_LANGUAGE
          value: zh
```

（4）创建 Dashboard 服务，指令操作如下：

```
kubectl apply -f recommended.yaml
```

（5）查看 Dashboard 运行状态。

```
kubectl get pod -n kubernetes-dashboard
kubectl get svc -n kubernetes-dashboard
```

（6）将 master 节点也设置为 node 节点，可以运行 Pod 容器任务，命令如下：

```
kubectl taint nodes --all node-role.kubernetes.io/master-
```

（7）基于 Token 的方式访问，设置和绑定 Dashboard 权限，命令如下：

```
#创建 Dashboard 的管理用户
kubectl create serviceaccount dashboard-admin -n kube-system
#将创建的 Dashboard 用户绑定为管理用户
kubectl create clusterrolebinding dashboard-cluster-admin --clusterrole=cluster-admin --serviceaccount=kube-system:dashboard-admin
#获取刚刚创建的用户对应的 Token 名称
kubectl get secrets -n kube-system | grep dashboard
#查看 Token 的详细信息
kubectl describe secrets -n kube-system $(kubectl get secrets -n kube-system | grep dashboard |awk '{print $1}')
```

（8）通过浏览器访问 Dashboard Web，https://192.168.1.188:31001/，输入 Token 登录即可，如图 20-9 所示。

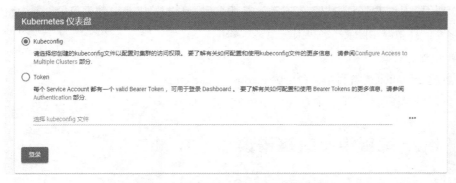

图 20-9　访问 Dashboard UI 登录界面

第 21 章 Kubernetes 配置故障实战

21.1 etcd 配置中心故障错误一

配置完毕却无法启动,如图 21-1 所示。

```
#[member]
ETCD_NAME=etcd3
ETCD_DATA_DIR="/data/etcd"
#ETCD_WAL_DIR=""
#ETCD_SNAPSHOT_COUNT="10000"
#ETCD_HEARTBEAT_INTERVAL="100"
#ETCD_ELECTION_TIMEOUT="1000"
ETCD_LISTEN_PEER_URLS="http://192.168.0.133:2380"
ETCD_LISTEN_CLIENT_URLS="http://192.168.0.133:2379,http://127.0.0.1:2379"
CD_MAX_SNAPSHOTS="5"
#ETCD_MAX_WALS="5"
#ETCD_CORS=""
#[cluster]
ETCD_INITIAL_ADVERTISE_PEER_URLS="http://192.168.0.133:2380"
# if you use different ETCD_NAME (e.g. test), set ETCD_INITIAL_CLUSTER value for this nam
ETCD_INITIAL_CLUSTER="etcd1=http://192.168.0.131:2380,etcd2=http://192.168.0.132:2380,etc
ETCD_INITIAL_CLUSTER_STATE="new"
ETCD_INITIAL_CLUSTER_TOKEN="etcd-cluster"
ETCD_ADVERTISE_CLIENT_URLS="http://192.168.0.133:2379"
#ETCD_DISCOVERY=""
#ETCD_DISCOVERY_SRV=""
```

图 21-1 etcd 启动服务报错

(1)原因分析:根据日志提示,启动 etcd 服务时,会初始化 etcd 数据目录/data/etcd/,而该目录不存在,自动创建目录提示权限被拒绝。

(2)解决方法如下:

① 创建/data/etcd/配置目录,命令为 mkdir -p /data/etcd/。

② 同时设置权限为可写,命令为 chmod 757 -R /data/etcd/。

21.2 etcd 配置中心故障错误二

部署 etcd 集群时,通常尝试启动 etcd1 时,报错信息如下:

```
Apr 12 01:06:49 Kubernetes-master etcd[3092]: health check for peer
618d69366dd8cee3
could not connect: dial tcp 10.0.0.122:2380: getsockopt: connection refused
Apr 12 01:06:49 Kubernetes-master etcd[3092]: health check for peer acd2ba924953b1ec
could not connect: dial tcp 192.168.0.143: 2380: getsockopt: connection
refused
Apr 12 01:06:48 Kubernetes -master etcd[3092]: publish error: etcdserver:
request timed out
```

（1）原因分析：etcd1 的配置文件/etc/etcd/etcd.conf 中的 etcd_INITIAL_CLUSTER_STATE 是 new，而在配置中 etcd_INITIAL_CLUSTER 写入了 etcd2、etcd3 的 IP:PORT，这时 etcd1 尝试连接 etcd2、etcd3，但是 etcd2、etcd3 的 etcd 服务此时还未启动。

（2）解决方法：启动 etcd2 和 etcd3 的 etcd 服务，再启动 etcd1。

21.3　Pod infrastructure 故障错误

Kubernetes Pod-infrastructure:latest 镜像下载失败，报错如下：

```
image pull failed for registry.access.redhat.com/rhel7/pod-infrastructure:
latest, this may be because there are no credentials on this request.  details:
(open /etc/docker/certs.d/registry.access.redhat.com/redhat-ca.crt: no
such file or directory)
Failed to create pod infra container: ImagePullBackOff; Skipping pod
"redis-master-jj6jw_default(fec25a87-cdbe-11e7-ba32-525400cae48b)": Back-
off pulling image "registry.access.redhat.com/rhel7/pod-infrastructure:
latest
```

报错原因：因为获取远程 Pod 镜像需要查找安全证书，所以需要从官网下载证书，但由于网络故障无法获取远程镜像。解决方法如下：

```
yum install *rhsm* -y
vim /etc/kubernetes/kubelet,pod 外网源如下：
KUBELET_POD_INFRA_CONTAINER="--pod-infra-container-image=registry.access
.redhat.com/rhel7/pod-infrastructure:latest"
可以更改为内网的私有源即可：
KUBELET_POD_INFRA_CONTAINER="--pod-infra-container-image=192.168.1.146:
5000/rhel7/pod-infrastructure:latest"
```

21.4　Docker 虚拟化故障错误一

在 Kubernetes 节点上维护管理 Docker 时，node 节点的 Docker 无法启动，报错信息如下：

```
systemctl start docker.service
```

```
Job for docker.service failed because the control process exited with error
code. See "systemctl status docker.service" and "journalctl -xe" for details.
```

(1) 错误原因：因为找不到 devicemap 存储块相关的库，导致无法启动 Docker，可以通过安装 devcie-map 相关的库解决问题。

(2) 解决方法如下：

```
yum install device-map* -y
```

21.5　Docker 虚拟化故障错误二

Kubernetes 客户端 node 节点，部署 Docker 服务，发现启动 Docker 容器时报错如下：

```
/usr/bin/docker-current: Error response from daemon: shim error: docker-runc
not installed on system.
```

(1) 原因分析：可能是 Docker 的版本 bug，Docker 软件安装不正确，导致无法启动。

(2) 解决方法：更换 yum 源，卸载当前 Docker，重新安装启动，Docker 容器可以正常运行。

21.6　Dashboard API 故障错误

当配置好 Dashborad Web UI 时，浏览器访问 UI：http://10.0.0.122:8080/ui，显示的是相关的提示信息，而不是 Kebernetes Dashboard 的页面。

(1) 原因分析：检查一下 master 端，重点检查 API server 接口问题。

(2) 解决方法：在 master 端修改/etc/kubernetes/apiserver 中的配置文件，删除 ServiceAccount，重新访问 Dashboard 界面如下：

```
KUBE_ADMISSION_CONTROL="--admission-control=NamespaceLifecycle,Namespace
Exists,LimitRanger,SecurityContextDeny,ResourceQuota"
```

21.7　Dashboard 网络访问故障错误

当配置好 Dashborad Web UI 后，浏览器访问 UI：http://10.0.0.122:8080/ui，显示的是相关的提示信息，而不是 Kebernetes Dashboard 的页面，修改 API server 也无法解决该问题，最终访问 UI 超时，显示如下信息：

```
Error: 'dial tcp 10.0.66.2:9090: getsockopt: connection timed out'
Trying to reach: 'http://10.0.66.2:9090/'
```

(1) 原因分析：远程服务器 10.0.66.2 的 9090 端口无法连接，应该属于网络的问题。

(2) 解决方法：重启 Kubernetes 集群相关服务即可，命令操作如下：

```
service etcd restart;service flanneld restart;service docker restart;
iptables -P FORWARD ACCEPT
```

通过 ifconfig 查看各个服务器的 IP 信息，同时确认 Flanneld 配置是否正确，最终解决问题，如图 21-2 所示。

图 21-2　Kubernetes Web 界面